Proceedings of the Ringberg Workshop

NEW TRENDS IN HERA PHYSICS 2005

PROCEEDINGS OF THE RINGBERG WORKSHOP

NEW TRENDS IN HERA PHYSICS 2005

edited by

G. Grindhammer
Max Planck Institute for Physics
Munich, Germany

B. A. Kniehl
G. Kramer
Hamburg University, Germany

W. Ochs
Max Planck Institute for Physics
Munich, Germany

World Scientific

NEW JERSEY • LONDON • SINGAPORE • BEIJING • SHANGHAI • HONG KONG • TAIPEI • CHENNAI

Published by

World Scientific Publishing Co. Pte. Ltd.
5 Toh Tuck Link, Singapore 596224
USA office: 27 Warren Street, Suite 401-402, Hackensack, NJ 07601
UK office: 57 Shelton Street, Covent Garden, London WC2H 9HE

British Library Cataloguing-in-Publication Data
A catalogue record for this book is available from the British Library.

NEW TRENDS IN HERA PHYSICS 2005
Proceedings of the Ringberg Workshop

ISBN 981-256-816-6

Printed in Singapore by Mainland Press

Preface

The international workshop entitled *New Trends in HERA Physics 2005* took place from 2 to 7 October 2005 at Ringberg Castle, which overlooks Lake Tegernsee in the foothills of the Bavarian Alps, one of the most picturesque locations to be found in the whole of Germany. The castle was built during the first half of the twentieth century by Duke Luitpold in Bavaria (Herzog Luitpold in Bayern), a member of the Wittelsbach family which ruled Bavaria for over 800 years, and his friend Friedrich Attenhuber, an all-round artist, architect, and interior decorator. The castle is entirely their creation, from the massive Renaissance-inspired exterior right down to the fittings and furniture, which, in every detail, were designed by Attenhuber himself and executed by native craftsmen. Attenhuber also painted every single picture exhibited in the castle. He found his models in the farmhouses around Lake Tegernsee. The castle embodies all trends of art and styles which dominated the first half of the last century, combined with local Alpine originality and the individual creative power of its constructors. According to the Duke's last will, the castle passed into the hands of the Max Planck Society after his death, in 1973. The castle was then transformed into a conference venue, where scientists can exchange their latest ideas and discuss problems with their colleagues from all over the world in beautiful surroundings and in a relaxed mountain atmosphere, high above the daily business activities.

This was the fifth event in a series of Ringberg workshops on HERA physics, which started in 1997 and continued in 1999, 2001, and 2003. In fact, at the end of these workshops, many participants expressed the opinion that this was a successful endeavour to bring theorists and experimentalists together in order to interpret the latest HERA data, and that it would be useful to organize a follow-up workshop in the same spirit.

On the occasion of the 2005 Ringberg workshop, forty-one experts of elementary-particle physics, both theorists and experimentalists, from twenty-seven universities and research institutions in nine countries congregated to present their latest results on the various aspects of HERA physics. Specifically, there were nineteen theorists and twenty-two experimentalists, the latter representing the H1, HERMES, and ZEUS collaborations at HERA. The topics included: proton structure function; polarized *ep* scattering; photoproduction of hadrons and jets; determination of the parton densities in the proton and the strong coupling α_s from HERA data alone; final states in deep-inelastic scattering, with special emphasis on hadrons, jets, resonances, and diquarks; heavy-flavour and charmonium production; elastic and diffractive *ep* scattering; new physics at HERA; future plans for HERA; and implications of HERA for LHC physics. We hope that the high-energy physics community will benefit from these proceedings, in which the ongoing efforts in understanding the nature of the strong interactions, with particular emphasis on HERA physics, are documented.

We wish to thank all our friends and colleagues who have contributed to these proceedings. We are indebted to the workshop secretary, Mrs. Rosita Jurgeleit, for her assistance before, during, and after the workshop and to Dr. Annette Holtkamp for her technical assistance in the editorial work. The local costs at Ringberg Castle and the costs for the publication and dissemination of these proceedings were covered in equal parts by the Deutsches Elektronen-Synchrotron at Hamburg and the Max-Planck-Institut für Physik at Munich, which we gratefully acknowledge.

Günter Grindhammer
Bernd A. Kniehl
Gustav Kramer
Wolfgang Ochs

Hamburg, January 2006

Contents

1 Proton Structure

PROTON STRUCTURE MEASUREMENTS AT HIGH Q^2 AND LARGE X

K. WICHMANN
Institute of Experimental Physics, Hamburg University
Luruper Chaussee 149,
22607 Hamburg, Germany
E-mail: kklimek@mail.desy.de

Proton structure measurements at high Q^2 and x, performed by the H1 and ZEUS Collaborations at the HERA accelerator, are reviewed. Neutral and charge current deep inelastic scattering cross sections and structure function measurements are presented. A new method employed to measure the NC cross section up to Bjorken-x values equal to one is described. The review also discusses the ZEUS-JETS PDF fits which include jet data giving direct access to the gluon densities. An estimate on the impact of HERA II data on the PDFs and their influence on the LHC measurements are mentioned.

1. Introduction

Precise knowledge of proton parton distribution functions (PDFs) is crucial for understanding proton structure. Additionally it is required for any cross-section calculation at hadron colliders for Standard Model (SM) physics as well as for the discovery of physics beyond the SM.

The PDFs are usually determined in global fits within the DGLAP formalism [1–4] at next-to-leading order (NLO) in QCD. These fits use data from many different experiments, with the inclusive cross-section data from deep inelastic scattering (DIS) experiments providing the major source of information. The wide kinematic range covered by the HERA DIS data [5–7], as well as their precision, has allowed the determination of PDFs across a broad range of phase space spanned by the fractional proton momentum, Bjorken-x, carried by the struck quark, and the negative squared four-momentum transfer Q^2. This paper reviews the measurements at high Q^2 and x performed by the H1 and ZEUS Collaborations at the HERA accelerator.

2. Cross Section Measurements and Structure Functions

The differential cross section for lepton-nucleon ($l^{\pm}N$) NC DIS is given in terms of structure functions by

$$\frac{d^2\sigma^{NC}(l^{\pm}N)}{dxdQ^2} = \frac{2\pi\alpha^2}{xQ^4}[Y_+F_2(x,Q^2) \mp Y_-xF_3(x,Q^2) - y^2F_L(x,Q^2)], \quad (1)$$

where α is the electromagnetic fine structure constant, y is the inelasticity parameter and $Y_{\pm} = 1 \pm (1-y)^2$. The longitudinal structure function, F_L , is significant only at large y and low Q^2 and is not discussed in this paper. The structure functions F_2 and xF_3 are directly related to quark distributions and their Q^2 dependence is predicted by perturbative QCD. For Q^2 up to around 1000 GeV^2 the charged lepton-hadron cross section is dominated by photon exchange and the F_2 structure function. For values of x lower than about 10^{-2} F_2 is sea-quark dominated and its Q^2 dependence is driven by the gluon density. That is why HERA can provide crucial information on gluon and quark distributions. The structure function F_3, which provides information on valence quark distributions, is dominated by a parity-violating term and thus makes a significant contribution only at large Q^2. The structure function F_3 is measured by subtracting the NC cross sections of electron-proton and positron-proton DIS. Fig.1 shows xF_3 at fixed values of Q^2 as a function of x. The measurements of the H1 [8] and ZEUS [9] Collaborations are in good agreement. The expectation of the SM, evaluated with the CTEQ6D [16] PDFs, gives a good description of the data. So far the biggest uncertainty comes from the smaller e^-p sample. With already about ten times more e^-p data at HERA II, more significant results are expected soon.

The charged-current (CC) interactions are sensitive to the flavor of the valence distributions at high x because their (LO) cross sections are given by

$$\frac{d^2\sigma^{CC}}{dxdQ^2}(e^+p) = \frac{G_F^2}{2\pi x}\frac{M_W^4}{(Q^2+M_W^2)^2}x[(\bar{u}+\bar{c}) + (1-y^2)(d+s)], \quad (2)$$

$$\frac{d^2\sigma^{CC}}{dxdQ^2}(e^-p) = \frac{G_F^2}{2\pi x}\frac{M_W^4}{(Q^2+M_W^2)^2}x[(u+c) + (1-y^2)(\bar{d}+\bar{s})], \quad (3)$$

where G_F is the Fermi coupling constant and M_W is the mass of the W-boson and the parton distributions u, d, s, c are functions of x and Q^2. The reduced cross sections, $\tilde{\sigma}$, are defined as

$$\tilde{\sigma} = \left[\frac{G_F^2}{2\pi x}\frac{M_W^4}{(Q^2+M_W^2)^2}\right]^{-1}\frac{d^2\sigma^{CC}}{dxdQ^2}.$$

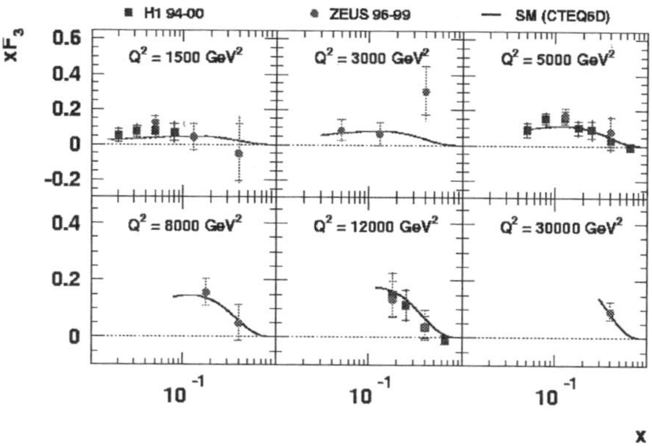

Figure 1. The structure function xF_3 as a function of x for different Q^2 values measured by the H1 (squares) and ZEUS (dots) Collaborations. The inner error bars show the statistical uncertainty, while the outer ones show the statistical and systematic uncertainties added in quadrature. Also shown are SM predictions using CTEQ6D PDFs.

The e^-p CC cross section gives information on the u valence quark at high x. The e^+p CC cross section gives information on the d valence quark at high x, and since the d valence quark distribution is poorly known these data are particularly valuable. Fig.2 shows the e^-p and e^+p reduced cross sections as a function of x, measured by H1 [8,10] and ZEUS [11], compared to SM predictions evaluated using CTEQ6D. The contributions from the PDF combinations $(u+c)$ and $\bar{d}+\bar{s}$ for e^-p and $(d+s)$ and $\bar{u}+\bar{c}$ for e^+p, obtained in the $\overline{\text{MS}}$ scheme from the ZEUS-S fit, are also shown. There is good agreement between the two measurements. The SM prediction evaluated using CTEQ6D gives a good description of the data.

3. High-x Measurement

Due to limitations in beam energies and measurement techniques a direct confrontation with data has not been possible to date for $x \to 1$. The highest measured points in the DIS regime are at $x = 0.75$ [12]. Data at higher x [13,14] exist, but they are in the resonance production region and cannot be easily interpreted in terms of parton distributions. The highest value of x so far for HERA structure function data is $x = 0.65$. A new method [15] was employed to measure the NC cross section up to Bjorken-x values equal to one with the ZEUS detector for $Q^2 > 576$ GeV^2. In most of the DIS events

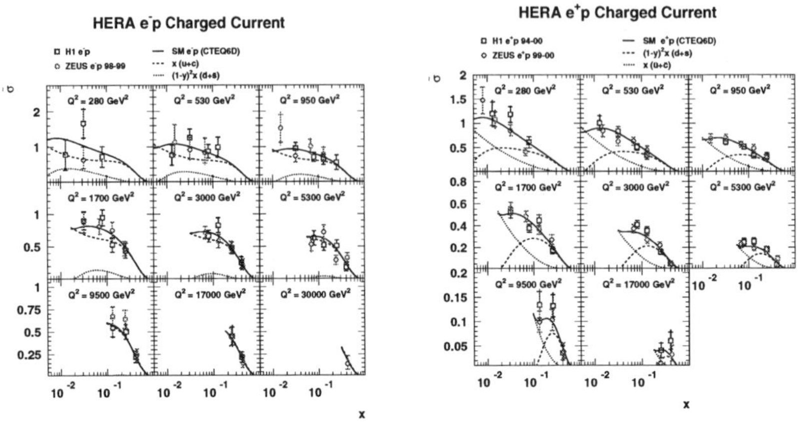

Figure 2. The reduced CC cross section, $\tilde{\sigma}$, as a function of x, for different values of Q^2 for e^-p (left) and e^+p (right) scattering measured by the H1 (squares) and ZEUS (circles) Collaborations. The inner error bars show the statistical uncertainty, while the outer ones show the statistical and systematic uncertainties added in quadrature. The SM expectations calculated using CTEQ6D PDF are shown as full line. The separate contributions from the PDF combinations $x(u+c)$ and $(1-y)^2x(\bar{d}+\bar{s})$ for e^-p, $(1-y)^2x(d+s)$ and $x(\bar{u}+\bar{c})$ for e^+p are indicated by dashed and dotted lines, respectively.

there is a jet present in the detector. As x increases, the jet is boosted in the forward direction and the jet angle decreases. When x is too high, a part of the jet is lost in the beam pipe. The value of x at which this occurs is Q^2 dependent: as Q^2 increases the x value for which jets are well contained increases. In this analysis for $Q^2 > 576$ GeV^2, the electron was scattered at large angles and was well contained in the detector. For the purpose of this analysis the electron and jet information were combined to allow a measurement of the differential cross section up to $x = 1$. Events are sorted into Q^2 and x bins to allow a measurement of the double differential cross section $d^2\sigma/dxdQ^2$. The value Q^2 was calculated using electron information only and the jet energy and angle were used to calculate x for events with a well reconstructed jet. Events with no jet reconstructed within the fiducial volume were assumed to come from high x and were collected in a bin with $x_{\text{Edge}} < x < 1$. These bins were generally large and the form of the PDF is not well known in this region and therefore an integrated cross section was calculated. Events with more than one high energy jet were discarded. The standard selection cuts ensured a background-free and well reconstructed DIS sample for e^+p and e^-p scattering with a good jet with transverse energy greater than 10 GeV. Good agreement between data and

MC simulation is observed for the kinematic, electron and jet variables for the full sample of events. Overall, a small excess of data over MC was observed for zero jet events. Also an offset in the $E - p_Z$ distribution is seen, with the MC distribution slightly lower than the data, but generally the distributions agree well and there is no indication of background in the sample.

The measured Born level cross sections for e^+p are shown in Fig.3 and compared to SM expectations at NLO using the CTEQ6D PDFs. Similar results were obtained for the e^-p data. The double differential cross sections are represented by solid points, and generally agree well with the expectations. The cross section in the highest x bin is the integrated cross section. In this bin, the expected cross section is drawn as a horizontal line, while the measured cross section is displayed by the open symbol.

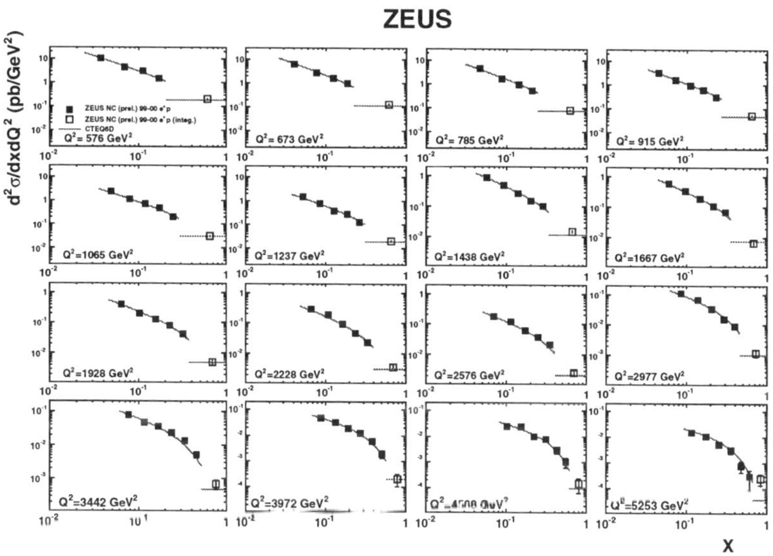

Figure 3. The double differential cross section for e^+p NC scattering (solid squares) and the integral of the double differential cross section divided by the bin width (open squares) compared to the Standard Model expectations evaluated using CTEQ6D PDFs (lines). The inner error bars show the statistical uncertainty, while the outer ones show the statistical and systematic uncertainties added in quadrature.

The measured double differential cross sections generally agree well with the expectations. For the highest x bins, which are in previously unmeasured kinematic ranges, the data has a tendency to lie above the expectations. These data are expected to have an impact on the extraction of the PDFs at the highest values of x, and via sum rules, also on the PDFs at smaller x. Due to the limited statistics the present measurement has a modest impact on the PDFs but the method is developed and HERA II will have significant impact.

4. Jet Data in Global Fits

The gluon density contributes only indirectly to the inclusive DIS cross sections. However it makes a direct contribution to jet cross sections through boson-gluon and quark-gluon scattering. Tevatron high-E_T data [17,18] have been used to constrain the gluon in the fits of MRST [19,20] and CTEQ [16]. However, these data suffer from very large correlated systematic uncertainties from a variety of sources. In the ZEUS-JETS fit [21], ZEUS NC e^+p DIS inclusive jet cross sections [22] and direct photoproduction dijet cross sections [23] have been used to constrain the gluon. The jet cross sections used in this analysis are shown in Fig.4 together with the SM predictions evaluated using the ZEUS-JETS parton densities. The data and predictions agree well.

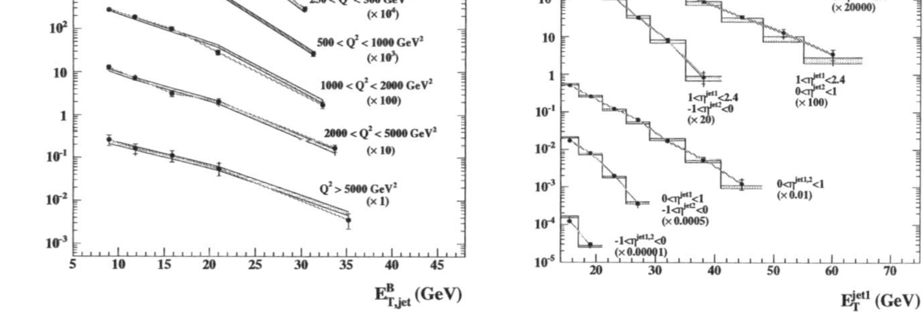

Figure 4. ZEUS-JETS fit compared to ZEUS DIS inclusive jet data (left) and photoproduction dijet data (right), respectively. Each cross section has been multiplied by the scale factor in brackets to aid visibility.

These jet data together with ZEUS data on NC and CC e^+p and e^-p DIS inclusive cross sections [9,11,24–27] were used as inputs to an NLO QCD DGLAP analysis in order to determine the PDFs. The predictions for the jet cross sections are calculated to NLO in QCD and are used in the fit rigorously, rather than approximately as in previous fits [16,19,20]. The details of the fit and the fitting procedure are described in detail elsewhere in these proceedings [28]. The valence, sea and gluon PDFs are compared for the ZEUS-JETS fit and the previous ZEUS-S global PDF analysis in Fig.5. There is good agreement between the ZEUS PDF extractions. In the figure there is also a comparison of the MRST and CTEQ PDFs to the ZEUS-JETS PDFs. Considering the size of the uncertainties on each PDF set, they are compatible with the ZEUS PDFs.

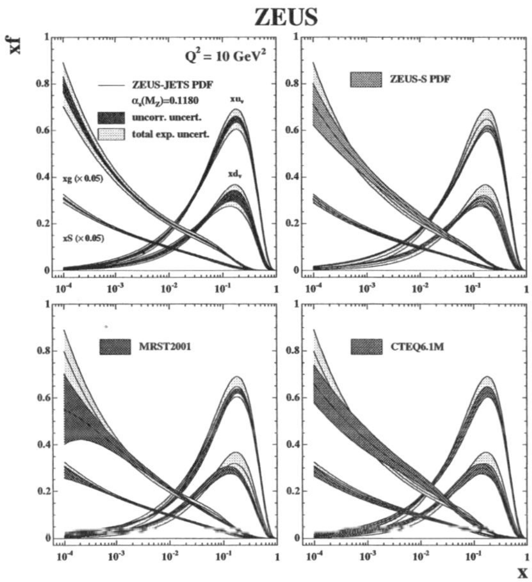

Figure 5. (a) PDFs extracted from the ZEUS-JETS fit. (b) PDFs extracted from the ZEUS-JETS fit compared to ZEUS-S PDFs. (c) PDFs extracted from the ZEUS-JETS fit compared to MRST2001 PDFs. (d) PDFs extracted from the ZEUS-JETS fit compared to CTEQ6.1M PDFs. The total experimental uncertainty bands are shown for each PDF set.

The use of ZEUS data alone eliminates the uncertainty from heavy-target corrections required in global analyses in which the νFe and μD

fixed-target data, together with isospin-symmetry constraints between u and d in the proton and neutron, have been used for determining the valence distributions. It also avoids the difficulties that can sometimes arise from combining data sets from several different experiments, thereby allowing a rigorous statistical treatment of the PDF uncertainties. In the ZEUS-JETS fit the lower Q^2 NC inclusive cross-section data determine the low-x sea and gluon distributions and the high-Q^2 NC and CC inclusive cross sections determine the valence distributions. The jet cross-section data constrain the mid- to high-x gluon density. The gluon and sea distributions are as well determined as the corresponding distributions in the global fits at low x since for all the fits these distributions are determined by the HERA inclusive NC data. The sea distributions at high x are constrained to be similar to those of the ZEUS-S global fit by the choice of the parametrisation. The uncertainties of the gluon have been reduced by the addition of the ZEUS jet data. In Fig.6 the uncertainty of the gluon distribution for fits with and without the jet data are compared.

The shapes of the PDFs are not changed significantly by the addition of jet data. The decrease in the uncertainty on the gluon distribution is significant. In the mid-x range, over the full Q^2 range a decrease in uncertainty by a factor of about two is found.

5. Prospects of HERA Measurements

PDF uncertainties from current global fits are, in general, limited by (irreducible) experimental systematics. In contrast, those from fits to HERA data alone, are largely limited by the statistical precision of existing measurements. Therefore, the impact of future data from HERA is likely to be most significant in fits to HERA data only. HERA II is now running efficiently and is expected to provide a substantial increase in luminosity. This will allow more precise measurements of cross sections that are currently statistically limited, in particular, the high-Q^2 NC and CC data and high-Q^2 and/or high-E_T jet data. In addition to the simple increase in luminosity, recent studies [29] have shown that future jet cross-section measurements, in kinematic regions optimized for sensitivity to PDFs, should have a significant impact on the gluon uncertainties. The effect on the PDF uncertainties, of both the higher precision expected from HERA II and the possibility of optimized jet cross-section measurements, has been estimated in a new QCD fit [29] ("HERA-II projection" fit). For the high-Q^2 inclusive data, a total integrated luminosity of 700 pb^{-1} was assumed, equally di-

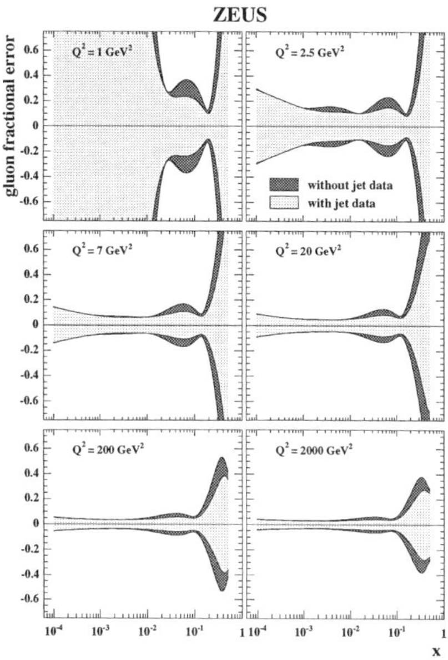

Figure 6. The total experimental uncertainty on the gluon PDF for the ZEUS-JETS fit (central error bands) compared to the total experimental uncertainty on the gluon PDF for a fit not including the jet data (outer error bands). The uncertainties are shown as fractional differences from the central values of the fits, for various values of Q^2. The total experimental uncertainty includes the statistical, uncorrelated and correlated systematic uncertainties and normalization, for both fits.

vided between e^+ and e^- . For the jet data, an integrated luminosity of 500 pb^{-1} was assumed. The central values and systematic uncertainties were taken from the published data in each case. In addition to the assumed increase in precision of the measurements, a set of optimized jet cross sections were also included, for forward dijets in γp collisions, as defined in a recent study [29], assuming luminosity of 500 pb^{-1}. The increased statistical precision of the high Q^2 data, as assumed in the HERA II projected fit, gives a significant improvement in the valence uncertainties over the whole range of x. For the sea quarks, a significant improvement in the uncertainties at high-x is also observed. In contrast, the low-x uncertainties are not visibly reduced due to the fact that the data constraining the low-x region tends to be at lower-Q^2, which are already systematically limited. This is also the reason why the low-x gluon uncertainties are not significantly reduced. However, the mid-to-high-x gluon, which is constrained by the jet data, is much improved in the HERA II projected fit. About half of the observed reduction in the gluon uncertainties is due to the inclusion of the simulated optimized jet cross sections.

The improvement to the high-x partons, observed in the HERA II projection will be particularly relevant for high-scale physics at the LHC. The prediction using the HERA II projected PDF shows a marked improvement at high jet transverse energy.

References

1. V.N. Gribov and L.N. Lipatov, *Sov. J. Nucl. Phys.* **15**, 438 (1972)
2. L.N. Lipatov, *Sov. J. Nucl. Phys.* **20**, 438 (1975)
3. G. Altarelli and G. Parisi, *Nucl. Phys.* **B20**, 298 (1977)
4. Yu.L. Dokshitzer, *Sov. Phys. JETP* **46**, 641 (1977)
5. ZEUS Collab., S. Chekanov et al., *Phys. Rev.* **D67**, 012007 (2003)
6. H1 Collab., C. Adloff et al., *Eur. Phys. J.* **C21**, 33 (2001)
7. H1 Collab., C. Adloff et al., *Eur. Phys. J.* **C30**, 1 (2003)
8. H1 Collab., C. Adloff et al., *Eur Phys J.* **C30**, 1 (2003)
9. ZEUS Collab., S. Chekanov et al., *Eur. Phys. J* **C28**, 175 (2003)
10. H1 Collab., C. Adloff et al., *Eur Phys J.* **C19**, 269 (2001)
11. ZEUS Collab., S. Chekanov et al., *Eur. Phys. J* **C32**, 16 (2003)
12. BCDMS Collab., A.C. Benvenuti et al., *Phys. Lett.* **B223**, 485 (1989)
13. L.W. Whitlow et al., *Phys. Lett.* **B282**, 475 (1992)
14. M. Osipenko et al., *Phys. Rev.* **D67**, 092001 (2003)
15. ZEUS Collab., S. Chekanov et al., 'Neutral current deep inelastic scattering measurements at high x', paper submitted to LP05 and EPS05 conferences
16. J. Pumplin et al., *JHEP* **0207**, 012 (2001)
17. DØ Collab., B. Abbott et al., *Phys. Rev. Lett.* **82**, 2451 (1999)
18. CDF Collab., T. Affolder et al., *Phys. Rev.* **D64**, 032001 (2001)
19. A.D. Martin et al., *Eur. Phys. J.* **C23**, 73 (2002)
20. A.D. Martin et al., *Eur. Phys. J.* **C28**, 455 (2002)
21. ZEUS Collab., S. Chekanov et al., *Eur. Phys. J.* **C42**, 1 (2005)
22. ZEUS Collab., S. Chekanov et al., *Phys. Lett.* **B547**, 164 (2002)
23. ZEUS Collab., S. Chekanov et al., *Eur. Phys. J.* **C23**, 615 (2002)
24. ZEUS Collab., S. Chekanov et al., *Eur. Phys. J.* **C21**, 443 (2001)
25. ZEUS Collab., S. Chekanov et al., *Eur. Phys. J.* **C12**, 411 (2000)
26. ZEUS Collab., S. Chekanov et al., *Phys. Lett.* **B539**, 197 (2002)
27. ZEUS Collab., S. Chekanov et al., *Phys. Rev.* **D70**, 052001 (2004)
28. A.M. Cooper-Sarkar, these proceedings
29. C. Gwenlan, A. Cooper-Sarkar, C. Targett-Adams, hep-ph/0509220

ELECTROWEAK PHYSICS AT HERA

J. MEYER

Deutsches Elektronen Synchrotron DESY
Notkestrasse 85, 22607 Hamburg, Germany
E-mail: joachim.meyer@desy.de

Results obtained at HERA concerning electroweak physics in deep inelastic electron-proton scattering are reviewed. At HERA I electroweak unification has been demonstrated. In a global fit to the measured proton structure functions QCD and electroweak parameters are determined simultaneously. The availability of polarised electrons at HERA II allows to determine the chirality properties of NC and CC interactions at high spacelike momentum transfers.

1. Results from the HERA I phase

1.1. *NC and CC cross sections*

The electron proton collider HERA operates since 1993 at center-of-mass energies above 300 GeV. Already the very first data accumulated in 1993 allowed to observe the charged current (CC) deep inelastic scattering (DIS) process at high spacelike momentum transfers Q^2 and established the W propagator effect [1]. Figure 1 shows the then measured CC cross section converted to an equivalent fixed target neutrino nucleon scattering cross section. The cross section clearly deviates from a linear rise indicating the finite mass in the spacelike W propagator.

In the HERA I phase (1993 - 2000) both experiments, H1 and ZEUS, each accumulated more than 100 pb^{-1} in e^+p and e^-p scattering. This amount of luminosity allowed to study electroweak effects extending into kinematic regions where the four momentum transfer Q exceeds the masses of the gauge bosons W and Z. Figure 2 shows the cross sections [2,3] differential in Q^2 for NC and CC processes in e^+p and e^-p scattering. Whereas the NC cross sections exceeds the CC cross section at low values of Q^2 by several orders of magnitude they become similar in the region where $Q^2 \sim M_W^2, M_Z^2$. This is the first direct demonstration of electroweak unification in spacelike DIS scattering. Figure 2 also shows the sensitivity of the cross sections to the lepton charge. In NC scattering the cross section difference arises from the Z exchange resulting in different signs of the $\gamma - Z$ interference term. The difference in CC scattering results from dependence on the quark type (u or d -type) and also the chirality dependence. All data sets agree well with the Standard Model (SM) predictions. Since it is evident that the exploration of the high Q^2 region is still statistics limited, the high luminosity phase at HERA II will allow to test this region with much higher precision.

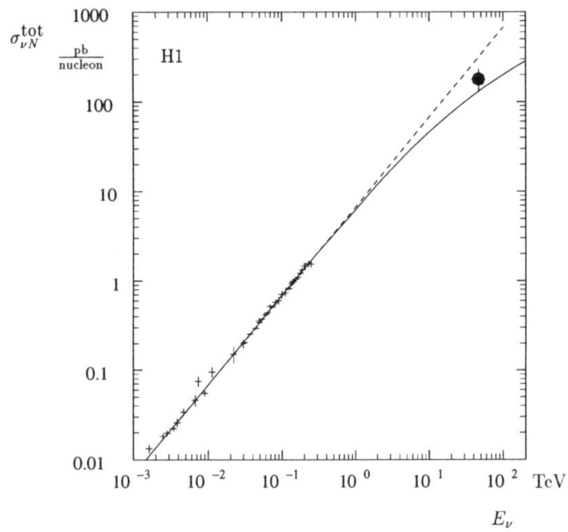

Figure 1. CC cross sections as function of fixed target lepton energy. Early 1993 HERA data and neutrino data are shown.

1.2. *Combined Electroweak-QCD Analysis*

Using the deep inelastic e^+p and e^-p charged and neutral current scattering cross sections derived from HERA I data, H1 performed a combined electroweak and QCD analysis [4] in order to determine electroweak parameters accounting for their correlation with parton distributions. The basis for this combined EW-QCD analysis is the H1 QCD analysis, the so-called H1PDF2000 [2]. In that analysis the cross sections were fitted describing the parton distributions by 10 free parameters while fixing the electroweak parameters to their SM values. In the new combined EW-QCD analysis in addition the electroweak parameters are fitted.

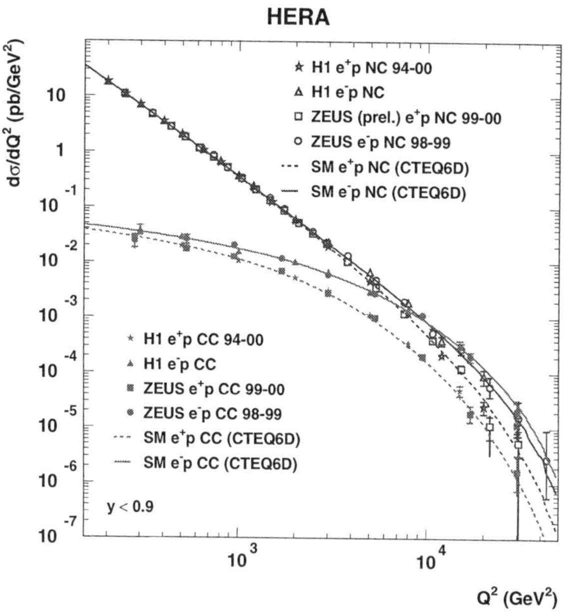

Figure 2. Differential cross section for NC and CC e^+p and e^-p scattering.

The additional free parameters depend on the choice of the electroweak renormalization scheme. The On-Mass-Shell scheme (OMS) uses the weak boson masses as fundamental parameters. In the Modified-On-Mass-Shell scheme (MOMS) the Fermi constant G_F is taken as fundamental parameter instead of M_W.

In the MOMS scheme the CC cross section reads

$$\frac{d^2\sigma^{CC}(e^\pm p)}{dxdQ^2} = \frac{G_F^2}{2\pi x}\left[\frac{M_{\text{prop}}^2}{M_{\text{prop}}^2 + Q^2}\right]^2 \phi_{CC}^{\pm}(x, Q^2) \tag{1}$$

with $\phi_{CC}^{\pm}(x, Q^2)$ describing the structure function dependence. Here G_F determines the normalization and the W mass entering the formula is the propagator mass M_{prop}. In the first fit performed in the MOMS scheme the parameters G_F and M_{prop} are considered as independent. The result on G_F and M_{prop} in this 12 parameter EW-QCD fit is shown in Figure 3(upper). When in a second fit the well measured Fermi constant is fixed to its PDG value the propagator mass is found to be

$$M_{\rm prop} = 82.87 \pm 1.82_{\rm exp} \, {}^{+0.30}_{-0.16}\Big|_{\rm model} \, {\rm GeV} \, . \tag{2}$$

Here the first error is experimental and the second corresponds to uncertainties due to input parameters and model assumptions. The propagator mass is within errors consistent with the W mass as measured at LEP and the Tevatron.

Within the OMS scheme the Fermi constant G_F is expressed by

$$G_F = \frac{\pi\alpha}{\sqrt{2} M_W^2 \left(1 - \frac{M_W^2}{M_Z^2}\right)} \frac{1}{1 - \Delta r} \, , \tag{3}$$

where $\alpha \equiv \alpha(Q^2 = 0)$ is the fine structure constant and the term Δr contains the electroweak radiative corrections, sensitive to the Higgs and top masses. In the OMS scheme the W mass is determined from the normalization of the CC cross section, in this case being however sensitive to the parameters (M_H, M_t) in the electroweak radiative corrections. Fixing $M_H = 120\,{\rm GeV}$ gives for M_t and M_W the result shown in Figure 3(lower). Fixing M_t to its world average value of 178 GeV leads to

$$M_W = 80.786 \pm 0.205_{\rm exp} \, {}^{+0.048}_{-0.029}\Big|_{\rm model} \pm 0.025_{\delta m_t} - 0.084_{\delta M_H} \pm 0.033_{\delta(\Delta r)} \, {\rm GeV} \, . \tag{4}$$

Here, in addition to the experimental and model uncertainties, three other error sources are considered, the uncertainty on the top quark mass $\delta m_t = 4.3\,{\rm GeV}$, a variation of the Higgs mass from 120 GeV to 300 GeV and the uncertainty of higher-order terms.

At HERA the NC interactions at high Q^2 receive contributions from γ and Z exchange. Thus the NC data can be used to extract the weak couplings of up- and down-type quarks to the Z boson. At high Q^2 and high x, where the NC $e^{\pm}p$ cross sections are sensitive to these couplings, the up- and down-type quark distributions are dominated by the light u and d quarks. Therefore, this measurement can be considered to determine the light quark couplings. The CC cross section data help disentangle the up and down quark distributions. In this analysis the vector and axial-vector couplings v_u-a_u-v_d-a_d of u and d quarks are treated as free parameters. The results of the EW-QCD fit are shown in Figure 4. The measured couplings are in good agreement with the SM expectations. The results are compared in Figure 4 with similar results obtained from the combined LEP data [5]. The HERA results resolve the sign ambiguity inherent in the LEP results.

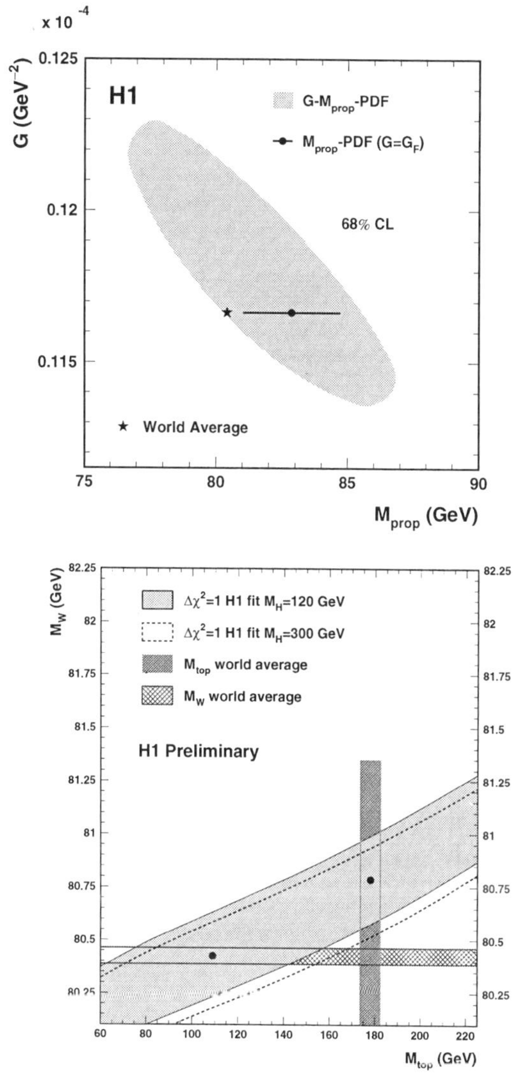

Figure 3. Upper figure : Result of the fit to G_F and $M_{\rm prop}$ (MOMS scheme). Lower figure : Result of the fit to M_t and M_W (OMS scheme).

2. Results from HERA II

The HERA accelerator and the experiments have undergone major modifications in the years 2000-2002 in order to achieve significantly higher

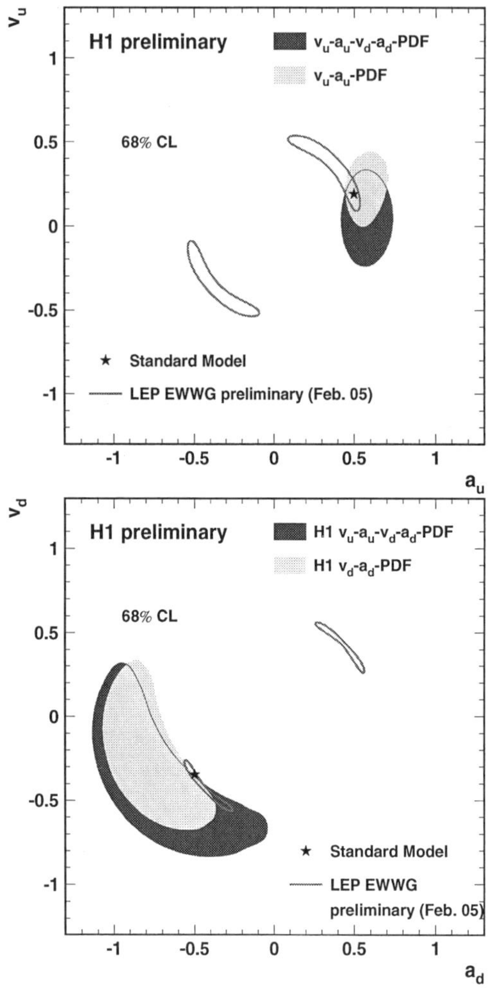

Figure 4. Results on the weak NC Z couplings to u (upper fig.) and d (lower fig.) quarks. The results from LEP are also shown.

luminosities. It is planned to accumulate around 700 pb^{-1} per experiment in the HERA II phase (up to 2007). This will significantly reduce the statistical errors in the high Q^2, region which is important for electroweak physics studies. Another important aspect for electroweak physics is the availability of longitudinally polarised leptons in the HERA II phase.

2.1. *Polarised CC cross sections*

The first physics analysis making use of the lepton polarisation is the measurement of the CC cross sections as a function of the degree of lepton polarisation. The double differential CC cross section for collisions of polarised leptons with unpolarised protons can be written as

$$\frac{d^2\sigma_{CC}^{e^\pm p}}{dxdQ^2} = [1 \pm P_e]\frac{G_F^2}{2\pi x}[\frac{M_W^2}{Q^2+M_W^2}]^2\Phi_{CC}^{\pm}, \tag{5}$$

Here, P_e is the degree of longitudinal polarisation defined as $P_e = (N_R - N_L)/(N_R + N_L)$ with $N_R(N_L)$ being the number of right (left) handed polarised leptons in the beam. The cross section has a linear dependence on the lepton polarisation and vanishes for lefthanded positrons and for righthanded electrons due to the V-A structure of the CC interaction.

Both collaborations used the data accumulated in 2003-2005 with polarised e^+ and e^- and combined these with the unpolarised data taken in the HERA I phase. The luminosities analysed up to now with polarised beams are relatively small (typically 10-30 pb^{-1} per charge and polarisation state) and the degree of polarisation is only 30-40 %. Nevertheless, both collaborations presented in their preliminary analyses for the 2005 summer conferences [6,7] clear evidence for the V-A structure of high Q^2 CC interactions.

Figure 5 shows the CC cross sections for e^+p and e^-p scattering in the region $Q^2 > 400\,\text{GeV}^2$ and $y < 0.9$ as function of the degree of longitudinal polarisation P. The data clearly exhibit the linear dependence with P and are compatible with vanishing cross sections for left(right)-handed positrons(electrons). Fitting a linear dependence to the e^+p data results in an extrapolated cross section of $\sigma_{CC}^{e^+}(P=-1) = 0.2 \pm 1.8(stat) \pm 1.6(syst)$ pb, consistent with the SM prediction of zero.

The ZEUS collaboration measured also polarisation dependences of differential CC cross sections [7]. Figure 6 shows these as function of the kinematic variables x, y and Q^2 for the case of e^-p scattering. As expected, all cross sections scale linearly with polarisation.

2.2. *Polarised NC cross sections*

The dependence of NC cross sections on the polarisation is more involved due to the contribution of γ and Z exchange in the amplitudes and the different chirality couplings. The bulk of the cross sections is dominated

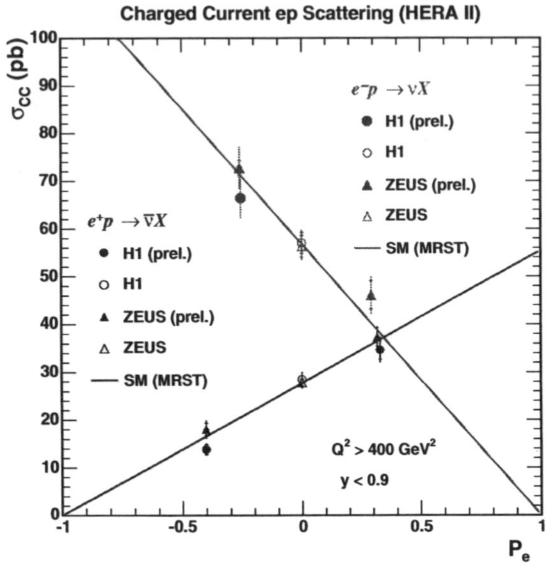

Figure 5. Total CC cross section for e^+p and e^-p scattering as function of the degree of longitudinal lepton polarisation P.

by γ exchange and thus independent of P and equal for e^+p and e^-p scattering. Only at highest Q^2 sensitivity to polarisation shows up due to the onset of the $\gamma - Z$ interference. Figure 7 shows the ZEUS results [7] for the dependence of the NC cross section on Q^2 for two polarisation states. The bottom part of the figure shows the cross section ratio for the two polarisations. Although the polarisation dependence is not observed conclusively, it is well compatible with the expected SM behaviour.

Summary

Many aspects of electroweak physics have been studied at HERA. In the HERA-I phase with unpolarised leptons luminosities $\sim 100\,\text{pb}^{-1}$ per experiment have been analysed. The measured NC and CC cross sections provide clear evidence for electroweak unification as predicted in the SM. In global fits to the structure functions electroweak parameters (M_{prop}, M_W and the vector and axial-vector couplings of the Z to up and down quarks) are determined. All results are in agreement with the SM predictions.

The newly available longitudinal lepton polarisation enhances the potential of electroweak measurements at HERA significantly. Already the

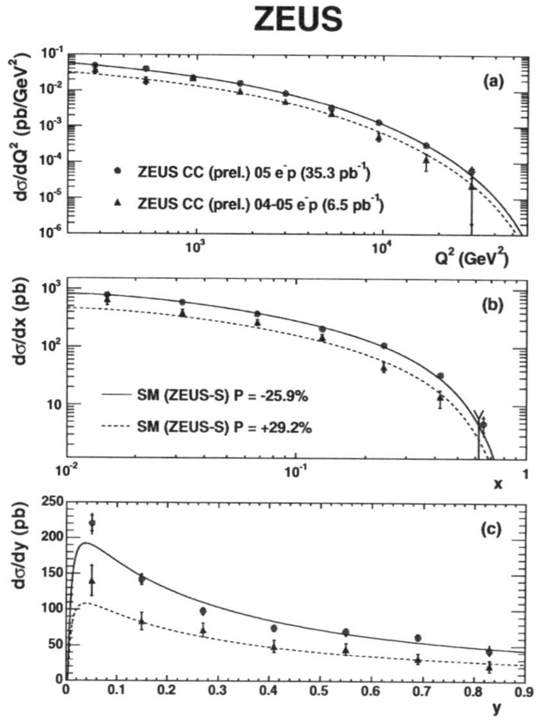

Figure 6. Differential CC e^-p cross sections for two polarisation states

very first analyses based on relatively low luminosity samples give encouraging results. The highlight here is the demonstration of the V-A structure of the charged current interaction at the very high spacelike momentum transfers never exploited before in deep inelastic scattering.

Since all results at high Q^2 are presently statistics limited the high luminosity expected to be accumulated until the end of HERA operation promises exciting physics results to come.

Acknowledgments

I like to thank the organisers of this workshop for the very stimulating atmosphere at Ringberg.

References

1. T. Ahmed et al. [H1 Collaboration], *Phys. Lett.* **B324**, 241 (1994).

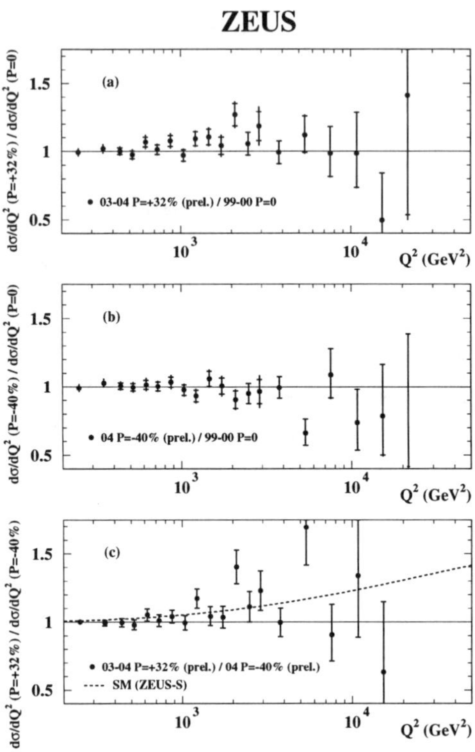

Figure 7. Differential NC e^+p cross sections for two polarisation states normalized to the unpolarised one a), b). In c) the cross section ratio for the two polarisation states is shown.

2. C. Adloff et al. [H1 Collaboration] *Eur.Phys.J.* **C30**, 1 (2003).
3. S. Chekanov et al. [ZEUS Collaboration] *Eur.Phys.J* **C32**, 1 (2003). S. Chekanov et al. [ZEUS Collaboration] *Phys.Rev.* **D70**, 052001 (2004).
4. A. Aktas et al. [H1 Collaboration] Accepted by *Phys. Lett.*, hep-ex/0507080 (2005).
5. LEP and SLD Electroweak Working Groups, http://lepewwg.web.cern.ch/LEPEWWG/plots/winter2005/.
6. H1 Collaboration, contributed paper LP2005 (2005), Uppsala, abstract 388. H1 Collaboration, contributed paper ICHEP04(2004), Beijing, abstracts 756 and 758.
7. ZEUS Collaboration, contributed papers LP2005 (2005), Uppsala, abstracts 254 and 255.

INCLUSIVE LOW Q^2 MEASUREMENTS AT HERA

V. LENDERMANN
Kirchhoff-Institut für Physik, Universität Heidelberg
Im Neuenheimer Feld 227, 69120 Heidelberg Germany
E-mail: victor@kip.uni-heidelberg.de

Inclusive *ep* scattering measurements at low virtualities of the exchanged boson, Q^2, allow precision tests of perturbative QCD at high gluon densities, as well as studies of the transition from the perturbative to non-perturbative QCD domain. Measurements in the transition region require special experimental approaches due to the limited detector acceptance. The current status and results of low Q^2 measurements at HERA are summarised.

1. Introduction

Inclusive measurements of *ep* scattering are the main source for our knowledge of proton structure. Over several decades they have played a decisive role in the development of Quantum Chromodynamics (QCD). So far the greatest kinematic coverage, over five magnitudes in the Bjorken scale variable x and in the modulus of the four-momentum transfer squared Q^2, is reached by the H1 and ZEUS experiments at HERA. Their inclusive DIS data [1,2,3,4,5] have shown that the Q^2 evolution of the proton structure function $F_2(x, Q^2)$ is well described by perturbative QCD (pQCD) in the range of $Q^2 \gtrsim 2 - 3\,\text{GeV}^2$. The data reach $2 - 3\%$ in precision for Q^2 values up to $\sim 100\,\text{GeV}^2$.

At $Q^2 \lesssim 2 - 3\,\text{GeV}^2$ the transition takes place into a domain in which non-perturbative effects dominate and the assumption of asymptotic freedom is no longer valid. A proper treatment of the transition from the soft to the hard QCD regime can thus improve our understanding of quark confinement. The description of the transition region remains a challenge for the theory and a field for phenomenological models.

The recent measurements performed at $Q^2 \lesssim 2 - 3\,\text{GeV}^2$ are presented in the following section. Afterwards, studies of the F_2 behaviour and comparisons to models are discussed. In the last section, extractions of the longitudinal structure function F_L are described.

2. Measurements in the Transition Region

Acceptance of the main H1 and ZEUS detectors is limited to $Q^2 \gtrsim 2\,\mathrm{GeV}^2$, therefore special experimental techniques are necessary to access the transition region. One way is to use special low Q^2 devices [7] mounted close to the outgoing lepton beam directon. However, the region $0.8 \lesssim Q^2 \lesssim 2\,\mathrm{GeV}^2$ is not reached via these devices since the respective angular range of the scattered lepton is complicated by the instrumentation of the main calorimeters.

This region is covered by data collected in special runs with the interaction vertex shifted in the direction of the proton beam [8,9,10]. In such an experimental configuration the scattered lepton is detected at higher polar angles[a] in the main detector, thus gathering events at lower Q^2 values.

In Fig. 1 the reduced cross section

$$\sigma_r = \frac{Q^4 x}{2\pi\alpha^2}\frac{d^2\sigma}{dxdQ^2} = F_2(x,Q^2) - \frac{y^2}{Y_+}F_L(x,Q^2)\ , \qquad (1)$$

with the inelasticity $y = Q^2/(xs)$ and $Y_+ = 1+(1-y)^2$, is shown. The ZEUS Beam Pipe Tracker (BPT) measurements [7], the preliminary H1 results of running in 1999 with the standard vertex position and in 2000 with the vertex shifted by 70 cm [10] are shown together with the fixed target data from NMC [11]. The HERA inclusive data in the transition domain reach a precision of $3-4\%$. The predictions of an extrapolated fractal model fit [12] and the ALLM97 parametrisation [13] are also displayed. All predictions are in good agreement with the data.

The low Q^2 measurements are further extended towards higher x values making use of events with hard photon radiation. The cross section for radiative processes becomes sufficiently large for distinct experimental configurations, in which the photon is emitted either nearly collinear with the electron beam (Initial State Radiation, ISR) or nearly collinear with the scattered lepton (Final State Radiation, FSR), or both the lepton and the photon are detected under finite polar angles nearly back-to-back in azimuth (QED Compton process, QEDC). Two of these topologies, ISR and QEDC, are used for measurements, as discussed below.

2.1. *Cross Section Measurement Using ISR Data*

The new H1 ISR analysis uses the shifted vertex data collected in 2000. Contrary to the previous HERA measurements [14,15,16], the emitted photon

[a]The polar angle is measured w.r.t. the proton beam direction. A higher polar angle means a lower scattering angle for the outgoing lepton.

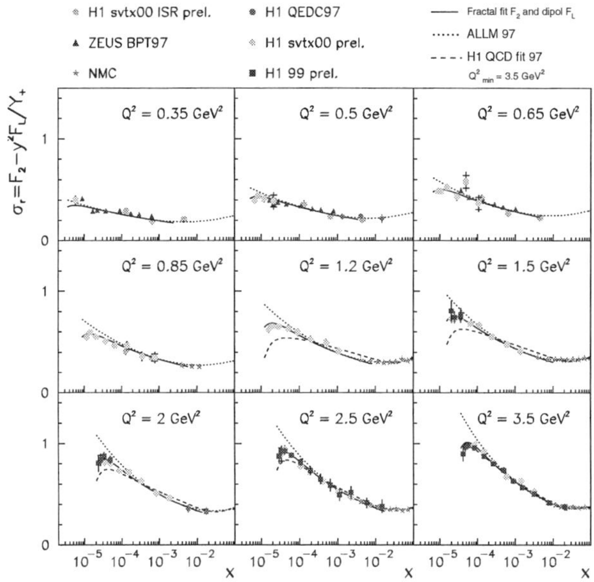

Figure 1. Reduced cross section measurements at $Q^2 \lesssim 3.5\,\text{GeV}^2$ by H1 (preliminary), ZEUS and NMC compared with the ALLM97 parametrisation and with a calculation based on the fractal fit to F_2 and F_L from the dipole model.

is not tagged explicitly in the new H1 ISR analysis [17]. Instead, its energy is inferred from the longitudinal momentum imbalance:

$$2E_\gamma = 2E_e^0 - (E - p_z)_e - (E - p_z)_h \ , \tag{2}$$

where E_e^0 is the electron beam energy, $(E - p_z)_e$ is the measured difference between the energy and the longitudinal momentum of the scattered electron and $(E - p_z)_h$ is the same quantity for the hadronic final state. As in the other recent H1 low Q^2 measurements, the Backward Silicon Tracker (BST) is used to identify the scattered electron and to reduce the contamination by neutral particle backgrounds.

The new data cover the region $0.35 \lesssim Q^2 \lesssim 0.85\,\text{GeV}^2$ and $10^{-4} \lesssim x \lesssim 5 \cdot 10^{-3}$. They are in a good agreement with the other measurements, as shown in Fig. 1.

2.2. F_2 *Measurement Using QED Compton Scattering*

The experimental signature of QEDC events is an approximately back-to-back azimuthal configuration of the outgoing electron and photon, both

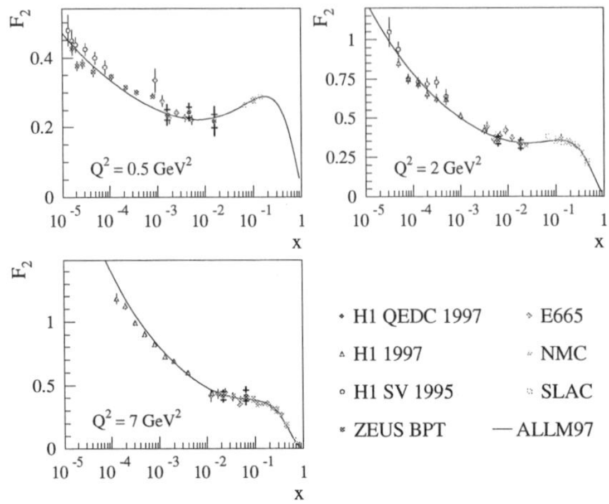

Figure 2.
F_2 measurements from QED Compton scattering by H1 (closed circles), compared with other measurements at HERA and fixed target experiments. The solid line depicts the ALLM97 parametrisation.

found in the main detector. In this configuration their transverse momenta nearly balance such that it is possible to reconstruct the event kinematics for very small values of Q^2.

The first F_2 measurement in inelastic QEDC scattering is performed by H1 [18] using data from the 1997 running period. A prominent background to inelastic QEDC scattering arises from inclusive DIS events in which one particle from the hadronic final state (typically a π^0) fakes the outgoing photon. At high y, where the hadronic final state lies mostly in the backward region, this process dominates the QEDC signature, hampering a clean QEDC event selection. For this reason the measurement is restricted to relatively low y values: $y \lesssim 0.006$.

At so low y the variables x and y cannot be determined solely from the measured electron and photon four-momenta, since their resolution deteriorates as $1/y$. Hence for the kinematic reconstruction the Σ-method is employed, which also uses information from the hadronic final state. As low y values correspond to small polar angles of the final state hadrons, one of the main challenges for the analysis is the reconstruction of the total momentum of the hadronic final state correcting for losses beyond the forward acceptance of the detector. This necessitates a good simulation of hadronisation processes at low Q^2 and low invariant masses of the hadronic final state W. The simulation of the hadronic final state down to the resonance region is performed using the SOPHIA program [19].

The F_2 values measured in QED Compton scattering are shown in Fig. 2 as a function of x at fixed Q^2 and are compared with other HERA [1,7,8] and fixed target [11,20,21] data. The QEDC analysis extends the kinematic range

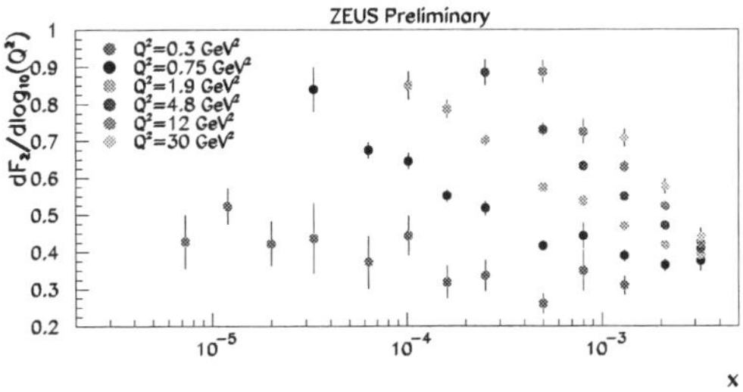

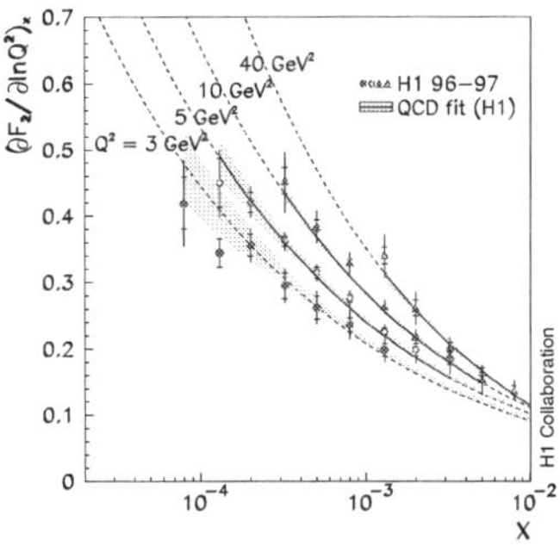

Figure 3. Derivatives $\partial F_2/\partial \ln Q^2|_x$ as functions of x for different Q^2.

of HERA at low Q^2 towards higher x values, thus complementing standard inclusive and shifted vertex measurements. The measurement is consistent with the results of fixed target experiments in the overlapping region. The data are well described by the ALLM97 parametrisation [13].

3. Interpretation of the Data

A principal feature of the HERA data is the dramatic rise of F_2 at low x driven by the gluon evolution. This rise questions the validity of the DGLAP [6] approach in the region of high parton densities, on which the current pQCD fits to the data are based. While in the DGLAP formalism only the $\ln Q^2$ terms are summed, the subleading terms involving powers of $\alpha_s \ln(1/x)$ may become large as x decreases. This may require a different summation scheme, such as BFKL [22] or CCFM [23], or non-linear corrections to the pQCD expansion [24]. The non-linear effects may lead to gluon-gluon absorption which would tame the growth of F_2 at low x.

High precision F_2 data at very low x are necessary in order to search for deviations from the DGLAP evolution and signs of saturation. As low x can only be reached at low Q^2 due to kinematical correlation, it is the low Q^2 data which are used for these studies.

An important quantity for searches of a new gluon dynamics is the derivative $\partial F_2(x,Q^2)/\partial \ln Q^2|_x$ describing scaling violations. The local derivatives measured by H1 [1] and ZEUS are shown in Fig. 3 as functions of x for different Q^2. They exhibit a continuous rise towards low x down to the lowest Q^2 without an indication of a change in the dynamics. The derivatives are well described by the pQCD calculations for $Q^2 \gtrsim 3\,\mathrm{GeV}^2$.

In the double asymptotic limit, the DGLAP evolution equation can be solved analytically and F_2 is expected to rise approximately as a power of x towards low x. A power behaviour is also predicted in BFKL. A

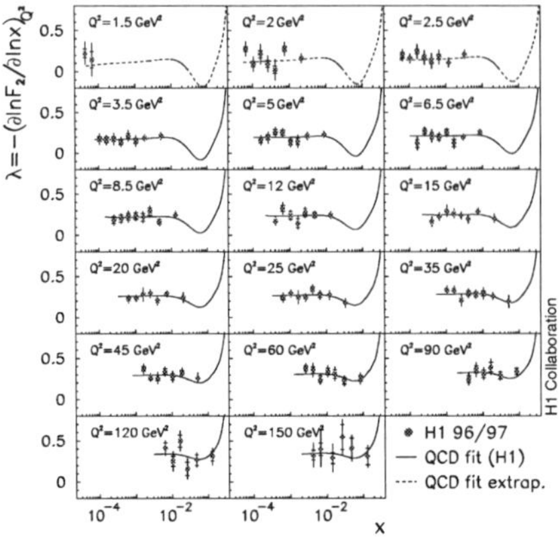

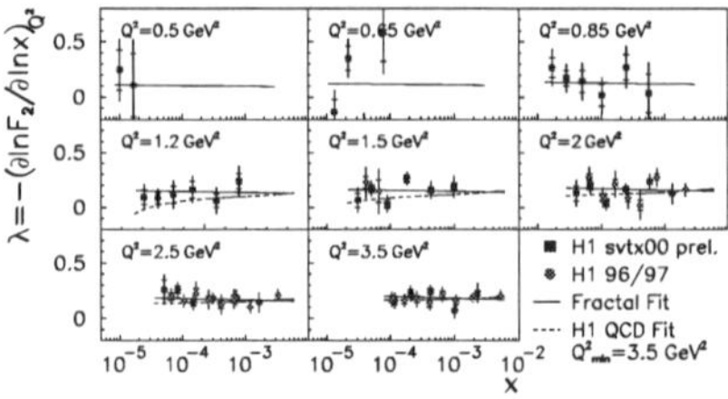

Figure 4. Derivatives $\partial F_2/\partial \ln x|_{Q^2}$ as functions of x for different Q^2.

damping of this rise would indicate the presence of novel QCD effects. A relevant observable for the investigation of the dynamics of this growth is the derivative $\lambda = -\partial F_2(x, Q^2)/\partial \ln x|_{Q^2}$.

The high precision of the present F_2 data allowed H1 to measure this quantity locally [10,25], as shown in Fig. 4. The measurements are consistent with no dependence of λ on x for $x < 0.01$. The monotonic rise of F_2 persists down to the lowest x measured at HERA, and no evidence for a change of this behaviour is found.

This suggests that F_2 can be parameterised by $F_2 = c(Q^2) \cdot x^{-\lambda(Q^2)}$. The results, obtained by fitting the present data at fixed Q^2 values are shown in Fig. 5. The left plot presents $\lambda(Q^2)$ values obtained separately from the H1 and ZEUS data. The extension of the x range of the H1 shifted vertex data achieved by including the ISR data allowed an improved extraction of λ. The highest precision can probably be reached by combining H1, ZEUS and fixed target data, as shown in the right plot.

The coefficient $c(Q^2) \approx 0.18$ and the logarithmic dependence of λ on Q^2 for $Q^2 \gtrsim 2 - 3\,\mathrm{GeV}^2$ are in accord with pQCD predictions. In contrast, at lower Q^2 the behaviour is changing to a weaker dependence compatible with reaching, as $Q^2 \to 0$, a constant consistent with the soft pomeron intercept $\alpha_{I\!P} - 1 = 0.08$, which is expected from the energy dependence of soft hadronic interactions [28]. The change takes place at distance scales of $\sim 0.3\,\mathrm{fm}$ and can be interpreted as being related to a transition from partonic to hadronic degrees of freedom. This change is the major challenge for the theory which must interpolate between the two x dependences.

For many approaches, colour dipoles appear to provide useful degrees of freedom in order to treat low x inclusive and diffractive data [29]. Among

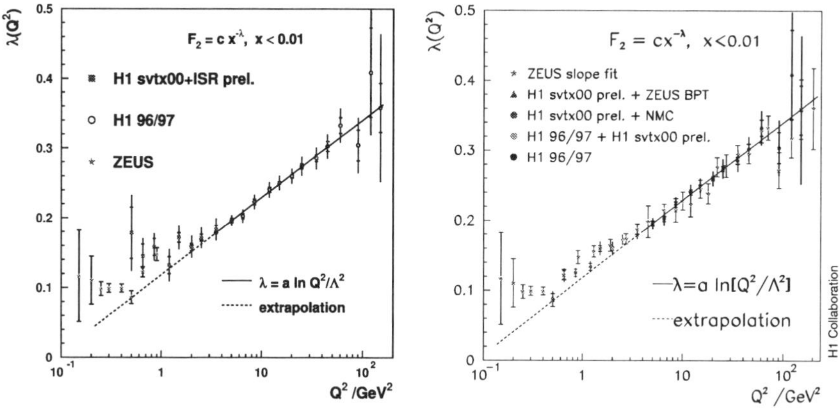

Figure 5. Selected HERA results of the λ extraction from low x data.

several dipole models, the saturation model[30] has been a subject of much debate. Using just a few parameters, the model successfully describes the transition region at low x via the saturation of the colour dipole–proton cross section at large transverse sizes of the dipole. The extended model[31] includes the gluon DGLAP evolution in order to describe the scaling violations in the perturbative region (see Fig. 6 left).

The saturation model predicts a geometric scaling of the γ^*p cross section in the small x region[32]. This means that the cross section σ_{γ^*p} should be a function of only one dimensionless variable $\tau = Q^2R_0^2(x)$, where the "saturation radius" $R_0(x)$ decreases with decreasing x. The HERA and fixed target data at $x < 0.01$ exhibit this behaviour, as shown in Fig. 6 right. This scaling is a manifestation of the presence of an internal saturation scale, $Q_s(x) \sim 1/R_0(x)$, characterising dense partonic systems.

However, despite the success of the saturation model and its appeal from a theoretical point of view, there is no compelling evidence for the saturation at low x yet, as other models are also able to describe the data with a similar presicion.

4. Extraction of F_L

The proton structure function F_L describes the exchange of longitudinally polarised photons. It imposes a constraint on the otherwise highly uncertain behaviour of the gluon distribution function in the proton at low Q^2. Though the gluon density is obtained in pQCD analyses of DIS data via the derivative $\partial F_2/\partial Q^2$, its determination at low Q^2 and low x suffers from

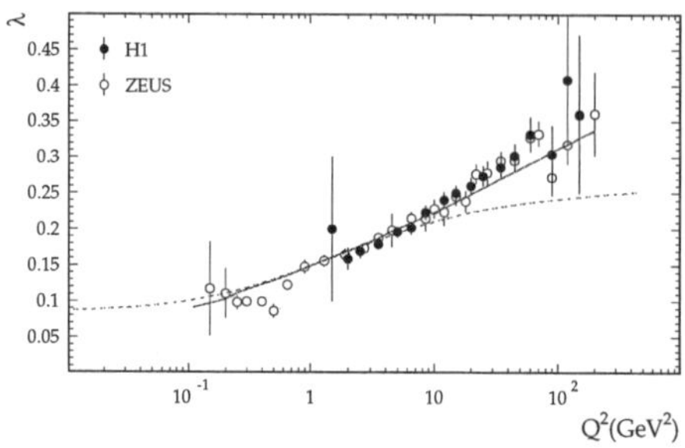

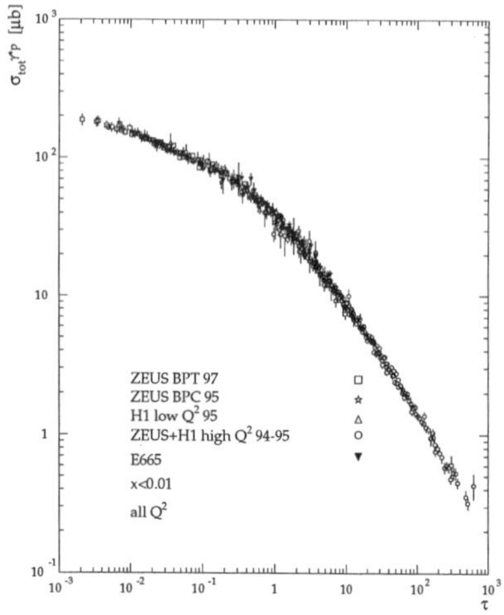

Figure 6. Left: the original (dotted line) and the extended (solid line) saturation model are shown against λ values extracted from the low x HERA data. Right: data on $\sigma_{\gamma^* p}$ from the region $x < 0.01$ as a function of the scaling variable $\tau = Q^2 R_0^2(x)$.

non-perturbative effects becoming significant.

A direct determination of F_L needs cross section values measured at different y values for the same x and Q^2. This can be achieved at lower ep centre-of-mass energies, e.g. by performing dedicated runs at lower proton beam energies. The possibility of taking such runs at the end of the HERAII running period is currently being discussed [33].

Alternatively, events with tagged Initial State Radiation can be employed, as presented by ZEUS [16]. The ISR process is interpreted as the inclusive ep scattering at a reduced electron beam energy. The emitted photon is detected in the luminosity monitor. The result of the ZEUS analysis of the data taken during 1996 and 1997 running periods are shown in Fig. 7. Much more statistics is required in order to achieve a sufficient accuracy of the F_L measurement in ISR.

A higher accuracy is achieved using indirect methods of F_L extraction, which are based on the analysis of the reduced cross section behaviour at high y values. The data of the minimum bias 1999 [26] and shifted vertex 2000 [10] runs are used by H1 [27] to extract F_L by two methods: the "derivative method" and the new "shape method".

In the derivative method, F_L is extracted from the partial derivative of the cross section on y at fixed Q^2

$$\left.\frac{\partial \sigma_r}{\partial \ln y}\right|_{Q^2} \approx \left.\frac{\partial F_2}{\partial \ln y}\right|_{Q^2} - \frac{2y^2(2-y)}{Y_+^2} F_L \tag{3}$$

which is dominated by the F_L-dependent term at high y.

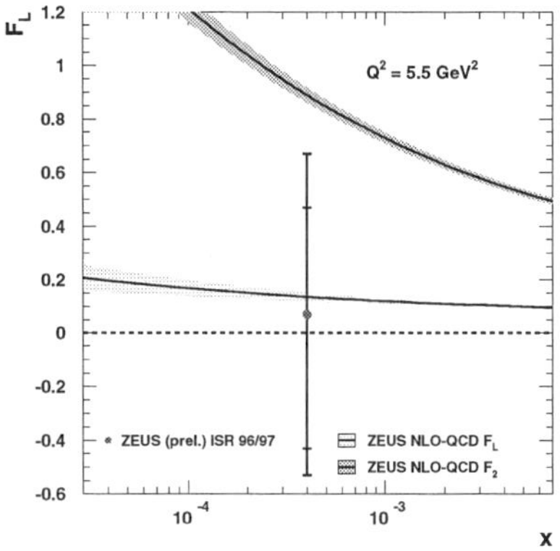

Figure 7. ZEUS ISR result for F_L plotted for $x = 4 \cdot 10^{-4}$ and $Q^2 = 5.5\,\text{GeV}^2$. The lighter (yellow) band shows the prediction for F_L from the ZEUS NLO QCD fit [3]. The darker (blue) band shows the prediction for F_2, which is the maximum possible value of F_L.

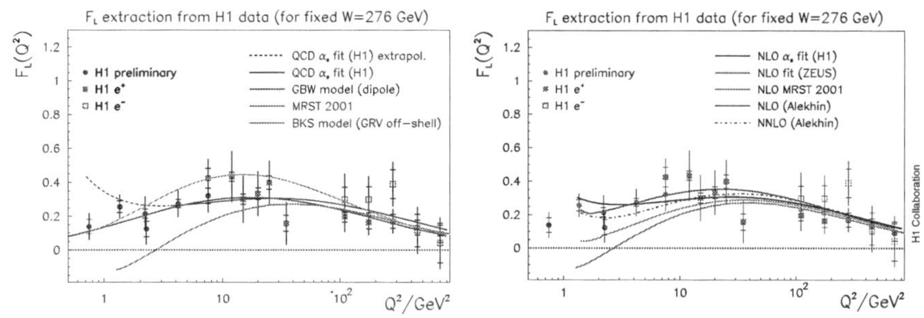

Figure 8. Q^2 dependence of $F_L(x, Q^2)$ at fixed $y = 0.75$, extracted from the H1 data. The lines represent various phenomenological models, as well as pQCD fits.

The shape method exploits the shape of σ_r in a given Q^2 bin. The shape is driven at high y by the kinematic factor y^2/Y_+ (eq. 1), and to a lesser extent by $F_L(x, Q^2)$ which is considered to be constant: $F_L = F_L(Q^2)$. Based on the analysis of the rise of F_2 towards low x, the reduced cross section is fitted by

$$\sigma_{r,\text{fit}} = cx^{-\lambda} - \frac{y^2}{Y_+} F_L \,, \tag{4}$$

and F_L is determined from the fit for different Q^2 bins. The errors obtained turn out to be significantly smaller than those from the derivative method. However, the x-dependence cannot be extracted.

The results for a fixed $y = 0.75$ ($W = 276\,\text{GeV}$) are presented in Fig. 8, in which an overview of all current H1 data in the Q^2 range $0.75 \leq Q^2 \leq 700\,\text{GeV}^2$ is given. The data are compared with higher order QCD fits. The H1 data favour a positive, not small F_L at low Q^2.

References

1. C. Adloff *et al.* [H1 Collaboration], *Eur. Phys. J.* C **21**, 33 (2001).
2. S. Chekanov *et al.* [ZEUS Collaboration], *Eur. Phys. J.* C **21**, 443 (2001).
3. S. Chekanov *et al.* [ZEUS Collaboration], *Phys. Rev.* D **67**, 012007 (2003).
4. C. Adloff *et al.* [H1 Collaboration], *Eur. Phys. J.* C **30**, 1 (2003).
5. S. Chekanov *et al.* [ZEUS Collaboration], *Phys. Rev.* D **70**, 052001 (2004).
6. V. N. Gribov and L. N. Lipatov, *Sov. J. Nucl. Phys.* **15**, 438 and 675 (1972);
   L. N. Lipatov, *Sov. J. Nucl. Phys.* **20**, 94 (1975);
   Y. L. Dokshitzer, *Sov. JETP* **46**, 641 (1977);
   G. Altarelli and G. Parisi, *Nucl. Phys.* B **126**, 298 (1977).
7. J. Breitweg *et al.* [ZEUS Collaboration], *Phys. Lett.* B **487**, 53 (2000).
8. C. Adloff *et al.* [H1 Collaboration], *Nucl. Phys.* B **497**, 3 (1997).
9. M. Derrick *et al.* [ZEUS Collaboration], *Z. Phys.* C **69**, 607 (1996).
10. H1 Collaboration, Contributed paper to EPS 2003, Aachen, Abstract **082**.
11. M. Arneodo *et al.* [New Muon Collaboration], *Nucl. Phys.* B **483**, 3 (1997).
12. T. Laštovička, *Eur. Phys. J.* C **24**, 529 (2002).
13. H. Abramowicz and A. Levy, DESY–97–251, hep–ph/9712415.
14. T. Ahmed *et al.* [H1 Collaboration], *Z. Phys.* C **66**, 529 (1995).
15. M. Derrick *et al.* [ZEUS Collaboration], *Z. Phys.* C **69**, 607 (1996).
16. ZEUS Collaboration, Contributed paper to EPS 2003, Aachen, Abstract **502**.
17. H1 Collaboration, Contrib. paper to ICHEP 2004, Beijing, Abstract **5-0170**.
18. A. Aktas *et al.* [H1 Collaboration], *Phys. Lett.* B **598**, 159 (2004).
19. A. Mücke *et al.*, *Comput. Phys. Commun.* **124**, 290 (2000).
20. L. W. Whitlow *et al.*, *Phys. Lett.* B **282**, 475 (1992).
21. M. R. Adams *et al.* [E665 Collaboration], *Phys. Rev.* D **54**, 3006 (1996).
22. E. A. Kuraev, L. N. Lipatov and V. S. Fadin, *Sov. JETP* **44**, 443 (1976);
    E. A. Kuraev, L. N. Lipatov and V. S. Fadin, *Sov. JETP* **45**, 199 (1977);
    I. I. Balitsky and L. N. Lipatov, *Sov. J. Nucl. Phys.* **28**, 822 (1978).
23. M. Ciafaloni, *Nucl. Phys.* B **296**, 49 (1988);
    S. Catani, F. Fiorani and G. Marchesini, *Phys. Lett.* B **234**, 339 (1990);
    S. Catani, F. Fiorani and G. Marchesini, *Nucl. Phys.* B **336**, 18 (1990);
    G. Marchesini, *Nucl. Phys.* B **445**, 49 (1995).
24. L. V. Gribov, E. M. Levin and G. M. Ryskin, *Phys. Rept.* **100**, 1 (1983);
    A. H. Mueller and J. Qiu, *Nucl. Phys.* B **268**, 427 (1986).
25. C. Adloff *et al.* [H1 Collaboration], *Phys. Lett.* B **520**, 183 (2001).
26. H1 Collaboration, Contributed paper to EPS 2001, Budapest, Abstract **799**.
27. H1 Collaboration, Contrib. paper to ICHEP 2004, Beijing, Abstract **5-0161**.
28. A. Donnachie and P. V. Landshoff, *Phys. Lett.* B **296**, 227 (1992);
    A. Donnachie and P. V. Landshoff, *Z. Phys.* C **61**, 139 (1994).
29. G. Shaw, these proceedings.
30. K. Golec-Biernat and M. Wüsthoff, *Phys. Rev.* D **59**, 014017 (1999);
    K. Golec-Biernat and M. Wüsthoff, *Phys. Rev.* D **60**, 114023 (1999).
31. J. Bartels, K. Golec-Biernat, H. Kowalski, *Phys. Rev.* D **66**, 014001 (2002).
32. A. M. Stasto, K. Golec-Biernat, J. Kwiecinski, *Phys. Rev. Lett.* **86**, 596 (2001).
33. J. Feltesse, these proceedings; R. Thorne, these proceedings.

RESUMMED PERTURBATIVE EVOLUTION AT HIGH ENERGY

R. D. BALL

School of Physics, University of Edinburgh
Edinburgh EH9 3JZ, Scotland

G. ALTARELLI

CERN, Department of Physics, Theory Division
CH-1211 Genève 23, Switzerland
and
Dipartimento di Fisica "E.Amaldi", Università Roma Tre
INFN, Sezione di Roma Tre
via della Vasca Navale 84, I-00146 Roma, Italy

S. FORTE

Dipartimento di Fisica, Università di Milano
INFN, Sezione di Milano, Via Celoria 16, I-20133 Milan, Italy

We present a small x resummation for the GLAP anomalous dimension and its corresponding dual BFKL kernel, which includes all the available perturbative information and nonperturbative constraints.

An accurate perturbative determination of the hard partonic cross-sections and of the anomalous dimensions which govern parton evolution is necessary for the precise extraction of parton densities. As is well known, such high-order perturbative calculations display classes of terms containing large logarithms, which ultimately signal the breakdown of perturbation theory. Because these terms are scale–dependent and in general non universal, lack of their inclusion can lead to significant distortion of the parton densities in some kinematical regions, thereby leading to loss of accuracy if parton distributions extracted from deep-inelastic scattering (DIS) or the Drell-Yan (DY) processes are used at the LHC.

Logarithmic enhancement of higher order perturbative contribution may take place when more than one large scale ratio is present. In DIS and DY this happens in the two opposite limits when the centre-of-mass energy of the partonic collision is close to the threshold for the production of the final state, or much higher than the characteristic scale of the process. These correspond respectively to the small x and large x kinematical regions, where $0 \leq x \leq 1$ is defined in terms of the invariant mass of the non-leptonic final state $\frac{(1-x)Q^2}{x}$. The corresponding perturbative contributions are respectively enhanced by powers of $\ln \frac{1}{x}$ and $\ln(1-x)$, or, equivalently, in the space of Mellin moments, by powers of $\frac{1}{N}$ and $\ln N$, where $N \to 0$ moments dominate as $x \to 0$ while $N \to \infty$ moments dominate as $x \to 1$. Here we will be concerned only with the small x (and thus small N) region.

In the DGLAP evolution equation one resums collinear logarithms first, resulting at LO and NLO in a good description of many data sets, and in particular the HERA F_2 data at perturbative Q^2 and values of x as low as 10^{-4} [1–3]. However in the singlet (gluonic) channel the fixed order splitting functions contain small x logarithms also, which at NNLO [4] begin to destabilise the perturbative expansion, so that a further resummation is needed if the evolution is to be reliable at small x. Small x logarithms may be resummed by using the BFKL equation. However the fixed order kernels of this equation, currently known to NLO [5] are perturbatively unstable for all realistic values of α_s, and are thus by themselves useless for reliable calculations [6–8]. This is because they contain collinear logarithms, which must be resummed even at LO if reliable predictions are to be obtained [9]. This collinear resummation of the BFKL kernel restores longitudinal momentum conservation [10], and leads ultimately to a stable expansion.

Two approaches to the simultaneous resummation of these two classes of logs have recently reached the stage where their phenomenological application can be envisaged. In the duality (ABF) approach [10–16] one concentrates on the problem of obtaining an improved anomalous dimension (splitting function) for DIS which reduces to the ordinary perturbative result at large N (large x), while including resummed BFKL corrections at small N (small x), determined through the BFKL kernel. The CCSS approach [17,18,20,21] is built up within the BFKL framework, by improving the whole hierarchy of subleading kernels in the collinear region consistently with the renormalization group. The BFKL quation is then solved and a perturbative splitting function extracted numerically.

Here we will briefly review the theoretical underpinnings of the duality approach, and then compare phenomenological results obtained in both approaches.

1. High Energy Duality

In the ABF approach one constructs an improved anomalous dimension (splitting function) for DIS which reduces to the ordinary perturbative result at large N (large x) given by:

$$\gamma(N,\alpha_s) = \alpha_s\gamma_0(N) \ + \ \alpha_s^2\gamma_1(N) \ + \ \alpha_s^3\gamma_2(N) \ \ \ldots. \tag{1}$$

while including resummed BFKL corrections at small N (small x) which are determined by the aforementioned BFKL kernel $\chi(M,\alpha_s)$:

$$\chi(M,\alpha_s) = \alpha_s\chi_0(M) \ + \ \alpha_s^2\chi_1(M) \ + \ \ldots, \tag{2}$$

which is the Mellin transform of the angular averaged kernel K with respect to $t = \ln\frac{k^2}{k_0^2}$. The main theoretical tool which enables this construction is the duality relation between the kernels χ and γ

$$\chi(\gamma(N,\alpha_s),\alpha_s) = N, \tag{3}$$

which is a consequence of the fact that the solutions of the BFKL and DGLAP equations coincide at leading twist [8, 10, 19]. Further improvements are obtained exploiting the symmetry under gluon interchange of the BFKL gluon-gluon kernel and through the inclusion of running coupling effects.

By using duality, one can construct a more balanced expansion for both γ and χ, the "double leading" (DL) expansion, where the information from χ is used to include in γ all powers of α_s/N and, conversely, γ is used to improve χ by all powers of α_s/M. A great advantage of the DL expansion is that it resums the collinear poles of χ at $M = 0$, enabling the imposition of the physical requirement of momentum conservation $\gamma(1,\alpha_s) = 0$, whence, by duality:

$$\chi(0,\alpha_s) = 1. \tag{4}$$

This procedure eliminates in a model independent way the alternating sign poles $+1/M, -1/M^2, \ldots$.. that appear in $\chi_0, \chi_1, \ldots$. These poles make the perturbative expansion of χ unreliable even in the central region of M: e.g., $\alpha_s\chi_0$ has a minimum at $M = 1/2$, while, at realistic values of α_s, $\alpha_s\chi_0 + \alpha_s^2\chi_1$ has a maximum.

At this stage, while the poles at $M = 0$ are eliminated, those at $M = 1$ remain, so that the DL expansion is still not finite near $M = 1$. The resummation of the $M = 1$ poles can be accomplished by exploiting the collinear-anticollinear symmetry, as in the CCSS approach [17, 18, 21]. In Mellin space, this symmetry implies that at the fixed-coupling level the kernel χ for evolution in $\ln \frac{s}{kk_0}$ must satisfy $\chi(M) = \chi(1-M)$ order by order in perturbation theory. This symmetry is however broken by the DIS choice of variables $\ln \frac{1}{x} = \ln \frac{s}{Q^2}$ and by the running of the coupling. In the fixed coupling limit the kernel χ_{DIS}, dual to the DIS anomalous dimension, is related to the symmetric one χ_σ through the implicit equation [5]

$$\chi_{\text{DIS}}(M + 1/2\chi_\sigma(M)) = \chi_\sigma(M). \tag{5}$$

Hence, the $M = 1$ poles can be resummed by performing the double-leading resummation of $M = 0$ poles of χ_{DIS}, determining the associated χ_σ through eq. (5), then symmetrizing it, and finally going back to DIS variables by using eq. (5) again in reverse. Using the momentum conservation eq. (4) and eq. (5), it is easy to show that $\chi_\sigma(M)$ is an entire function of M, with $\chi_\sigma(-1/2) = \chi_\sigma(3/2) = 1$, and thus necessarily has a minimum at $M = 1/2$. Through this procedure one obtains order by order from the DL expansion a symmetrized DL kernel χ_{DIS}, and its corresponding dual anomalous dimension γ. The kernel χ_{DIS} has to all orders a minimum and satisfies a momentum conservation constraint $\chi_{\text{DIS}}(0) = \chi_{\text{DIS}}(2) = 1$.

The final ingredient of the ABF approach is a treatment of the running coupling corrections to the resummed terms. Indeed, their inclusion in the resummed anomalous dimension greatly softens the asymptotic behavior near $x = 0$. Hence, the dramatic rise of structure functions at small x, which characterized resummations based on leading–order BFKL evolution and is ruled out phenomenologically, is replaced by a much milder rise. This requires a running coupling generalization of the duality equation (3), which is possible noting that in M space the running coupling $\alpha_s(t)$ becomes a differential operator, since $t \to d/dM$. Hence, the BFKL evolution equation for double moments $G(N, M)$, which is an algebraic equation at fixed coupling, becomes a differential equation in M for running coupling. In the ABF approach, one solves this differential equation analytically when the kernel is replaced by its quadratic approximation near the minimum. The solution is expressed in terms of an Airy function if the kernel is linear in α_s [12], for example in the case of $\alpha_s\chi_0$, or of a Bateman function in the more general case of a nonlinear dependence on α_s [16] as is the case for the DL kernels. The final result for the improved anomalous dimension is given in terms of the DL expansion plus the "Airy" or "Bateman" anomalous dimension, with the terms already included in the DL expansion subtracted away.

For example, at leading DL order, i.e. only using $\gamma_0(N)$ and $\chi_0(M)$, the improved anomalous dimension is

$$\gamma_I^{NL}(\alpha_s, N) = \left[\alpha_s\gamma_0(N) + \alpha_s^2\gamma_1(N) + \gamma_s(\frac{\alpha_s}{N}) - \frac{n_c\alpha_s}{\pi N}\right] + \gamma_A(c_0, \alpha_s, N) - \frac{1}{2} + \sqrt{\frac{2}{\kappa_0\alpha_s}[N - \alpha_s c_0]}. \quad (6)$$

The terms within square brackets give the LO DL approximation, i.e. they contain the fixed–coupling information from γ_0 and (through γ_s) from χ_0. The "Airy" anomalous dimension $\gamma_A(c_0, \alpha_s, N)$ contains the running coupling resummation, i.e. it is the exact solution of the running coupling BFKL equation which corresponds to a quadratic approximation to χ_0 near $M = 1/2$. The last two terms subtract the contributions to $\gamma_A(c_0, \alpha_s, N)$ which are already included in γ_s and γ_0. In the limit $\alpha_s \to 0$ with N fixed, $\gamma_I(\alpha_s, N)$ reduces to $\alpha_s\gamma_0(N) + O(\alpha_s^2)$. For $\alpha_s \to 0$ with α_s/N fixed, $\gamma_I(\alpha_s, N)$ reduces to $\gamma_s(\frac{\alpha_s}{N}) + O(\alpha_s^2/N)$, i.e. the leading term of the small x expansion. Thus the Airy term is subleading in both limits. However, if $N \to 0$ at fixed α_s, the Airy term replaces the leading singularity of the DL anomalous dimension, which is a square root branch cut, with a simple pole, located on the real axis at rather smaller N, thereby softening the small x behaviour. The quadratic approximation is sufficient to give the correct asymptotic behaviour up to terms which are of subleading order in comparison to those included in the DL expression in eq. (6).

The running coupling resummation procedure can be applied to a symmetrized kernel, which possesses a minimum to all orders, and then extended to next-to-leading order [15, 16]. This entails various technical complications, specifically related to the nonlinear dependence of the symmetrized kernel on α_s, to the need to include interference between running coupling effects and the small x resummation, and to the consistent treatment of next-to-leading log Q^2 terms, in particular those related to the running of the coupling.

2. Results

Even though the basic underlying physical principles of the ABF and CCSS approaches are close, there are technical differences in the construction of the symmetrized DL kernel, in the derivation from it of an anomalous dimension and associated splitting function, and in the inclusion of running coupling effects. Therefore, we will compare results for the resummed fixed-coupling χ kernel (BFKL characteristic function), then the corresponding fixed-coupling splitting functions, and finally the running coupling splitting functions which provide the final result in both approaches. In order to assess the phenomenological impact on parton evolution we will finally compare the convolution of the splitting function with a "typical" gluon distribution.

In fig. 1 we show the solution $\chi_{\rm DIS}$ of eq. (5) for the symmetrized NLO DL kernel. The pure Lx and NLx (BFKL) and next-to-leading $\ln Q^2$ (DGLAP) are also shown. All curves are determined with frozen coupling ($\beta_0 = 0$), and with $n_f = 0$, in order to avoid complications related to the diagonalization of the DGLAP anomalous dimension matrix and to the choice of scheme for the quark parton distribution. Also shown is the corresponding resummed kernel in the RGI CCSS approach. The resummed ABF and CCSS results are very close, the main difference being due to the fact that at small M the ABF result coincides by construction with NLO DGLAP, whereas for very small M (i.e. large x) the CCSS result reduces to LO DGLAP. Because of the underlying symmetry this small difference is also seen in the anticollinear region $M \sim 1$, though the two curves coincide by construction at the momentum conservation points $M = 0$ and $M = 2$. In comparison to DGLAP, the resummed kernels have a minimum, related to the fact that both collinear and anticollinear logs are resummed. In comparison to BFKL, which has a minimum at LO but not NLO, the resummed kernels always have a perturbatively stable minimum, characterized by a lower intercept than leading–order BFKL: specifically, when $\alpha_s = 0.2$, $\lambda \sim 0.25$ instead of $\lambda \sim 0.5$. This corresponds to a softer small x rise of the associated splitting function.

The fixed–coupling resummed splitting functions up to NLO are shown in fig. 2, along with the unresummed DGLAP splitting functions up to NNLO. For $n_f = 0$ the NLO DGLAP splitting function has the property that it vanishes at small x — this makes it relatively straightforward to combine not just LO DGLAP but also NLO DGLAP with the NLLx resummation. In the ABF approach the splitting function is the inverse Mellin transform of the anomalous dimension obtained using duality eq. (3) from the symmetrized DL χ kernel. Hence, the LO and NLO resummed result respectively reproduce all information contained in the LO and NLO χ and γ kernel with the additional constraint of collinear-anticollinear symmetry. Both the ABF LO and NLO results are shown in fig. 2. The CCSS NLx+LO and NLx+NLO curves are also shown for comparison.

In comparison to unresummed results, the resummed splitting functions display the characteristic rise at small x of fixed-coupling leading-order BFKL resummation, though the small x rise is rather milder ($\sim x^{-0.25}$

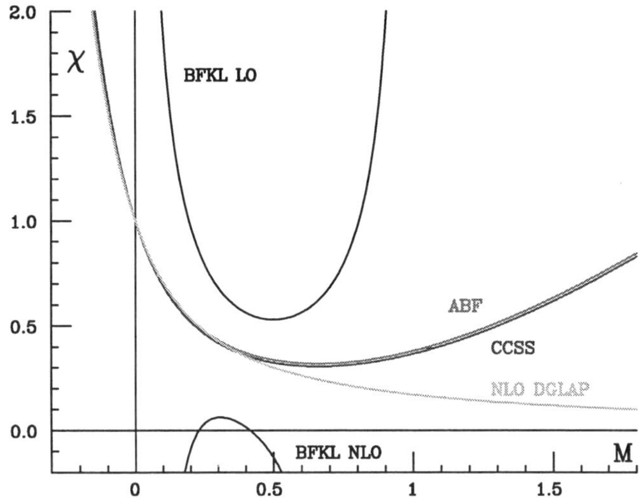

Figure 1. The kernel χ (BFKL characteristic function) for fixed coupling ($\beta_0 = 0$) $\alpha_s = 0.2$ and $n_f = 0$. The BFKL curves are the LO and NLO truncations of eq. (2), the DGLAP curve is the dual eq. (3) of the NLO anomalous dimension eq. (1), while the ABF and CCSS curves are respectively the solution $\chi_{\rm DIS}$ of eq. (5) and the solution for ω of a similar eqn. in the CCSS approach.

instead of $\sim x^{-0.5}$ for $\alpha_s = 0.2$). At large x there is good agreement between the resummed results and the corresponding LO (dashed) or NLO (solid) DGLAP curves. At small x the difference between the ABF LO and CCSS NLx+LO (dashed) curves is mostly due to the different way the symmetrization is implemented, as both approaches contain the same dominant small x terms. This difference is reduced when one compares the CCSS NLx+nLO with ABF NLO, and it might be taken as an estimate of the intrinsic ambiguity of the fixed coupling resummation procedure. At intermediate x the NLO resummed splitting functions is of a similar order of magnitude as the NLO DGLAP result even down to quite small x, but with a rather different shape, characterized by a dip at $x \sim 10^{-3}$. The unstable small x drop of the NNLO DGLAP result is a consequence of the unresummed $\frac{\alpha_s^3}{N^2}$ double pole in the NNLO anomalous dimension.

The running-coupling resummed splitting functions are displayed in fig. 3. Note that the unresummed curves are the same as in the fixed coupling case since their dependence on α_s is just through a prefactor of α_s^k, whereas in the resummed case there is an interplay between the running of the coupling and the structure of the small-x logs. All the resummed

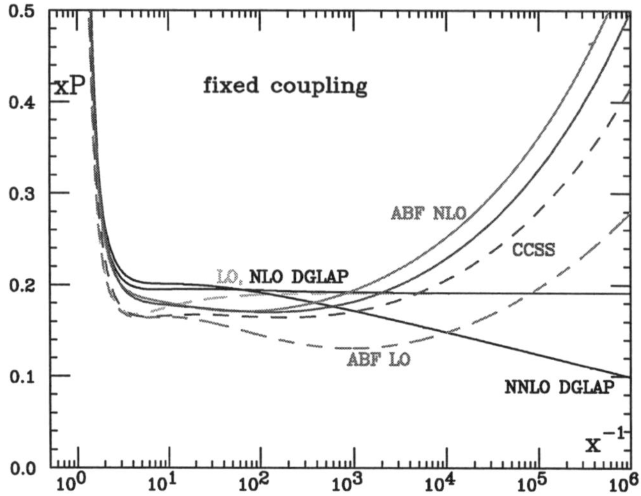

Figure 2. The fixed coupling ($\beta_0 = 0$) $xP_{gg}(x)$ splitting function, evaluated with $\alpha_s = 0.2$ and $n_f = 0$. The dashed curves are LO for DGAP, symmetrized LO DL for ABF, and NLx+LO for CCSS, while the solid curves are NLO and NNLO for DGLAP, symmetrized NLO DL for ABF and NLx+NLO for CCSS.

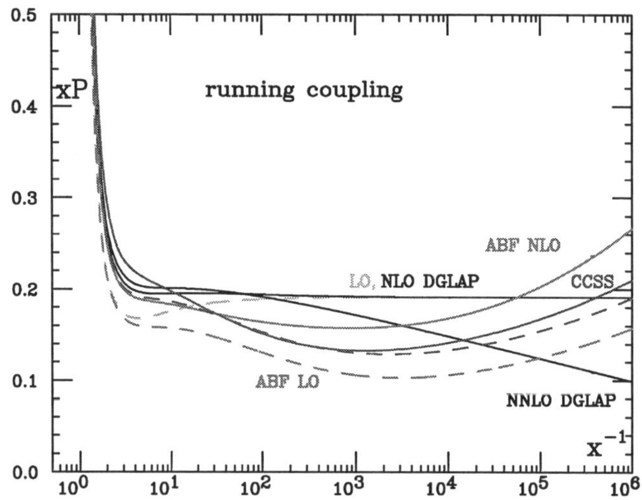

Figure 3. The running coupling $xP_{gg}(x)$ splitting function, evaluated with $\alpha_s = 0.2$ and $n_f = 0$. The various curves correspond to the same cases as in fig. 2.

curves display a considerable softening of the small x behaviour in comparison to their fixed-coupling counterparts, due to the softening of the leading small x singularity in the running-coupling case [12, 17]. As a consequence, the various resummed results are closer to each other than in the fixed-coupling case, and also closer to the unresummed LO and NLO DGLAP results. The resummed perturbative expansion appears to be stable, subject to moderate theoretical ambiguity, and qualitatively close to NLO DGLAP.

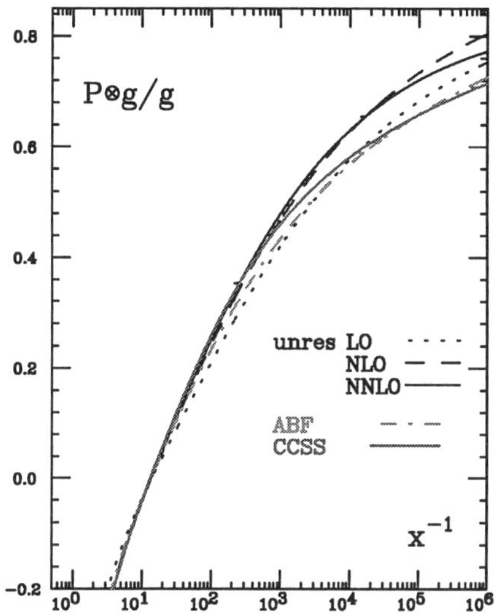

Figure 4. Convolution of resummed and fixed-order P_{gg} splitting functions with a toy gluon distribution, eq. (7), normalised to the gluon distribution itself, with $\alpha_s = 0.2$ and $n_f = 0$. The resummed ABF and CCSS curves are obtained using respectively the ABF NLO and CCSS NLx+NLO splitting function shown in fig. 3.

Finally, to appreciate the impact of resummation it is useful to investigate not only the properties of the splitting function, but also its convolution with a physically reasonable gluon distribution. We take the following toy gluon

$$xg(x) = x^{-0.18}(1-x)^5 \,, \tag{7}$$

and show in fig. 4 the result of its convolution with various splitting functions of fig. 3. The differences between resummed and unresummed results, and between the ABF and CCSS resummations are as expected partly washed out by the convolution, even though the difference between the unresummed LO and NLO DGLAP results is clearly visible. In particular, differences between the fixed-order and resummed convolution start to become significant only for $x \lesssim 10^{-2} - 10^{-3}$, even though resummation effects started to be visible in the splitting functions at somewhat larger x. However it should be clear from this figure why the structure function data from HERA were so well described by LO and NLO GLAP evolution [1–3]: significant deviations from GLAP will only be seen for smaller x and larger Q^2 than is accessible at HERA, but may well be important for many important processes at the LHC.

References

1. R.D. Ball and S. Forte, Phys. Lett. **B335**, 77 (1994).
2. R.D. Ball and S. Forte, Phys. Lett. **B336**, 77 (1994).
3. R.D. Ball and S. Forte, Phys. Lett. **B358**, 365 (1995).
4. S. Moch, J.A.M. Vermaseren and A. Vogt, Nucl. Phys. **B726**, 317 (2005).
5. V.S. Fadin and L.N. Lipatov, Phys. Lett. **B429**, 127 (1998).
6. R.D. Ball and S. Forte, hep-ph/9805315, 1998.
7. D.A. Ross, Phys. Lett. **B431**, 161 (1998).
8. R.D. Ball and S. Forte, Phys. Lett. **B465**, 271 (1999).
9. G.P. Salam, J.H.E.P. **9807**, 019 (1998).
10. G. Altarelli, R.D. Ball and S. Forte, Nucl. Phys. **B575**, 313 (2000).
11. G. Altarelli, R.D. Ball and S. Forte, Nucl. Phys. **B599**, 383 (2001).
12. G. Altarelli, R.D. Ball and S. Forte, Nucl. Phys. **B621**, 359 (2002).
13. G. Altarelli, R.D. Ball and S. Forte, Nucl. Phys. **B674**, 459 (2003).
14. G. Altarelli, R.D. Ball and S. Forte, hep-ph/0310016, 2003.
15. G. Altarelli, R.D. Ball and S. Forte, Nucl. Phys. Proc. Suppl. **135**, 163 (2004).
16. G. Altarelli, R.D. Ball and S. Forte, hep-ph/0512237, 2005.
17. M. Ciafaloni, D. Colferai and G.P. Salam, Phys. Rev. **D60**, 114036 (1999).
18. M. Ciafaloni, D. Colferai, G.P. Salam and A.M. Stasto, Phys. Lett. **B576**, 143 (2003).
19. R.D. Ball and S. Forte, Phys. Lett. **B405**, 317 (1997).
20. M. Ciafaloni and D. Colferai, hep-ph/0507106, 2005.
21. M. Ciafaloni, D. Colferai, G.P. Salam and A.M. Stasto, Phys. Rev. **D68**, 114003 (2003).

COLOUR DIPOLE PHENOMENOLOGY

G. SHAW*

Particle Physics Group,
School of Physics and Astronomy,
The University of Manchester,
Manchester M13 9PL, U.K.
E-mail: graham.shaw@manchester.ac.uk

Recent, precise HERA data on diffractive photoprocesses are analysed in the context of colour dipole models. The data on the deep inelastic structure function F_2 is used to determine the dipole cross section, and shown to prefer parameterisations which include saturation effects. The resulting dipole cross sections are then used to predict cross sections for DVCS and DDIS, which are in good agreement with experiment.

1. The colour dipole model

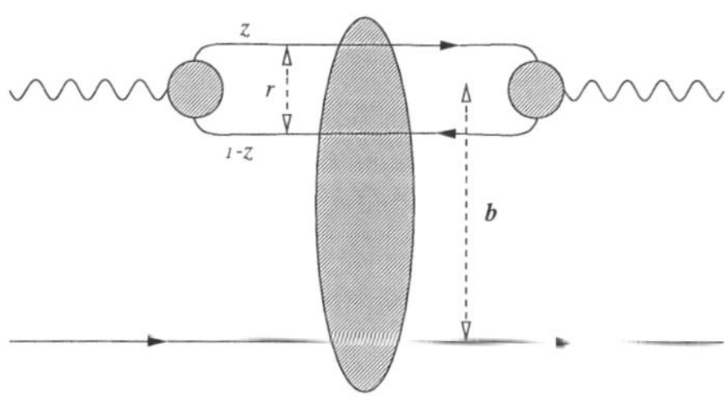

Figure 1. The colour dipole model for elastic Compton scattering $\gamma^* p \to \gamma^* p$ and Deeply Virtual Compton Scattering (DVCS) on protons, $\gamma^* p \to \gamma p$

In the colour dipole model [1,2], the forward amplitude for virtual Compton scattering is assumed to be dominated by the mechanism illustrated in Figure 1 in which the photon fluctuates into a $q\bar{q}$ pair of fixed transverse

*This work describes work carried out in collaboration with J. R. Forshaw.

separation r and the quark carries a fraction z of of the incoming photon light-cone energy. Using the optical theorem, this leads to

$$\sigma_{\gamma^* p}^{L,T} = \int dz\, d^2r\, |\psi_{L,T}(z,r,Q^2)|^2 \sigma(s^*,r) \tag{1}$$

for the total virtual photon-proton cross section, where $\psi_{L,T}(z,r,Q^2)$ are the appropriate spin-averaged light-cone wavefunctions of the photon and $\sigma(s^*,r)$ is the dipole cross section. As implied, the dipole cross section is usually assumed to be independent of z (see below), and is parameterized in terms of an energy variable s^* which depends on the model.

For small r, the light-cone photon wavefunctions are given by tree level QED expressions [1,3], which decay exponentially at large r, with typical r-values of order Q^{-1} at large Q^2 and of order m_f^{-1} at $Q^2 = 0$. However for large dipoles $r \geq 1$ fm, which are important at low $Q^2 \leq 4m_f^2$, a perturbative treatment is not really appropriate. Nonetheless, we still use the perturbative QED wavefunction, treating the quark mass as a parameter which can be used to adjust the wavefunction at large r-values. At high Q^2 where small dipoles dominate, the wavefunctions are insensitive to the quark mass.

The dipole cross section itself is described by models devised to incorporate various theoretical requirements, like colour transparency for small dipoles and soft hadronic behaviour for large dipoles. The parameters of the models are then determined by fitting to DIS structure function data, and the resulting dipole cross sections used to make predictions for other reactions, like DVCS (Deeply Virtual Compton Scattering) and especially DDIS (Diffractive Deep Inelastic Scattering), in order to test the model. Here we shall consider three such models. Since the details have been published elsewhere, we shall summarize their properties rather briefly.

The FS2004 Regge model This simple model[4] combines colour transparency for small dipoles $r < r_0$ with "soft pomeron" behaviour for large dipole sizes $r > r_1$ by assuming:

$$\begin{aligned} \sigma(x_m, r) &= A_H r^2 x_m^{-\lambda_H} \quad \text{for} \quad r < r_0 \quad \text{and} \\ &= A_S x_m^{-\lambda_S} \quad \text{for} \quad r > r_1, \end{aligned} \tag{2}$$

where

$$x_m = \frac{Q^2}{Q^2 + W^2}\left(1 + \frac{4m^2}{Q^2}\right). \tag{3}$$

For light quark dipoles, the quark mass m is a parameter in the fit, whilst for charm quark dipoles the mass is fixed at 1.4 GeV. In the intermediate region $r_0 \leq r \leq r_1$ we interpolate linearly between the two forms of (2).

If the boundary parameters r_0 and r_1 are kept constant then this parameterisation reduces to a sum of two powers, as might be predicted in a two pomeron approach, and can be thought of as an update of the original FKS Regge model [5] to accommodate the latest data (see below). It is plainly unsaturated, in that the dipole cross section obtained at small r-values and fixed Q^2 grows rapidly with increasing s (or equivalently with decreasing x) without damping of any kind.

The FS2004 Saturation model Saturation can be introduced into the above model by adopting a device previously utilized in [7]. Instead of taking r_0 to be fixed we now determine it to be the value at which the hard component is some fixed fraction of the soft component, i.e.

$$\sigma(x_m, r_0)/\sigma(x_m, r_1) = f \tag{4}$$

and treat f instead of r_0 as a fitting parameter. This introduces no new parameters compared to the Regge model. However, the scale r_0 now moves to lower values as x decreases, and the rapid growth of the dipole cross section at a fixed, small value of r begins to be damped as soon as r_0 becomes smaller than r. In this sense we model saturation, albeit crudely, with r_0 a saturation radius.

The CGC saturation model The above two models are closely related, allowing the introduction of saturation effects to be studied in a direct manner. In addition we shall consider the CGC dipole model of Iancu, Itakura and Munier [8], which purports to include the main features of the "Colour Glass Condensate" regime, and can be thought of as a more sophisticated version of the original "Saturation Model" of Golec-Biernat and Wüsthoff [9].

2. Analysis of the F_2 data

In this section, we shall fit the recent ZEUS data [10] in the kinematic range

$$0.045\,\text{GeV}^2 < Q^2 < 45\,\text{GeV}^2 \qquad x \leq 0.01\,. \tag{5}$$

This is identical to that used by Iancu et al. [8] to determine the parameters of the CGC saturation model, albeit neglecting charm. The corresponding

H1 data [11] can also be fitted, but it is necessary to float the relative normalization of the two data sets to achieve this, and we will not consider it here.

The best fit obtained with the FS2004 Regge model is shown as the dashed line in Figure 2. As can be seen, the quality of the fit is not good, corresponding to a χ^2/data point of 428/156. This is not just a failing of this particular parameterisation. We have attempted to fit the data with many other Regge inspired models, including the original FKS parameterizations [5,6], without success.

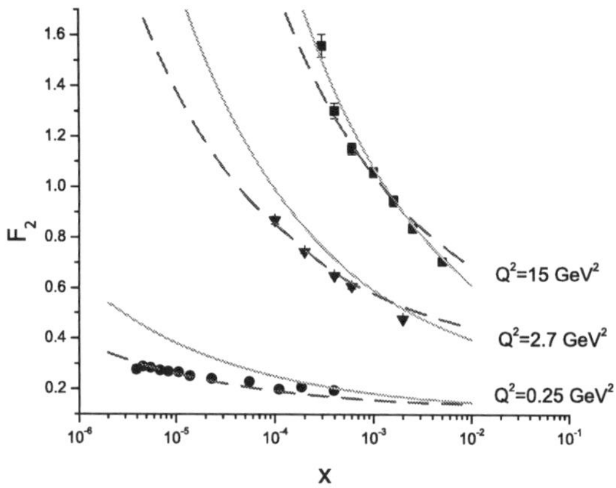

Figure 2. Comparison of the FS2004 Regge dipole fit (dashed line) to a subset of DIS data Also shown (solid line) is a fit of the same model to data in the restricted range $5 \times 10^{-4} < x < 10^{-2}$, extrapolated over the whole x-range $x < 0.01$.

A possible reason for this failing is suggested by the solid curve in Figure 2, which shows the result of fitting only to data in the x-range $5 \times 10^{-4} < x < 10^{-2}$, and then extrapolating the fit to lower x values, corresponding to higher energies at fixed Q^2. As can be seen, this leads to a much steeper dependence at these lower x-values (i.e. at higher energies) than is allowed by the data at all Q^2. An obvious way to solve this problem is by introducing saturation at high energies, to dampen this rise. This is confirmed by fitting with the FS2004 saturation model, which yields a χ^2/data point of 155/156. The resulting fit is shown in Figure 3, which also shows the very similar results previously obtained using the more sophisticated CGC model (no charm fit) [8].

It is clear from these results that the introduction of saturation into the model immediately removes the tension between the soft and hard components which is so disfavoured by the data. However, it is important to note that this conclusion relies on the inclusion of the data in the low Q^2 region: both the Regge and saturation models yield satisfactory fits if we restrict to $Q^2 \geq 2\,\mathrm{GeV}^2$, with χ^2/data point values of 78/86 and 68/86 respectively.

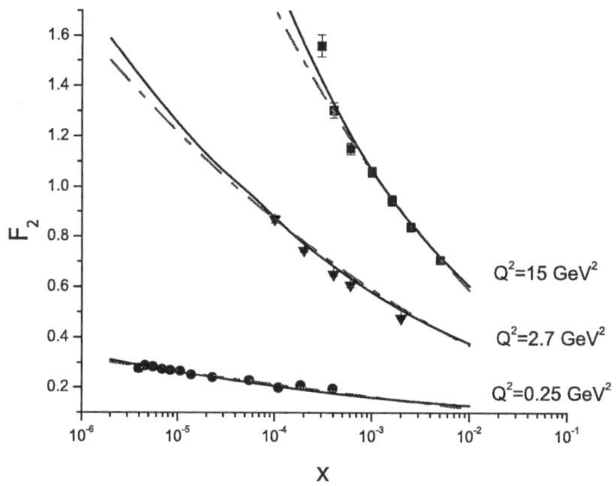

Figure 3. Comparison of the FS2004 saturation fit (solid line) to a subset of DIS data. Also shown (dot-dashed line) is the corresponding fit obtained using the CGC dipole model.

3. Predictions for other processes

One result of the above is that we now have well-determined parameterisations of the colour dipole cross section which, in the case of the FS2004, but not the CGC model, take account of charm. These can be used to yield predictions for other processes, in order to test the model against experimental data. Here we will compare these predictions with new, precise data on deeply virtual Compton scattering (DVCS) and, especially, diffractive deep inelastic scattering (DDIS). In doing so we stress that the photon wavefunctions and dipole cross sections used are precisely those determined in the previous section, without any adjustment of parameters. Predictions can also be obtained for vector meson production processes, but these will not be considered here because of the large uncertainties associated with the badly known vector meson wavefunctions.

3.1. *Deeply virtual Compton scattering (DVCS)*

In the colour dipole model, the imaginary part of the scattering amplitude for deeply virtual Compton scattering

$$\gamma^* + p \to \gamma + p \,, \tag{6}$$

at zero momentum transfer is given by

$$\mathrm{Im}\, A_{\mathrm{DVCS}}(W^2, Q^2, t=0) = \int dz\; \mathrm{d}^2 r\; \Psi_\gamma^{T*}(r, z, Q^2)\, \sigma(s, r)\, \Psi_\gamma^T(r, z, 0) \tag{7}$$

corresponding again to Figure 1. Thus dipole models provide predictions for the imaginary part of the forward amplitude with no adjustable parameters beyond those used to describe DIS. To calculate the forward cross section a correction for the contribution of the real part, estimated from dispersion relation arguments to be typically of order 10% [6], has to be included. Predictions for the measured total cross sections are then obtained by assuming an exponential t-distribution and integrating over t to obtain

$$\sigma_{L,T}(\gamma^* p \to \gamma p) = \frac{1}{B} \left. \frac{d\sigma^{T,L}}{dt} \right|_{t=0} , \tag{8}$$

where the value of the slope parameter B is taken from experiment. For this reason, the normalization is always somewhat uncertain.

In Figure 4, we compare the predictions of the FS2004 Regge and saturation models and the CGC model with the most recent data[13], using a fixed value $B = 6.02\ \mathrm{GeV}^{-2}$ for the slope parameter, which was measured in the same experiment to be $6.02 \pm 0.35 \pm 0.39$. Bearing in mind the normalization uncertainty, the agreement is excellent for all three models, although significant differences between the saturated and unsaturated models appear when the predictions are extrapolated to higher energies, as one would expect.

3.2. *Diffractive deep inelastic scattering(DDIS)*

We finally turn to the so-called diffractive deep inelastic scattering (DDIS) process

$$\gamma^* + p \to X + p \ , \tag{9}$$

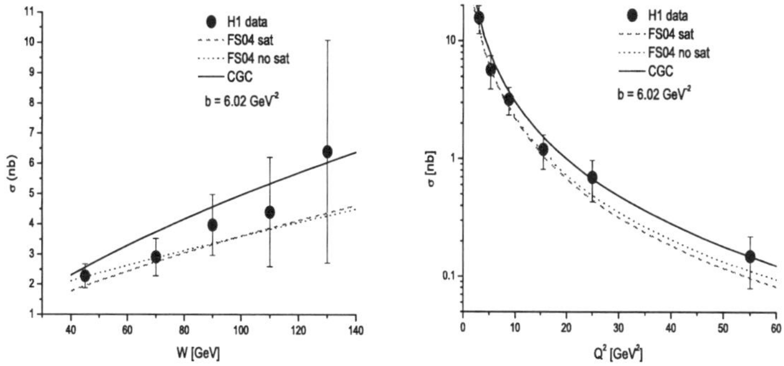

Figure 4. Comparison of the H1 DVCS data with the predictions of the FS2004 Regge model (dotted line), the FS2004 saturation model (dashed line) and the CGC saturation model (solid line). The plots show the W dependence at fixed $Q^2 = 8$ GeV^2 and the Q^2 dependence at fixed $W = 82$ GeV, respectively.

where the hadronic state X is separated from the proton by a rapidity gap. In this process, in addition to the usual variables x and Q^2 there is a third variable m_X^2, giving rise to a rich body of data. In practice, x and m_X^2 are often replaced by the variables $x_{I\!P}$ and β, given by

$$x_{I\!P} \simeq \frac{m_X^2 + Q^2}{W^2 + Q^2} \qquad \beta = \frac{x}{x_{I\!P}} \simeq \frac{Q^2}{m_X^2 + Q^2}. \tag{10}$$

in the diffractive limit $s >> Q^2, m_X^2$.

In the dipole model, the contribution due to quark-antiquark dipoles to the structure function $F_2^{D(3)}$ can be obtained from a momentum space treatment as described in [14,9]. However, if we are to confront the data at low values of β, corresponding to large invariant masses m_X, it is necessary to also include a contribution from the higher Fock state $q\bar{q}g$. We can estimate this contribution using an effective "two-gluon dipole" approximation due to Wüsthoff[14], as illustrated in Figure 5.

Again, the predictions obtained in this way involve no adjustment of the dipole cross sections and photon wavefunctions used to describe the F_2 data. However, we are again free to adjust the forward slope for inclusive diffraction, B, within the range acceptable to experiment, which means that the overall normalization, but not the energy dependence, of $F_2^{D(3)}$ is free to vary somewhat. We are also somewhat free to vary the value of α_s used to define the normalization of the model dependent $q\bar{q}g$ component, which is important at low values of β. Rather arbitrarily we take $\alpha_s = 0.1$.

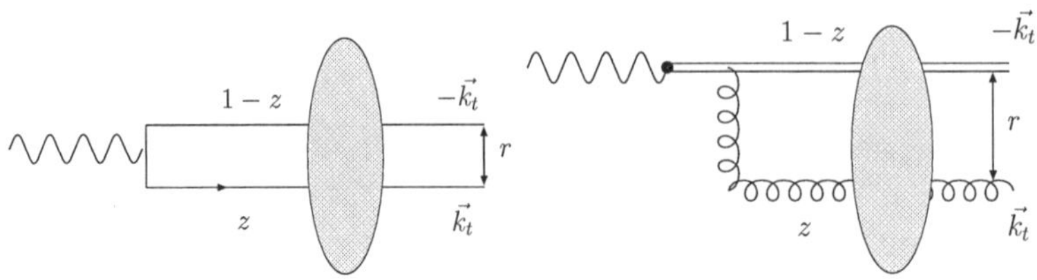

Figure 5. The $q\bar{q}$ and $q\bar{q}g$ contributions to $F_2^{D(3)}$.

In Figure 6 we compare the recent ZEUS LPS data [15] on the $x_{I\!P}$ dependence of the structure function $F_2^{D(3)}$ at various fixed Q^2 and β with the predictions obtained from the FS2004 Regge and saturation models and the CGC model. The agreement is in general good, except at the larger $x_{I\!P}$ values, corresponding to low energies, where there is clearly a significant non-diffractive contribution, neglected in the above predictions,

This problem is avoided in the LPC data [16], in which the non-diffractive contribution is explicitly removed in the data extraction. However this comes at the cost of including proton dissociation contributions, since the mass of the target fragments is only limited to below 2.3 GeV. The predictions of our three models are compared with this data in Figures 7 and 8, where the theory curves have been divided by 0.7 to allow for proton dissociation contributions, assumed to be a fixed fraction of the total, independent of the kinematic variables. This is obviously a crude approximation. Nonetheless, the agreement is remarkably good for the saturation models, except for the bins with both the lowest Q^2 values and the highest β values, corresponding to the lowest masses M_X, in the resonance region.

4. Conclusion

The colour dipole model offers a successful unified picture of diffractive photoprocesses over a wide range of Q^2, with a preference for dipole cross sections incorporating some form of saturation.

Acknowledgments

This research was supported in part by a UK Particle Physics and Astronomy Research Council grant number PPA/G/0/2002/00471.

References

1. N.N. Nikolaev and B.G. Zakharov, Z. Phys. **C49** (1991) 607; **C53** (1992) 331.

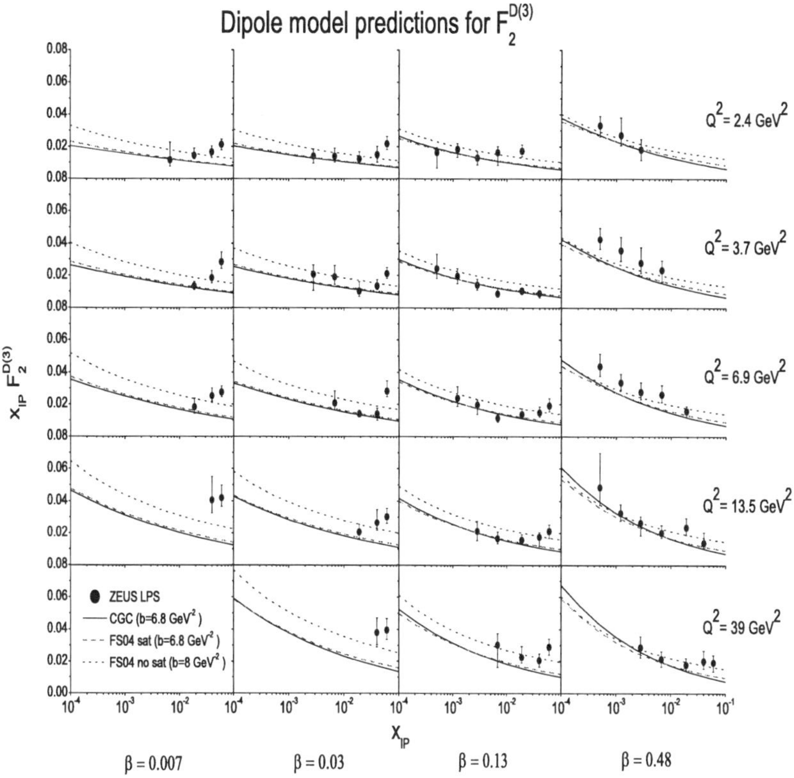

Figure 6. Predictions for the ZEUS LPS data: FS2004 Regge model (dotted line), the FS2004 saturation model (dashed line) and the CGC model (solid line)

2. A.H. Mueller, Nucl. Phys. **B415** (1994) 273; A.H. Mueller and B. Patel, Nucl. Phys. **B425** (1994) 471.
3. H.G. Dosch, T. Gousset, G. Kulzinger and H.J. Pirner, Phys. Rev. **D55** (1997) 2602.
4. J.R. Forshaw and G. Shaw, JHEP 0412 (2004) 052
5. J.R. Forshaw, G. Kerley and G. Shaw, Phys. Rev. **D60** (1999) 074012; J.R. Forshaw, G. Kerley and G. Shaw, Nucl. Phys. **A675** (2000) 80.
6. M.F. McDermott, R. Sandapen and G. Shaw, Eur. Phys. J. **C22** (2002) 665.
7. M. McDermott, L. Frankfurt, V. Guzey and M. Strikman, Eur. Phys. J. **C16** (2000) 641.

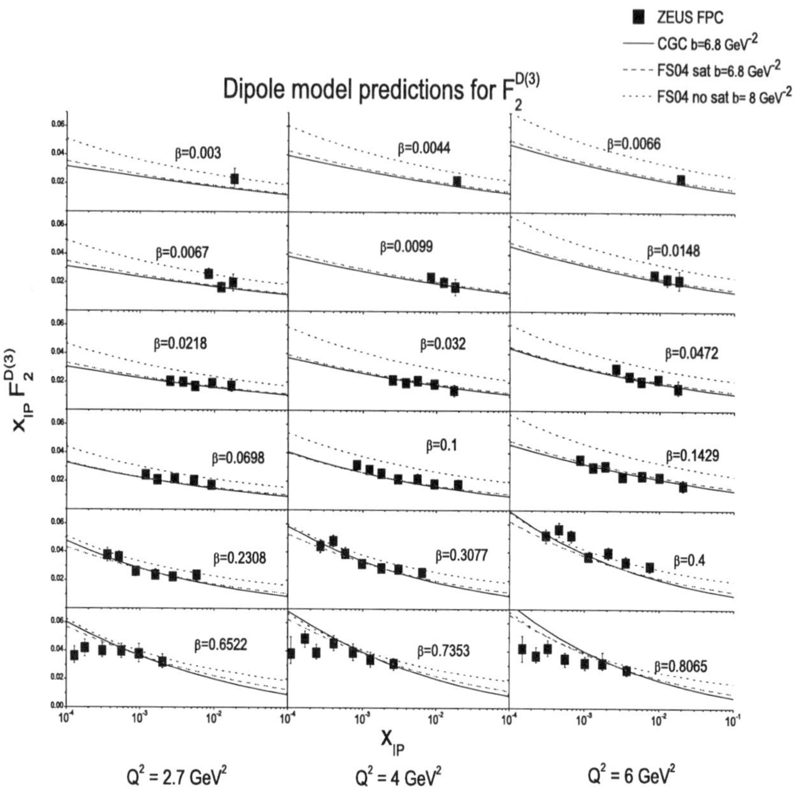

Figure 7. Predictions for the ZEUS FPC data at low Q^2: FS2004 Regge model(dotted line), the FS2004 saturation model (dashed line) and the CGC model (solid line)

8. E. Iancu, K. Itakura and S. Munier, Phys. Lett. **B590** (2004) 199.
9. K. Golec-Biernat and M. Wüsthoff, Phys. Rev. **D59** (1999) 014017; K. Golec-Biernat and M. Wüsthoff, Phys. Rev. **D60** (1999) 114023.
10. S. Chekanov et al., ZEUS Collab., Eur. Phys. J. **C21** (2001) 442.
11. C. Adloff et al., H1 Collab., Eur. Phys. J. **C21** (2001) 33.
12. H1 Collaboration: Proc. of the International Europhysics Conference on High Energy Physics, Aachen, July 2003.
13. A. Aktas et al., H1 Collab., Eur. Phys. **C44** (2005) 1.
14. M Wüsthoff, Physical Review **D56** (1997) 4311.
15. S. Chekanov et al., ZEUS Collab., Eur. Phys. J. **C38** (2004) 43.

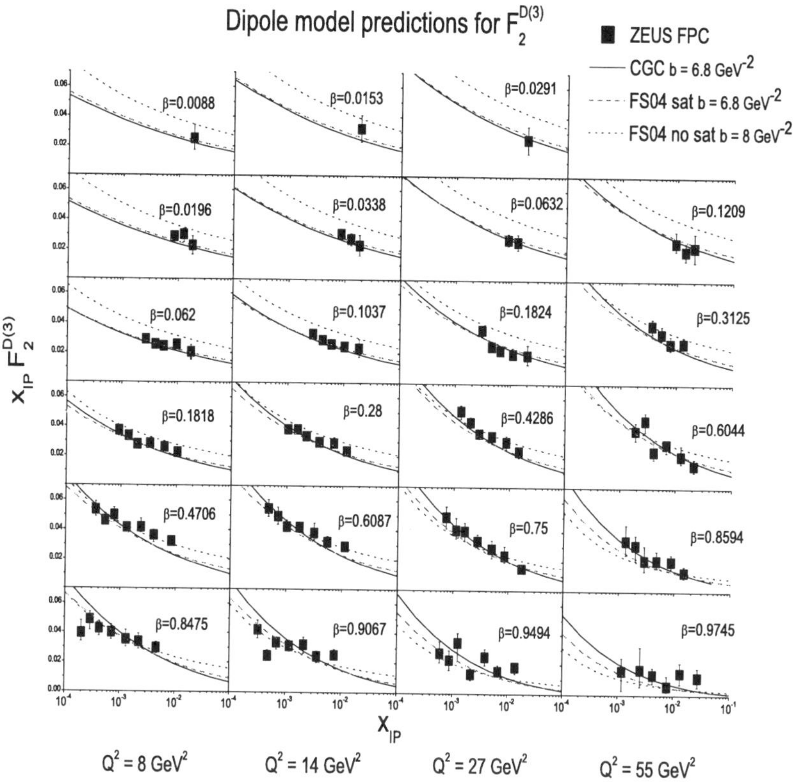

Figure 8. Predictions for the high Q^2 ZEUS FPC data: FS2004 Regge model(dotted line), the FS2004 saturation model (dashed line) and the CGC model (solid line)

16. S. Chekanov et al., ZEUS Collab., Nucl. Phys. **B713** (2005) 3.

2 Spin Physics

EXCLUSIVE REACTIONS AT HERMES

F. ELLINGHAUS*
University of Colorado,
Department of Physics,
Boulder, Colorado 80309-0390, USA
E-mail: Frank.Ellinghaus@desy.de

FOR THE HERMES COLLABORATION

Hard exclusive reactions are the tool to learn about generalized parton distributions, which provide a more complete parameterization of the nucleon than the ordinary parton distribution functions. Measurements performed by the HERMES collaboration of the exclusive production of photons, i.e., Deeply-Virtual Compton Scattering, and of mesons are briefly reviewed, focusing on new results which may allow the determination of the total angular momentum of quarks in the nucleon.

1. Introduction

Similar to the case of inclusive and semi-inclusive DIS, where the nucleon structure is described using **P**arton **D**istribution **F**unctions (PDFs), hard exclusive reactions can be expressed in terms of **G**eneralized **P**arton **D**istributions (GPDs) [1,2,3]. The PDFs and elastic nucleon **F**orm **F**actors (FFs) are included in the GPDs as limiting cases and moments of GPDs, respectively [2]. This is visualized in Fig. 1. While FFs derived in elastic scattering describe the transverse location of partons inside the nucleon and PDFs describe their longitudinal momentum distribution, GPDs are able to provide information on both at the same time. Thus exclusive reactions are able to give a certain 3–dimensional picture of the nucleon structure [4,5,6]. In particular, GPDs offer for the first time a possibility to determine the total angular momentum carried by the quarks in the nucleon [2].

Below the experimental status of measurements on the exclusive production of photons (DVCS) and mesons at HERMES is summarized. The data has been taken with polarized and unpolarized gas targets using the

*This work is supported in part by the US Department of Energy.

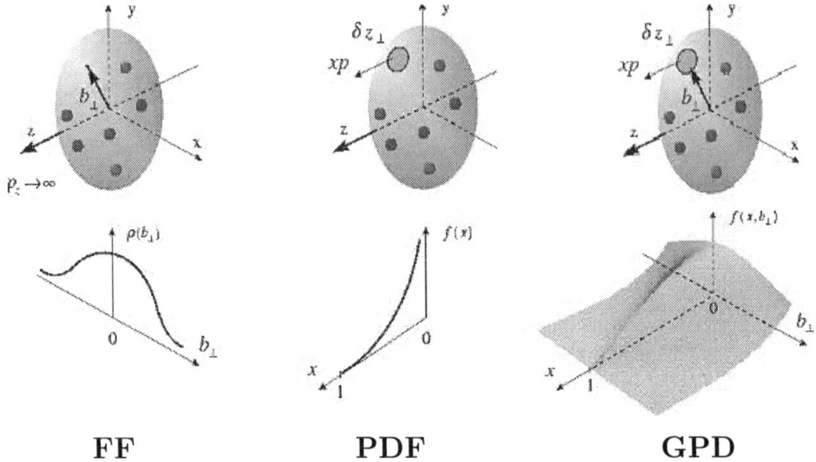

Figure 1. Probabilistic interpretation of form factors (FF), parton distribution functions (PDF) and generalized parton distributions (GPD). For details on figure see Ref. [5].

HERMES spectrometer [7] at the HERA electron/positron–proton collider at DESY, which offers longitudinally polarized 27.6 GeV electron and positron beams. The connection between the measured asymmetries and the four leading twist GPDs H, $\widetilde{H}$, E, $\widetilde{E}$ is given, especially focusing on the connection to the total angular momentum.

2. Deeply–virtual Compton scattering

The hard exclusive leptoproduction of real photons, e.g., $e\,p \to e\,p\,\gamma$, is called **D**eeply–**V**irtual **C**ompton **S**cattering (DVCS). DVCS amplitudes can be measured through the interference between the DVCS and **B**ethe–**H**eitler (BH) processes, in which the photon is radiated from a parton in the former and from the electron in the latter process. Both processes have an identical final state, i.e., they are indistinguishable, and thus their amplitudes (τ_{DVCS}, τ_{BH}) add coherently. For an unpolarized proton target, the cross section for the exclusive leptoproduction of photons reads [8]

$$\frac{d\sigma}{dx_B\, dQ^2\, d|t|\, d\phi} = x_B\, y^2 \frac{|\tau_{BH}|^2 + |\tau_{DVCS}|^2 + \overbrace{\tau_{DVCS}\, \tau_{BH}^* + \tau_{DVCS}^*\, \tau_{BH}}^{I}}{32\,(2\pi)^4\, Q^4\, \sqrt{1 + 4\, x_B^2\, M_p^2/Q^2}}, \tag{1}$$

where x_B represents the Bjorken scaling variable, $-Q^2$ the virtual–photon four–momentum squared, y the fraction of the incident lepton energy carried by the virtual photon and t and M_p are the Mandelstam variable and the proton mass, respectively. The azimuthal angle ϕ is defined as the angle between the lepton scattering plane, and the photon production plane made up correspondingly by the virtual and the real photon. At leading twist, the interference term I is given by [8]

$$I = \pm \frac{\sqrt{32}}{t\,Q\,x_B} \frac{m\,e^6}{\sqrt{1-x_B}} \left[\cos\phi \frac{1}{\sqrt{\epsilon(1-\epsilon)}} \mathcal{R}eM^{1,1} - P_l \sin\phi \sqrt{\frac{1+\epsilon}{\epsilon}} \mathcal{I}mM^{1,1} \right], \quad (2)$$

where +(-) denotes a negatively (positively) charged lepton beam with polarization P_l, e represents the electron charge and ϵ is the ratio of the number of longitudinal to transverse initial photons in the DVCS process. The ϕ–dependence of a cross section asymmetry with respect to the charge (spin) of the lepton beam predominantly singles out the $\cos\phi$ ($\sin\phi$) amplitude of the interference term, which provides information about the real (imaginary) part of the DVCS amplitude $M^{1,1}$. The DVCS amplitude [9]

$$M^{1,1} = F_1\,H_1 + \frac{x_B}{2 - x_B}(F_1 + F_2)\,\tilde{H}_1 - \frac{t}{4M^2} F_2\,E_1 \quad (3)$$

is given by a linear combination of the so–called Compton form factors (CFFs) H_1, $\tilde{H}_1$ and E_1, together with the Dirac and Pauli form factors F_1 and F_2, respectively. The CFFs are convolutions of the respective twist–2 GPD H, $\tilde{H}$ or E and of the hard scattering amplitude. While the above holds true only for an unpolarized target, other linear combinations of GPDs can be accessed in a similar way using a longitudinally or transversely polarized proton target.

The event selection at HERMES requires events with exactly one photon and one charged track, identified as the scattered lepton, with $Q^2 > 1$ GeV^2. The photon is identified by detecting an energy deposition in the preshower scintillator and in the calorimeter without an associated charged track. The recoiling proton is not detected. Exclusive events are identified by the fact that the missing mass of the reaction $ep \rightarrow e\gamma X$ corresponds to the proton mass. Due to the finite energy resolution the exclusive sample is defined as $-1.5 < M_x < 1.7$ GeV, as determined from signal–to–background studies using a Monte-Carlo simulation. A recoil detector, presently being assembled in the target area, will help to more clearly identify exclusive reactions and thus to reduce the underlying background to less than 1%.

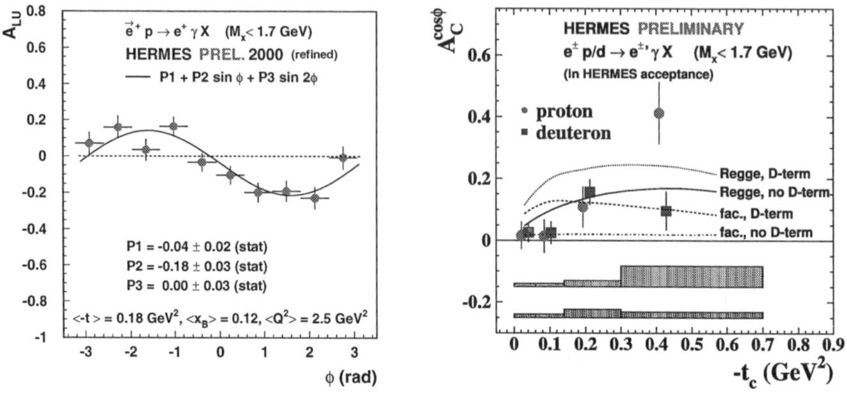

Figure 2. Left panel: Beam–spin asymmetry A_{LU} for the hard electroproduction of photons off the proton [10] as a function of the azimuthal angle ϕ for the exclusive sample. The solid curve shows the result of the indicated fit with the values given in the plot. Right panel: The $\cos\phi$ amplitude of the beam–charge asymmetry on hydrogen and on unpolarized or spin–averaged polarized deuterium [12] as a function of $-t$ at average values of $\langle x_B \rangle = 0.10$ and $\langle Q^2 \rangle = 2.5$ GeV^2. The GPD model calculations use a factorized or a Regge–inspired t–dependence with or without a D-term contribution.

2.1. *Access to GPD H via beam-charge asymmetry, "ordinary (charged)" and "neutral" beam-spin asymmetry*

The beam–spin asymmetry (BSA) and the beam–charge asymmetry (BCA) as a function of ϕ are calculated as

$$A_{LU}(\phi) = \frac{1}{<|P_l|>} \frac{\overrightarrow{N}(\phi) - \overleftarrow{N}(\phi)}{\overrightarrow{N}(\phi) + \overleftarrow{N}(\phi)}, \qquad A_C(\phi) = \frac{N^+(\phi) - N^-(\phi)}{N^+(\phi) + N^-(\phi)}, \quad (4)$$

with the normalized yields $\overrightarrow{N}$ ($\overleftarrow{N}$) or N^+ (N^-) using a beam with positive (negative) helicity or a positron (electron) beam, respectively. The BSA on the proton as a function of ϕ is shown in the left panel of Fig. 2 for the exclusive sample [10]. The results from the indicated fit show the $\sin\phi$ dependence for the BSA as expected from Eqn. 2. Note that similar results for the BSA have also been obtained using heavier targets (D, Ne) [10].

The expected predominant $\cos\phi$ behavior of the BCA (see Eqn. 2) has also been observed at HERMES [11]. The right panel of Fig. 2 shows the $\cos\phi$ amplitudes of the BCA on deuterium and hydrogen as a function of $-t$ which are derived from a fit to the BCA in each $-t$ bin [12]. For both targets, the signal only becomes sizeable for larger values of $-t$. While their slightly different behavior at large $-t$ values can be due to incoherent scattering

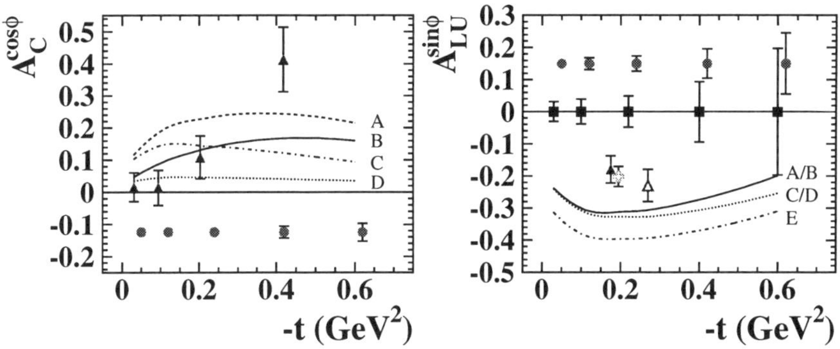

Figure 3. The $\cos\phi$ amplitude of the BCA (left panel, closed triangles) from HERMES [12] and the $\sin\phi$ amplitude of the BSA (right panel) from the 96/97 HERMES data [13] (open triangle), from the 2000 HERMES data [10] (closed triangle), and from CLAS [14] (open cross) as a function of $-t$. The CLAS result was achieved using an electron beam and is thus multiplied by -1. Statistical and systematic uncertainties are added in quadrature. The parameter sets for the GPD models are given in the table. The predictions from the models A, B, C and D are used for comparison with the BCA (left panel). The BSA is insensitive to the D-term contribution and thus yields the same result for the models A/B and C/D (right panel). See text for explanation on the projections (squares,circles).

on the neutron in the deuteron, the approximately 40% contribution of coherent scattering on the deuteron in the first $-t$ bin does not result in a different asymmetry when compared to the proton target, within the uncertainties.

These results together with the ones on the BSA from the 96/97 data [13] and from the CLAS collaboration [14] are shown in Fig. 3. Note that the smaller average $-t$ value of the 2000 data (see right panel) when compared to the result from the 96/97 data is due to the fact that events with small angles between the virtual and the real photon have been included for the analysis of the 2000 data. The model calculations shown in this plot and in the right panel of Fig. 2 are based on a GPD model described in Refs. [15,16]. Since at small values of x_B (t) the contribution from the CFFs $\tilde{H}_1$ (E_1) to the amplitude $M^{1,1}$ is suppressed (see Eqn. 3), the main contribution to the asymmetries in the kinematic region of HERMES originates from the CFF H_1 and thus it originates from the GPD H. Consequently the theoretical calculations shown in Fig. 2 and Fig. 3 were derived by only varying the model parameters for the GPD H in the underlying code [17] in order to calculate the asymmetries at the average kinematics of every bin. The model parameters (see table) differ with respect to including or neglecting the so–called D–term [18] in the GPD model and whether the t–dependence of the GPD H is calculated in either the simplest ansatz where the t–dependence factorizes from the t–independent part or in the Regge–motivated ansatz. In addition, the parameters b_{val} and b_{sea} in the profile function [19] have been varied. The BCA data appears to disfavor the model with the non–factorized t-dependence and the contribution from the so–called D–term while all models seem to overshoot the absolute size of the $\sin\phi$ amplitude of the BSA.

The extraction of the kinematic dependences of the BSA for the HERA Run I data on hydrogen (1996-2000) is almost completed, the projected statistical precision is shown in the same figure (squares). The additional statistics expected from a dedicated running on unpolarized hydrogen in the final two years of HERA (2006 – 2007) is also displayed (circles), provided that 1 fb^{-1} with each beam charge can be collected at a beam polarization of 40%. While up to now the BSA has always been measured with a positron beam at HERMES and an electron beam at CLAS, the question arises if there are advantages in having both charges and how to combine the data in the first place. The two beam charges do not only result in a different sign for the asymmetry but also the absolute values will be different. This is due to the fact that the numerator of the BSA does not only contain the (helicity dependent) $\sin\phi$ term from the interference term (see Eqn. 2), but also has a (helicity dependent) $\sin\phi$ contribution from the DVCS squared term in Eqn. 1. Despite the fact that the latter is a higher-twist contribution that is suppressed in comparison to the former, it is likely that it will increase the extracted $\sin\phi$ amplitude for one beam charge and diminish it for the other. Especially since kinematic dependences will be studied, the $\sin\phi$ amplitude of the DVCS squared term will not be negligible compared to the $\sin\phi$ amplitude of the interference term in every given kinematic bin. Thus the *ordinary* BSA (see Eqn. 4) will not give a clean access to the interference term, i.e., to linear combinations of GPDs, but will always be disturbed by a

Model	D–Term	b_{val}	b_{sea}	Ansatz t–dependence
A	Yes	1	∞	Regge
B	No	1	∞	Regge
C	Yes	1	∞	factorized
D	No	1	∞	factorized
E	Yes/No	1	1	factorized

term which is proportional to GPDs squared. In addition, the denominator of the BSA should be dominated by the (helicity independent) BH squared term (see Eqn. 1) when compared to the helicity independent part of the interference term, i.e., the $\cos\phi$ term (see Eqn. 2). However, it is also here likely that this (beam charge dependent) $\cos\phi$ term will alter the *ordinary (charged)* BSA since depending on the beam charge it is added or subtracted to/from the BH squared term.

The solution is to define a new cross section asymmetry, which one might call the *neutral* BSA:

$$A^0_{LU}(\phi) = \frac{(d\overrightarrow{\sigma}^+ + d\overleftarrow{\sigma}^-) - (d\overleftarrow{\sigma}^+ + d\overrightarrow{\sigma}^-)}{(d\overrightarrow{\sigma}^+ + d\overleftarrow{\sigma}^-) + (d\overleftarrow{\sigma}^+ + d\overrightarrow{\sigma}^-)}, \tag{5}$$

Apart from the fact that this asymmetry turns out to be independent of the beam charge (*neutral*), it has the important advantage that its $\sin\phi$ amplitude stems solely from the interference term since the (higher twist) $\sin\phi$ dependence of the DVCS squared term drops out. Thus it gives clean access to linear combinations of GPDs, while differences between the ordinary and the neutral BSA can in principle be used to determine the size of these higher twist contributions. It can be concluded that by using the upcoming HERMES data (see Fig.3) for a measurement of this new neutral BSA, together with a measurement of the BCA, one should be able to put serious constraints on the GPD H.

2.2. *Access to GPDs H and $\widetilde{H}$ via longitudinal target-spin asymmetry*

The longitudinal target-spin asymmetry (LTSA) utilizes an unpolarized (U) beam and a longitudinally polarized (L) target. It is calculated as

$$A_{UL}(\phi) = \frac{1}{<|P_T|>}\frac{(N^{\leftarrow\Leftarrow}(\phi) + N^{\rightarrow\Leftarrow}(\phi)) - (N^{\leftarrow\Rightarrow}(\phi) + N^{\rightarrow\Rightarrow}(\phi))}{(N^{\leftarrow\Leftarrow}(\phi) + N^{\rightarrow\Leftarrow}(\phi)) + (N^{\leftarrow\Rightarrow}(\phi) + N^{\rightarrow\Rightarrow}(\phi))} \tag{6}$$

where P_T is the target polarization and $\leftarrow$ ($\rightarrow$) and $\Leftarrow$ ($\Rightarrow$) denote the orientation of the beam and target spin with respect to the beam direction, respectively. The LTSA should be dominated by a $\sin\phi$ amplitude which eventually is most sensitive to a linear combination of the imaginary parts of H and $\widetilde{H}$. The HERMES results [20] on the $\sin\phi$ and $\sin 2\phi$ amplitudes of the LTSA on the proton and on the deuteron as a function of $-t$ are shown in Fig. 4 for the exclusive sample. While both deuteron amplitudes are slightly negative but essentially compatible with zero, it comes as a surprise that the $\sin 2\phi$ amplitude on the proton is not suppressed with respect to

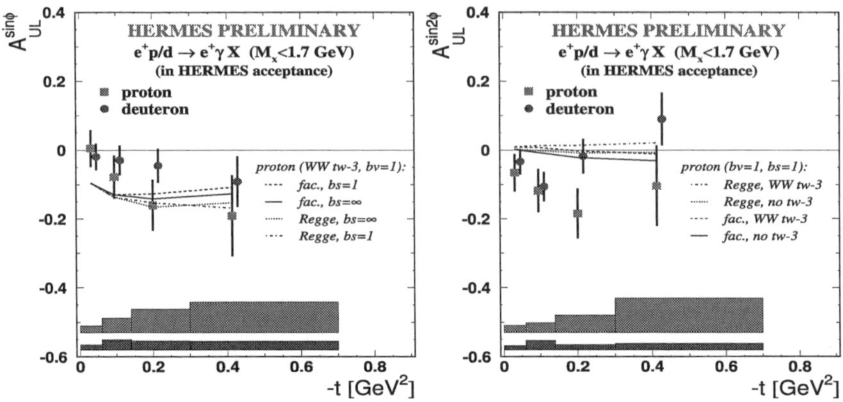

Figure 4. The $\sin\phi$ and the $\sin 2\phi$ amplitude of the longitudinal target-spin asymmetry on hydrogen and deuterium as a function of $-t$ [20].

the non-zero leading-twist $\sin\phi$ amplitude. No difference between the two targets can be seen in the first $-t$-bin where an about 40% contribution from coherent processes on the deuteron is expected. The proton and deuteron results are systematically different for both amplitudes of the asymmetry at higher $-t$ bins, which indicates that the asymmetries on protons and neutrons are different. The results of the model calculations shown in the figures are for the proton case. They are achieved in the same way as described above and show little variation when compared to the statistical error of the data. While the model calculations are in agreement with the data for the $\sin\phi$ amplitude, the (twist-3) $\sin 2\phi$ amplitude of the proton is unexpectedly large when compared to the calculations, as already discussed above. It should be noted here that the model only includes the Wandzura-Wilczek part of the twist–3 contribution but not the genuine twist-3 term.

2.3. *Access to GPDs H and E via transverse target-spin asymmetry*

As described above, HERMES will be able to provide sufficient data to largely constrain the GPD H in the kinematic region of HERMES. It is natural to ask to what extent the GPD E can be accessed, which is the other important GPD necessary in order to determine J_q, the total orbital angular momentum of the quarks in the nucleon [2]. For an unpolarized target the contribution from the GPD E is suppressed with respect to H (see Eqn. 3), but this is different for transverse target polarization [9,21]. Using an unpolarized beam (U) and a transversely (T) polarized target,

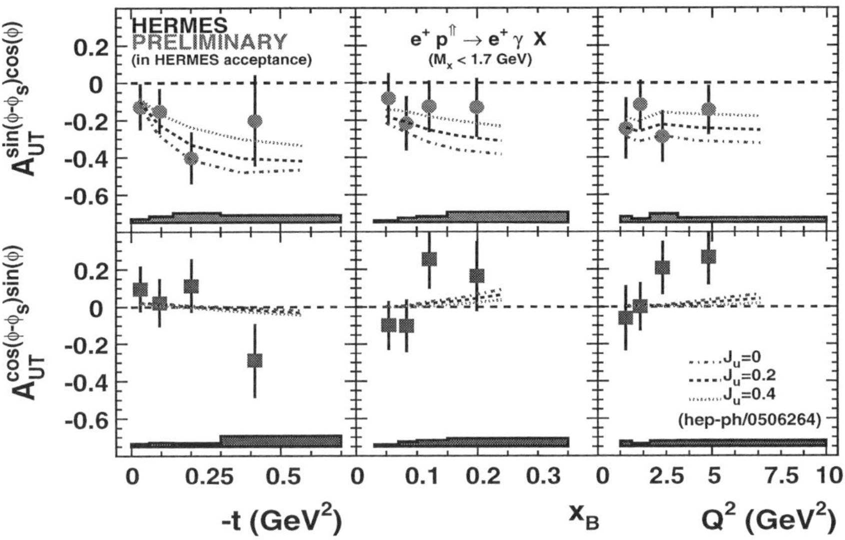

Figure 5. The $\sin(\phi - \phi_S)\cos\phi$ and the $\cos(\phi - \phi_S)\sin\phi$ amplitude of the transverse target-spin asymmetry for the hard electroproduction of photons off the proton as a function of $-t$, x_B and Q^2 in comparison to theoretical predictions from Ref. [22].

the transverse target-spin asymmetry (TTSA) can be written as [22]:

$$A_{UT}(\phi, \phi_s) = \frac{1}{|P_T|} \cdot \frac{d\sigma(\phi, \phi_S) - d\sigma(\phi, \phi_S + \pi)}{d\sigma(\phi, \phi_s) + d\sigma(\phi, \phi_S + \pi)} \propto \tag{7}$$

$$\mathrm{Im}[F_2\mathcal{H}_1 - F_1\mathcal{E}_1]\sin(\phi - \phi_S)\cos\phi + \mathrm{Im}[F_2\widetilde{\mathcal{H}}_1 - F_1\xi\widetilde{\mathcal{E}}_1]\cos(\phi - \phi_S)\sin\phi,$$

where ϕ_s denotes the azimuthal angle of the target polarization vector with respect to the lepton scattering plane.

From 2002 to 2005, HERMES collected data on a transversely polarized hydrogen target. The data set up to 2004 has been analyzed and the results are shown in Fig. 5. They agree with the model calculations [22] shown in the same figure. Since J_u is a parameter in the model for the GPD E, the model calculations confirm that the $\sin(\phi - \phi_s)\cos\phi$ amplitude of the TTSA is sensitive to J_u and thus to E, as expected from Eqn. 7. The calculations of $A_{UT}^{\cos(\phi-\phi_s)\sin\phi}$ on the other hand are almost insensitive to variations on the input values for the GPD model of E, but can in principle be used in order to access the GPDs $\widetilde{H}$ and $\widetilde{E}$. It is interesting to note that the results of the model calculations are largely insensitive to all model parameters but J_u. Taking into account that the GPD H and therefore the available theoretical models will be well constrained by the upcoming

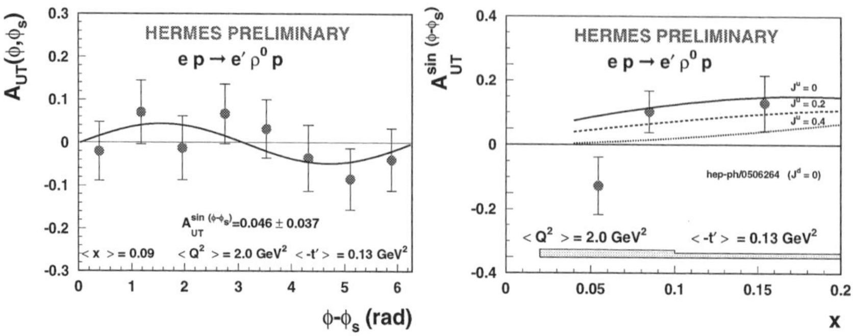

Figure 6. Left panel: Transverse target–spin asymmetry A_{UT} for the hard electroproduction of ρ^0 mesons off the proton as a function of the angle $\phi - \phi_S$. Right panel: The $\sin(\phi - \phi_S)$ amplitude of the transverse target-spin asymmetry as a function of x_B in comparison to model calculations from Ref. [22].

HERMES data, and that the data shown here is less than half of the data taken on the transversely polarized hydrogen target, a model-dependent extraction of J_u with a reasonable precision seems to be possible.

3. Access to GPDs H and E via exclusive ρ^0 production

Besides the TTSA in DVCS, the only other known reaction to access the GPD E on a proton target is the exclusive electroproduction of ρ^0 vector-mesons, $ep \to ep\rho^0$, with the initial proton being transversely polarized. The TTSA is defined as

$$A_{UT}(\phi - \phi_S) = \frac{d\sigma(\phi - \phi_S) - d\sigma(\phi - \phi_S + \pi)}{d\sigma(\phi - \phi_S) + d\sigma(\phi - \phi_S + \pi)} = A_{UT}^{\sin(\phi-\phi_S)} \cdot \sin(\phi - \phi_S). \tag{8}$$

In a strong simplification $A_{UT}^{\sin(\phi-\phi_S)}$ is proportional to the GPD ratio E/H.

Almost half of the HERMES data taken on transversely polarized hydrogen has been analyzed. Events are selected if they contain exactly onc positron with $Q^2 > 1$ GeV^2 and exactly two oppositely charged hadrons. Under the assumption that the two hadrons are pions, the calculated invariant mass of the two hadron system has to be within 0.6 and 1.0 GeV while it has to be above 1.04 GeV if calculated under the assumption that both hadrons are kaons. The exclusivity of the reaction is ensured by requiring the missing energy ΔE of the reaction to be below 0.6 GeV. The TTSA as a function of $\phi - \phi_S$ is shown in the left panel of Fig. 6. The fit shows the expected predominant sinusoidal behavior. The $\sin(\phi - \phi_S)$ amplitude of the TTSA as a function of x_B is shown in the right panel of the same figure. While the data agrees with the model calculations [22] it has to be noted that the model calculations are for longitudinal photons only, for which factorization is proven. The data on the other hand is not yet separated into its longitudinal and transverse parts, which are approximately of equal size at HERMES kinematics. Thus the sensitivity of this TTSA is smaller than the one in DVCS but might still provide additional constraints.

4. Summary

The full HERMES data set on unpolarized targets in addition to the data already taken on the transversely polarized hydrogen target will allow for a model-dependent extraction of the total angular momentum of the u-quark in the nucleon with reasonable precision.

References

1. D. Müller et al., Fortschr. Phys. **42** (1994) 101.
2. X. Ji, Phys. Rev. Lett. **78** (1997) 610, Phys. Rev. **D55** (1997) 7114.
3. A.V. Radyushkin, Phys. Rev. **D56** (1997) 5524.
4. M. Burkardt, Phys. Rev. **D62** (2000) 071503; Erratum–ibid. **D 66** (2002) 119903.
5. A.V. Belitsky and D. Müller, Nucl. Phys. **A711** (2002) 118.
6. J.P. Ralston and B. Pire, Phys. Rev. **D66** (2002) 111501.
7. HERMES Coll., K. Ackerstaff et al., Nucl. Instr. and Meth. **A417** (1998) 230.
8. M. Diehl et al., Phys. Lett. **B411** (1997) 193.
9. A.V. Belitsky, D. Müller and A. Kirchner, Nucl. Phys. **B629** (2002) 323.
10. HERMES Coll., F. Ellinghaus, Proc. of the 15th Intern. Spin Physics Symposium, Upton, New York, AIP Conf. Proc. **675** (2002) 303, hep-ex/0212019.
11. HERMES Coll., F. Ellinghaus, Nucl. Phys. **A711** (2002) 171.
12. F. Ellinghaus for the H1, ZEUS and HERMES Coll., hep-ex/0410094.
13. HERMES Coll., A. Airapetian et al., Phys. Rev. Lett. **87** (2001) 182001.
14. CLAS Coll., S. Stepanyan et al., Phys. Rev. Lett. **87** (2001) 182002.
15. M. Vanderhaeghen, P.A.M. Guichon and M. Guidal, Phys. Rev. **D60** (1999) 094017.
16. K. Goeke, M.V. Polyakov and M. Vanderhaeghen, Prog. Part. Nucl. Phys. **47** (2001) 401.
17. M. Vanderhaeghen, Private Communication, 2001.
18. M.V. Polyakov and C. Weiss, Phys. Rev. **D60** (1999) 114017.
19. I.V. Musatov and A.V. Radyushkin, Phys. Rev. **D61** (2000) 074027.
20. HERMES Coll., M. Kopytin, Proc. of the 13th Intern. Workshop on Deep Inelastic Scattering, Madison, Wisconsin, AIP Conf. Proc. **792** (2005) 424.
21. M. Diehl, Phys. Rept. **388** (2003) 41.
22. F. Ellinghaus, W.-D. Nowak, A.V. Vinnikov and Z. Ye, hep-ph/0506264.

TRANSVERSE SPIN EFFECTS IN SINGLE AND DOUBLE HADRON ELECTROPRODUCTION AT HERMES

B. ZIHLMANN

University of Gent
Dept. Subatomic Physics
Proeftuinstraat 86
B-9000 Gent, Belgium

In the first phase of the HERA II running period the HERMES collaboration is taking data with a transversely polarized Hydrogen target. It was installed during the HERA shutdown in the fall of 2000. The transverse target extends the spin physics program of HERMES to the topic of transversity. It allows to investigate the distribution of transversely polarized quarks in a transversely polarized nucleon. In the data evidence is found not only for a non-zero signal on transversity but also on the Sivers mechanism that can be related to non-zero orbital angular momenta of the quarks in a transversely polarized nucleon.

1. Introduction

Deep inelastic scattering (DIS) of leptons off a nucleon is an excellent tool to investigate the structure of the nucleon. The HERMES experiment uses the 27.6 GeV electron or positron beam from the HERA machine. In combination with a fixed gaseous transversely polarized Hydrogen target[1] the center of mass energy is about $\sqrt{s} = 7$ GeV. The scattered lepton is detected together with some of the produced hadrons in the DIS process in the HERMES spectrometer [2]. The tracking detectors in combination with the spectrometer magnet allow the momentum determination with a resolution between 1.5% and 2.6% depending on the momentum of the particle. For particle identification (PID) four different detectors are available. A lead-glass calorimeter, a pre-shower scintillator hodoscope, a transition radiation detector (TRD) and a dual radiator ring imaging Cherenkov detector (RICH). The latter is a mandatory prerequisite for a clear separation of the different hadronic particle types pions, kaons and protons[3].

2. Helicity amplitudes

In the handbag diagram, shown on the left hand side in fig.1, the DIS process is factorized into a hard part and a soft part. The hard scattering of the photon-quark vertex can be calculated accurately in QED while the soft part retains the unknown information about the structure of the target and has to be parameterized. This lower part of the handbag diagram can be represented in terms of helicity amplitudes dependent on the helicities of the quark and proton propagators. The amplitude is then given in the helicity basis as $A_{\Lambda\lambda,\Lambda'\lambda'}$ where Λ, λ, Λ' and λ' are the nucleon and quark helicity in the initial and final state, respectively. Out of the 16 possible

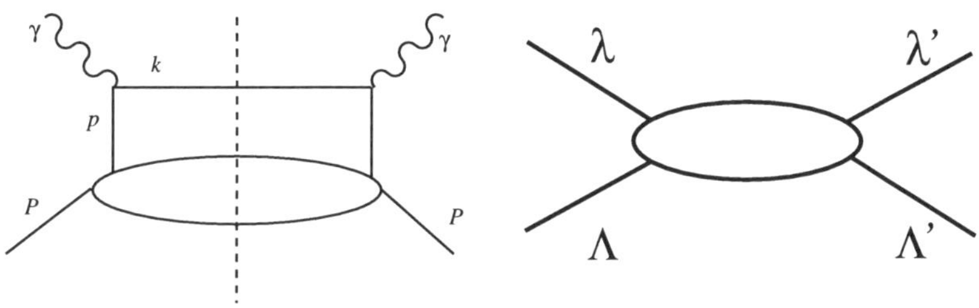

Figure 1. Handbag diagram left and quark-nucleon helicity amplitude right

helicity amplitudes only three are independent and possibly non-zero after applying helicity conservation ($\Lambda + \lambda = \Lambda' + \lambda'$) and parity conservation ($A_{\Lambda\lambda,\Lambda'\lambda'} = A_{-\Lambda-\lambda,-\Lambda'-\lambda'}$). These three amplitudes are:

$$A_{++,++};\ A_{+-,+-};\ A_{+-,-+}\ .$$

The first two of these amplitudes conserve the helicity of the nucleon and quark lines where in the last one $A_{+-,-+}$ the helicity of the quark is flipped as $\lambda \to -\lambda'$ and hence the proton helicity $\Lambda \to -\Lambda'$ in order to conserve the total helicity of the system. These amplitudes can be related to quark distributions with the following relation:

$$f(x) = f_+(x) + f_-(x) \sim Im(A_{++,++} + A_{+-,+-}), \tag{1}$$

$$\Delta f(x) = f_+(x) - f_-(x) \sim Im(A_{++,++} - A_{+-,+-}), \tag{2}$$

$$\delta f(x) = f_\uparrow(x) - f_\downarrow(x) \sim Im(A_{+-,-+}). \tag{3}$$

$f(x)$ is the well known momentum distribution of quarks in the nucleon and $\Delta f(x)$ is the difference of parallel to anti-parallel longitudinally polarized quarks with respect to the longitudinally polarized nucleon. $\delta f(x)$

describes the distribution difference of transversely polarized quarks parallel minus anti-parallel to the parent transversely polarized nucleon. $f_+(x)$ is the distribution of quarks which are polarized parallel to the parent nucleon and $f_-(x)$ is the distribution of quarks which are polarized opposite to the parent nucleon. In the third distribution $\delta f(x)$ the transversity basis ($|\uparrow\rangle$ and $|\downarrow\rangle$) is used for the quark distributions. This is necessary to obtain a probabilistic interpretation of the quark distributions. Using the relations $|\uparrow\rangle = \frac{1}{\sqrt{2}}(|+\rangle + i \mid -\rangle)$ and $|\downarrow\rangle = \frac{1}{\sqrt{2}}(|+\rangle - i \mid -\rangle)$ the helicity flip amplitude can be expressed in this transversity basis and the quark transversity distribution becomes

$$\delta f(x) = f_\uparrow(x) - f_\downarrow(x) \sim Im(A_{\uparrow\uparrow,\uparrow\uparrow} - A_{\uparrow\downarrow,\uparrow\downarrow}). \tag{4}$$

In the case of a gluon-nucleon helicity amplitude a helicity flip amplitude cannot exist due to helicity conservation. The change of the gluon helicity by two units can not be compensated by a helicity change of the nucleon. As a consequence there is no gluonic contribution to $\delta f(x)$ in the spin 1/2 nucleon.

3. Transversity in lepton scattering

In DIS the virtual photon couples to the quark in the nucleon thereby probing its structure. This process conserves helicity. Hence the inclusive DIS process where only the scattered lepton is detected is not suitable to measure transversity. In the handbag diagram this is equivalent to the fact that the helicity of the quark line coupling to the photons does not change. An additional process is needed to take care of a quark helicity flip. Such a possible process is described by the Collins fragmentation function $H_1^\perp(z, (-z\mathbf{k_T})^2)$ [4]. This chiral odd function is the difference of the probability density of a transversely polarized struck quark with its initial spin parallel to the parent nucleon minus anti-parallel to the parent nucleon. The detected final state hadron is a given type. k_T is the transverse momentum of the struck quark. Consequently one has to measure semi-inclusive deep inelastic scattering (SIDIS) processes to get access transversity. Consequently one measures the convolution of the transversity distribution function times the Collins fragmentation function.

3.1. *Single spin azimuthal asymmetry*

In SIDIS the observable that allows to access the product of this distribution function (DF) times the fragmentation function (FF) is a single spin azimuthal asymmetry (SSA). This asymmetry is formed between the SIDIS cross section on the two transverse target polarization states. It is calculated as a function of two angles, the azimuthal angle ϕ between the lepton scattering plane and the hadron production plane and the angle ϕ_S between the lepton scattering plane and the nucleon polarization axis. The kinematics of this process is shown in fig. 2. The signature of the Collins

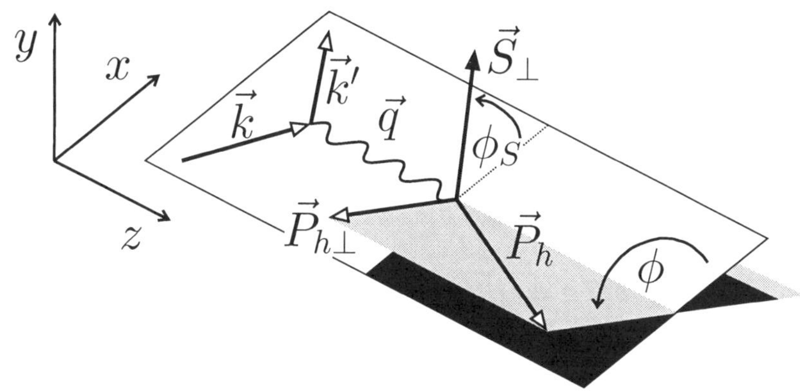

Figure 2. SIDIS on a transversely polarized nucleon target. ϕ_S is the angle between the nucleon polarization axis and the lepton scattering plane (LSP). The angle ϕ is between the LSP and the hadron production plane

mechanism is a $\sin(\phi+\phi_S)$ modulation of the cross section which enters the SSA. There is another combination of DF and FF on leading twist that can generate a SSA in SIDIS. This is the so called Sivers mechanism [5] where the Sivers DF $f_{1T}^{\perp}(x)$ describes the transverse momentum distribution of quarks in a transversely polarized nucleon. It appears in the cross section in combination with the unpolarized FF. The signature of this contribution to the cross section is a $\sin(\phi - \phi_S)$ modulation. The Collins and Sivers mechanism can be extracted from SSA when using a transversely polarized Hydrogen target. In this case the full angular range of ϕ_S from 0 and 2π is covered by the experimental setup. The SSA of an unpolarized lepton beam with a transversely polarized Hydrogen target is given as

$$\begin{aligned}A_{UT}(\phi,\phi_S) &= \frac{1}{S_\perp}\frac{N^{\uparrow}(\phi,\phi_S)-N^{\downarrow}(\phi,\phi_S)}{N^{\uparrow}(\phi,\phi_S)+N^{\downarrow}(\phi,\phi_S)} \\ &\propto \sin(\phi+\phi_S)\frac{\sum_q e_q^2\cdot\delta q(x)\otimes H_1^{\perp q}(z,(zk_T)^2)}{\sum_q e_q^2\cdot q(x)D_1^q(z)} \\ &\quad+\sin(\phi-\phi_S)\frac{\sum_q e_q^2\cdot f_{1T}^{\perp q}(x,p_T^2)\otimes D_1^q(z)}{\sum_q e_q^2\cdot q(x)D_1^q(z)} \\ &\quad+\dots\quad . \end{aligned} \tag{5}$$

Here $N^{\uparrow(\downarrow)}(\phi,\phi_S)$ are the yields of SIDIS events with upward and downward polarized protons, respectively. These values are calculated as function of the two azimuthal angles ϕ and ϕ_S. The sums are over all quark flavors weighted by their charge squared. The product of the DF and FF functions indicated by the symbol $\otimes$ are convolution integrals over the intrinsic transverse momenta of the quarks before (p_T) and after (k_T) the absorption of the virtual photon. In order to obtain the Collins and Sivers moments in an unbiased way the two sine modulations are extracted simultaneously from the data by fitting the asymmetries in a two dimensional grid of the two azimuthal angles ϕ and ϕ_S.

3.2. *Extracting Collins and Sivers*

The current available statistics obtained with the HERMES detector using the 27.6 GeV lepton beam from HERA and the transversely polarized Hydrogen target allows us to calculate the asymmetry A_{UT} in a grid of 12 times 12 bins in ϕ and ϕ_S ranging from 0 to 2π. These asymmetries are then fitted with the form

$$\begin{aligned}A_{UT}^{\pi^\pm}(\phi,\phi_S) &= 2\cdot\langle\sin(\phi+\phi_S)\rangle_{UT}^{\pi^\pm}\cdot\frac{B(\langle y\rangle)}{A(\langle x\rangle,\langle y\rangle)}\cdot\sin(\phi+\phi_S) \\ &\quad+2\cdot\langle\sin(\phi-\phi_S)\rangle_{UT}^{\pi^\pm}\cdot\sin(\phi-\phi_S) \\ &\quad+c_3\cdot\sin(2\phi-\phi_S)+c_4\cdot\sin(\phi_S)+c_5. \end{aligned} \tag{6}$$

In addition to the two different angular dependences of Sivers and Collins moments this fit also takes into account possible contributions from subleading twist cross section terms indicated by $\sin(2\phi-\phi_S)$ and $\sin(\phi_S)$. The factors $A(\langle x\rangle,\langle y\rangle)$ and $B(\langle y\rangle)$ take into account the different y dependence of the Collins and Sivers moments. Here $\langle x\rangle$ is the average Bjorken scaling variable and $\langle y\rangle$ is the average fractional energy of the virtual photon within a given angular bin. The factor 2 is added to be consistent with the "Trento convention" [6].

As indicated by the index $\pi^{\pm}$ the extraction of the moments is done independently for positive and negative pions in the final state, respectively. Due to the quite different physical meaning of the Collins and Sivers mechanisms the results on their moments are discussed in the following separately.

3.3. *Collins moment*

The results for the Collins moment which is proportional to the product of the transversity distribution function and the Collins fragmentation function is shown in figure 3. The amplitude of the $\sin(\phi+\phi_S)$ moment is plotted

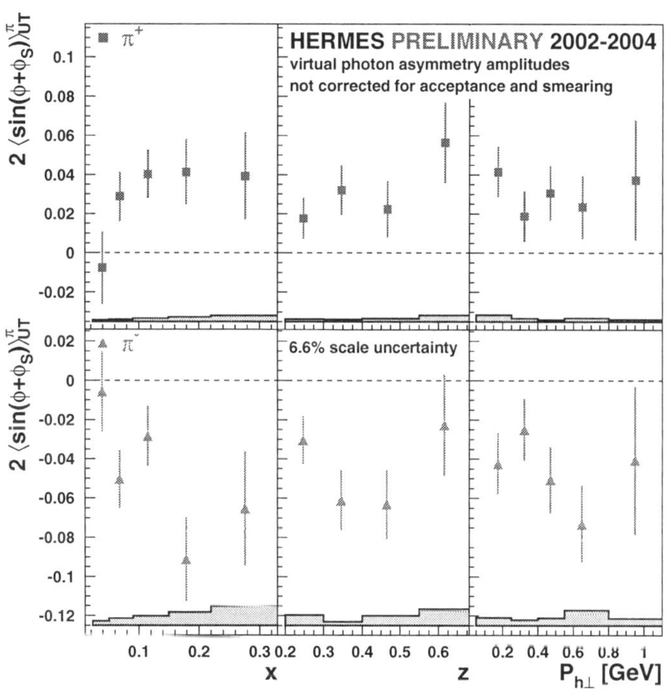

Figure 3. Virtual photon Collins moments as a function of x, z and $P_{h\perp}$. The 6.6% scale uncertainty is due to the target polarization.

as a function of the Bjorken scaling variable x, the fractional hadron energy z and the transverse momentum of the pion. We find a clear positive signal for the Collins moment in the case of a positive pion in the final state. Having negative pions in the final state we find a signal of opposite sign to that of positive pions. The absolute value is as large or even larger. This large value is somewhat surprising as one would expect, due to u-quark dominance, that the negative pions are suppressed similarly as in the longitudinal case. This behavior could imply that the unfavored FF where $u \to \pi^-$ is of similar magnitude as the favored FF. In order to extract from these moments the transversity distribution $\delta f(x)$ the Collins FF is needed. This FF can be measured at the Belle experiment, an e^-e^+ collider experiment. First results of the azimuthal asymmetries have been reported by the Belle collaboration [7]. This is the first step towards the extraction of the Collins FF.

First estimates to describe the HERMES data on the Collins moment are done using parameterizations for transversity and Collins functions in [8]. This study is based on the QCD factorization approach. They find that the unfavored Collins function has the same size as the favored one with opposite sign. However, higher-statistics data is needed to establish this.

3.4. *Sivers moment*

In case of the Sivers mechanism we look for a $\sin(\phi - \phi_S)$ moment in the SSA. It turns out that the amplitude for the $\sin(\phi - \phi_S)$ moments for positive pions are significantly different from zero and positive. This can be seen in figure 4 as a function of x, z and the transverse momentum the pion $P_{h\perp}$. The amplitude is of similar size as for the Collins moment. Since the Sivers mechanism involves the transverse momentum of unpolarized quarks in a polarized nucleon a non-zero signal implies non-vanishing orbital angular momenta of the quarks. While for the positive pions we observe a clear positive signal of the order of 0.045 the results for the negative pions are very small and compatible with zero. It was pointed out in ref. [9] that a non-vanishing SSA caused by a T-odd distribution function, such as the Sivers function, is related to final state interactions. This means that after the hard scattering process the struck quark interacts with the target remnant. In addition it was shown in ref. [10] that this T-odd function changes sign in hadron induced interactions like Drell-Yan. In such a case initial state interactions are taking place, such that the initial quark from one nucleon interacts with the spectator part of the opposite nucleon before it interacts with the quark of that nucleon.

The HERMES results for the Sivers moments were fitted using parame-

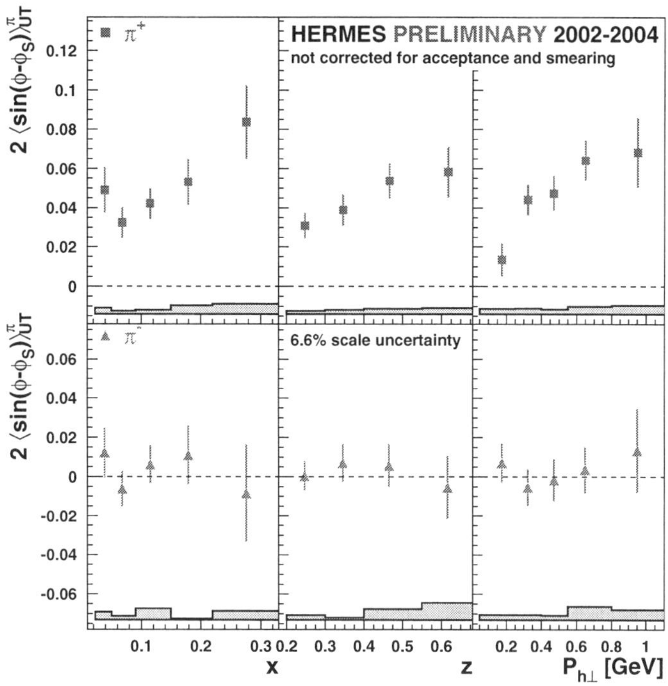

Figure 4. Virtual photon Sivers moments as a function of x, z and $P_{h\perp}$.

terizations for the distribution functions for the u and d quarks taking a Gaussian Ansatz for the p_T and k_T momentum distributions [11]. The conclusion therein is that the Sivers distribution functions for u and d quarks are of the same magnitude but with opposite sign. Such a behavior could explain the very small asymmetries measured in the COMPASS experiment using a transversely polarized deuterium target.

4. Two hadron SSA

Another way to access transversity is to measure SSA with two oppositely charged hadrons in the final state [12]. Assuming that the two hadrons inherit the dependence on the transverse momentum from the same initial struck quark, in relative quantities of the two hadrons like momentum and angle this dependence will cancel[13]. As a consequence the SSA will have no Sivers type contribution but only receive a contribution from transversity. However, the price to pay is - apart from the introduction of a new unknown FF, the interference fragmentation function (IFF) - the much lower statistics available. At leading twist the asymmetry with unpolarized beam and transversely polarized proton target is

$$A_{UT} \sim \sin(\phi_{R\perp} + \phi_S) \sin(\theta) \delta q(x) H_1^{\triangleleft}(z, M_{\pi\pi}) \qquad (7)$$

Here the angle $\phi_{R\perp}$ is the azimuthal angle between the lepton scattering plane and the plane defined by the two hadrons (see fig. 5). The third angle θ is the angle between the positively charged hadron and the direction of the sum of the two hadron momenta in the center of mass system of the two hadrons. In order to gain some understanding about the unknown IFF

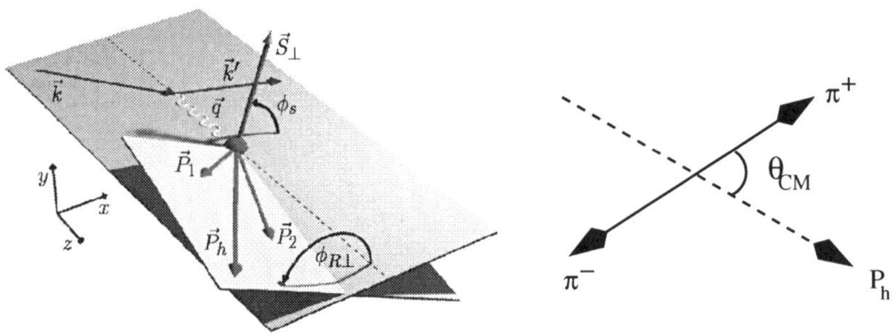

Figure 5. Two-hadron kinematics.

$H_1^{\triangleleft}$ its dependence on the invariant mass of the two hadrons in the final state is investigated using two different theoretical approaches. In a first approach [14] the dependence is estimated using the results from the pion-pion interaction phase shift analysis. The IFF is expanded using Legendre polynomials to separate the s-p and p-p wave interference of the two pions. Another approach uses a spectator model describing the unobserved part of the string break when the quark fragments into hadrons [15]. In this model the number of FF's reduces as the flavor dependence vanishes.

With the HERMES detector we observe a clear signal in the asymmetry A_{UT} with two hadrons in the final state as shown in figure 6. Also shown is the amplitude of the $\sin(\phi_{R\perp} + \phi_S)$ moment as a function of the invariant mass of the two pions in the final state. The signal is positive over the whole measured range. The mass of the ρ^0 meson is indicated by the dashed vertical line. No sign change is observed in this mass region.

The HERMES program with a transversely polarized Hydrogen target has just finished data taking. We recorded in the order of 10 million DIS events

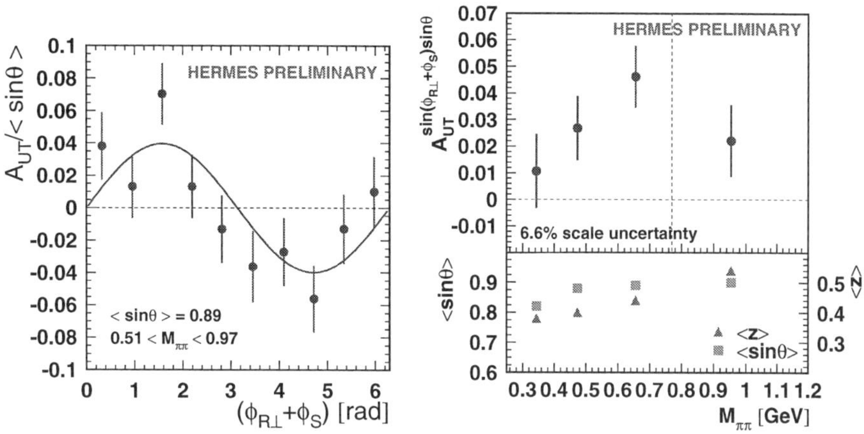

Figure 6. SSA for two hadrons in the final state integrated over the invariant mass range of 0.51 GeV to 0.97 GeV(left) and the extracted $\sin(\phi_{R\perp} + \phi_S)$ moment as a function of the invariant mass of the two pions

in total before data quality cuts. This represents more than double the statistics of the sample on which the results in this report are based.

References

1. A. Airapetian et al, Nucl. Instr. and Meth. A 540, 68(2005)
2. K. Ackerstaff et al., Nucl. Instr. and Meth. A 417, 230(1998)
3. N. Akopov et al., Nucl. Instr. and Meth. A 479, 511(2002)
4. J.C. Collins, Nucl. Phys. B 396, 161(1993)
5. D.W. Sivers, Phys. Rev. D 41, 83(1990)
6. A. Bacchetta et al., Phys. Rev. D 70, 117504(2004)
7. K. Abe et al., hep-ex/0507063
8. W. Vogelsang et al., hep-hp/0507266
9. S.J. Brodsky et al., Phys. Lett. B 530, 99(2002)
10. J. Collins, Phys. Lett. B 536, 43(2002)
11. M. Anselmino et al., hep-ph/0507181
12. J. Collins, Nucl. Phys. B 420, 182(2003)
13. A. Bacchetta, M. Radici, Phys. Rev. D 67, 094002(2003)
14. R. Jaffe et al., Phys. Lett. 80, 1166(1998)
15. M. Radici et. al., Phys. Rev. D65, 074031(2002)

PRESENT UNDERSTANDING OF THE NUCLEON SPIN STRUCTURE IN VIEW OF RECENT EXPERIMENTS*

A. METZ

Institut für Theoretische Physik II,
Ruhr-Universität Bochum, 44780 Bochum, Germany
E-mail: metza@tp2.ruhr-uni-bochum.de

The present understanding of the spin structure of the nucleon in view of recent experiments is briefly reviewed. The main focus is on parton helicity distributions, orbital angular momentum of partons as defined through generalized parton distributions, as well as single spin asymmetries and time-reversal odd correlation functions.

1. Introduction

The history of the non-trivial nucleon spin structure started already in 1933 with the discovery of the anomalous magnetic moment of the proton by Frisch and Stern [1]. This observation led to the important conclusion that the nucleon cannot be pointlike.

In the meantime the field has grown tremendously. This short review concentrates on the QCD spin structure of the nucleon which is usually quantified in terms of various parton distributions. In this context one is dealing with three kinds of parton distributions: (1) forward distributions (quark and gluon helicity distribution), (2) generalized parton distributions (GPDs) which contain information on the orbital angular momentum of partons, (3) transverse momentum dependent distributions (TMDs) which can lead to single spin asymmetries (SSAs). Related experiments are currently running at CERN, DESY, Jefferson Lab, and RHIC.

Many issues like the transversity distribution [2], parton distributions for $x \to 0, 1$, various sum rules, subleading twist etc. cannot be covered. For such topics the reader is referred to existing review articles (like, e.g., Refs. 3–6) and references therein.

*This work is part of the European Integrated Infrastructure Initiative Hadron Physics project under contract number RII3-CT-2004-506078.

2. Parton helicity distributions

2.1. *Quark helicity distribution*

Up to now our knowledge about the quark helicity distribution Δq has mostly come from inclusive lepton scattering off the nucleon. By measuring double spin asymmetries (polarized lepton beam and polarized target) one can extract the structure function $g_1(x, Q^2)$ which is given by

$$g_1^{p,n} = \frac{1}{9}\Delta\Sigma \pm \frac{1}{12}\Delta q_3 + \frac{1}{36}\Delta q_8 \,, \tag{1}$$

with the flavor combinations

$$\begin{aligned} \Delta\Sigma &= (\Delta u + \Delta\bar{u}) + (\Delta d + \Delta\bar{d}) + (\Delta s + \Delta\bar{s}) \,, \\ \Delta q_3 &= (\Delta u + \Delta\bar{u}) - (\Delta d + \Delta\bar{d}) \,, \\ \Delta q_8 &= (\Delta u + \Delta\bar{u}) + (\Delta d + \Delta\bar{d}) - 2(\Delta s + \Delta\bar{s}) \,. \end{aligned} \tag{2}$$

In the past many QCD analyses, using (slightly) different assumptions and different schemes, of polarised DIS data were performed. Information on the first moment of Δq_3 and Δq_8 from beta decay of the neutron and hyperons usually serves as an important independent constraint. The results of such QCD analyses can roughly be summarized as follows: while $\Delta\Sigma$ and Δq_3 are fairly well known, Δq_8 is not known with the same accuracy. In particular, this means that there still exists a considerable uncertainty for the distribution of strange quarks. Most importantly, however, inclusive DIS measurements do not permit to determine Δq and $\Delta\bar{q}$ separately.

At this point additional information can be obtained from semi-inclusive DIS where one extracts the double spin asymmetry

$$A^h \propto \frac{\sum_q e_q^2 \,\Delta q(x)\, D_q^h(z)}{\sum_q e_q^2 \,q(x)\, D_q^h(z)} \,. \tag{3}$$

Detecting one hadron h in the final state not only addresses the distribution of specific quark flavors (e.g., by looking at kaons one can learn something about the strange-quark distribution), but also makes it possible to separate the quark and antiquark distributions, since the fragmentation functions D_q^h, $D_{\bar{q}}^h$ put different weights on Δq and $\Delta\bar{q}$. The results for such an analysis from the HERMES Collaboration [7] are shown in Fig.1, where, in particular, it turned out that the data are consistent with a vanishing sea quark distribution for all three flavors. It has been claimed, however, that the extraction method used in Ref. 7 has some model dependence [8]. Recently, there has been quite some activity aiming at an entirely model-independent analysis of semi-inclusive DIS data [9,10].

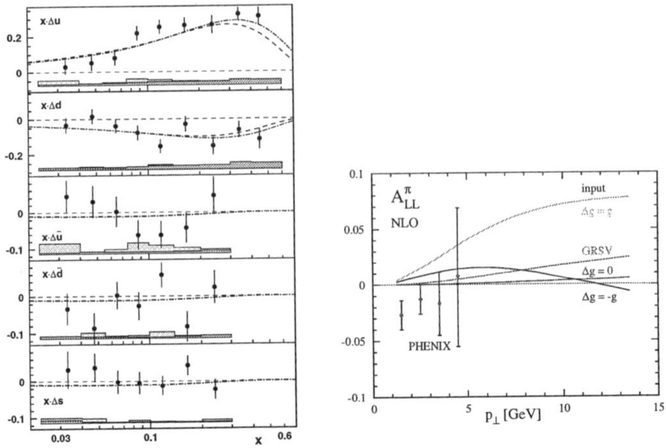

Figure 1. Quark helicity distributions at $\langle Q^2 \rangle = 2.5\,\mathrm{GeV}^2$ from HERMES [7] (left panel), and longitudinal double spin asymmetry for $\vec{p}\vec{p} \to \pi^0 X$ from PHENIX [21] (right panel).

Measuring a parity-violating SSA in $\vec{p}p \to W^\pm X$ can provide complementary information on the helicity distribution of the light flavors [11,12], where for W^+ production one has

$$A_L^{W^+} = \frac{\Delta u(x_1)\bar{d}(x_2) - \Delta\bar{d}(x_1)u(x_2)}{u(x_1)\bar{d}(x_2) + \bar{d}(x_1)u(x_2)}\,. \tag{4}$$

Since x_1 and x_2 are fixed by external kinematics one can disentangle the contributions of the different flavors. If, e.g., x_1 is large, then both $\Delta\bar{d}$ and $\bar{d}$ are small and one can extract the ratio $\Delta u/u$. On the other hand the ratio $\Delta\bar{d}/\bar{d}$ can be obtained if x_2 is large. Analogously, production of W^- allows one to measure $\Delta d/d$ and $\Delta\bar{u}/\bar{u}$. This method is rather clean and is planned to be exploited at RICH. Eventually, polarized neutrino DIS could be used to get additional information on Δs and $\Delta\bar{s}$ [13].

2.2. *Gluon helicity distribution*

In order to get information on the gluon helicity distribution Δg of the nucleon one studies lepton nucleon scattering as well as *pp*-collisions, where different final states are considered in both cases.

In the DIS measurements one tries to isolate the partonic subprocess of photon-gluon fusion (PGF), $\gamma g \to q\bar{q}$. From the experimental point of view, inclusive DIS represents the simplest reaction containing PGF. However, since it only enters through evolution, this process merely provides a rather indirect measurement of Δg. Because of the limited range in x

and Q^2, the currently available data put no strong constraint on Δg, even though a positive Δg is prefered in the analyses. This is in contrast to the situation of the unpolarized gluon distribution, where a lot of information is coming from unpolarized inclusive DIS which has been explored in a wide kinematical region.

A more direct measurement of the PGF process is possible by detecting high-p_T jets or hadron pairs in the final state (see, e.g., Refs. 3–6 and references therein). In this context a special role is played by the production of a pair of charmed mesons created through $\gamma g \to c\bar{c}$, because background processes like the QCD-Compton reaction $\gamma q \to gq$ are automatically suppressed without making specific kinematical cuts. To measure Δg via charm production is a central aim of the COMPASS Collaboration [14].

The only published numbers for Δg from such type of reactions are coming from the production of high-p_T hadron pairs. The measurements of the HERMES [15], COMPASS [16] and SMC [17] Collaborations, performed at different average values of x, yielded

$$\Delta g/g|_{\langle x\rangle=0.17} = 0.41 \pm 0.18\,(\text{stat}) \pm 0.03\,(\text{syst}) \quad (\text{from Ref. 15})\,, \tag{5}$$

$$\Delta g/g|_{\langle x\rangle=0.07} = -0.20 \pm 0.28\,(\text{stat}) \pm 0.10\,(\text{syst}) \quad (\text{from Ref. 17})\,, \tag{6}$$

$$\Delta g/g|_{\langle x\rangle=0.095} = 0.024 \pm 0.089\,(\text{stat}) \pm 0.057\,(\text{syst}) \quad (\text{from Ref. 16})\,. \tag{7}$$

Unfortunately, these data are still suffering from large statistical errors. While the SMC result was obtained in the DIS regime ($Q^2 > 1\,\text{GeV}^2$), HERMES and COMPASS used photoproduction which led to speculations about background contributions from resolved photons.

The second class of processes providing information on Δg are longitudinal double spin asymmetries in proton-proton collisions. To be specific, the following reactions are considered: prompt photon production ($\vec{p}\vec{p} \to \gamma X$), production of heavy flavors ($\vec{p}\vec{p} \to c\bar{c}X$, $b\bar{b}X$), jet production ($\vec{p}\vec{p} \to \text{jet}X$), as well as inclusive production of hadrons ($\vec{p}\vec{p} \to hX$). The processes have already been computed up to NLO in QCD. A detailed discussion of the advantages and drawbacks of the different reactions can be found in Refs. 12, 5, 18, 19, 20 and references therein. At RICH there are extensive ongoing activities in order to study the various channels for different kinematics.

The PHENIX Collaboration presented final data for inclusive production of neutral pions [21]. The asymmetry is shown in Fig.1 as function of the transverse momentum of the pion and compared to a NLO calculation [18]. Measuring A_{LL}^{π} with good statistics at higher values of $p_\perp$, where the sensitivity of the asymmetry to the gluon helicity is larger as compared to the low $p_\perp$ region, can already provide an important constraint on Δg.

3. Generalized parton distributions and orbital angular momentum

Knowing the helicity distributions is not sufficient to understand how the spin of the nucleon is decomposed. One also needs information on the orbital angular momentum of the partons. In 1996 it was shown [22] that generalized parton distributions (see, e.g., Refs. 23–26) can provide the pertinent information. GPDs appear in the description of hard exclusive processes like deep-virtual Compton scattering off the nucleon and meson production, where in both cases data have already been published (see Refs. 26, 27, 28 and references therein). Neglecting the scale dependence, GPDs are functions of three variables, x, ξ, t. While ξ and t, describing the longitudinal and total momentum transfer to the nucleon, are fixed by the external kinematics of an experiment, x is integrated over which complicates the extraction of the x-dependence of GPDs.

GPDs contain a vast amount of physics, and show several interesting properties which put strong constraints on models. They are related to forward parton distributions and nucleon form factors, obey the so-called polynomiality condition [29], and satisfy positivity bounds [30]. Moreover, they contain information on the shear forces partons experience in the nucleon [31]. In particular, they can provide a 3-dimensional picture of the nucleon [32].

Concerning the nucleon spin structure it is important that the total angular momentum (for longitudinal polarization) of quarks is related to the GPDs according to [22]

$$J_q^z = \frac{1}{2} \int_{-1}^{1} dx x \Big[H_q(x, \xi, t = 0) + E_q(x, \xi, t = 0) \Big] \,, \tag{8}$$

where $H_q(x, 0, 0) = q(x)$, while the GPD E_q has no relation to a normal forward distribution. For J_g^z an analogous formula holds. Knowing both the total angular momentum and the helicity of partons allows one to address the orbital angular momentum by means of the decomposition

$$\frac{1}{2} = \sum_q J_q^z + J_g^z = \sum_q \left[\frac{1}{2} \int_0^1 dx \Big(\Delta q(x) + \Delta \bar{q}(x) \Big) + L_q^z \right] + \Delta g + L_g^z \,. \tag{9}$$

(Note that also for a transversely polarized nucleon a decomposition like in (9) has been proposed [33].) Recently, Lattice QCD [34,35] as well as models and phenomenological parametrizations of GPDs [36,37,38] were used to estimate the orbital angular momentum of the quarks. Lattice data, e.g., result in a small contribution to the angular momentum if one sums over the quarks, but the uncertainties of these calculations are still large.

4. Single spin asymmetries

Single spin asymmetries are currently under intense investigation from both the experimental and theoretical point of view. For the process $p^{\uparrow}p \to \pi X$, e.g., Fermi-Lab [39] observed large transverse SSAs (up to 40%) at the *cm*-energy $\sqrt{s} = 20\,\text{GeV}$, and recent results from the STAR Collaboration [40] have shown that the effect survives at $\sqrt{s} = 200\,\text{GeV}$ (see Fig.2). Also for pion production in semi-inclusive DIS transverse SSAs have been measured by the HERMES [41] and COMPASS [42] Collaborations (see also Fig.2).

In general, SSAs are generated by so-called time-reversal odd (T-odd) correlation functions (parton distributions and fragmentation functions). They vanish in leading twist collinear factorization [43]. To get non-zero effects one has to resort to (collinear) twist-3 correlators [44,45] or to transverse momentum dependent functions [46,47,48]. There exist four T-odd leading twist TMD correlation functions, where the Sivers function $f_{1T}^{\perp}$ [49], describing the azimuthal asymmetry of quarks in a transversely polarized target, is the most prominent T-odd parton distribution. In the case of fragmentation the Collins function [50] (transition of a transversely polarized quark into an unpolarized hadron) has attracted a lot of interest, since in semi-inclusive DIS it gets coupled to the transversity distribution of the nucleon.

For A_N in *pp*-collisions both TMD twist-2 and collinear twist-3 correlators were used to describe the data as can be seen in Fig. 2. For the twist-2 analysis, very recently the invoked kinematics has been revisited carefully. As a result it turns out that the Collins mechanism actually cannot explain the data [51], while the Sivers mechanism could well do so [52]. In contrast to A_N, in semi-inclusive DIS at low transverse momentum of the detected hadron one can unambiguously select the Sivers mechanism shown in Fig. 2.

For quite some time it was believed that T-odd TMD distributions like the Sivers function should vanish because of T-invariance of the strong interaction [50], whereas T-odd fragmentation functions may well exist because of final state interactions [50,53]. However, in 2002 a simple spectator model calculation provided a non-zero SSA in DIS [54]. A reanalysis then revealed that in fact the Sivers function can be non-zero, but only if the Wilson-line ensuring color gauge invariance is taken into account in the operator defintion [55]. The presence of the Wilson line which can be process-dependent in turn endangers universality of TMD correlation functions [55,56,57]. This problem affects also the soft factor appearing in factorization formulae for transverse momentum dependent processes. The schematical structure of

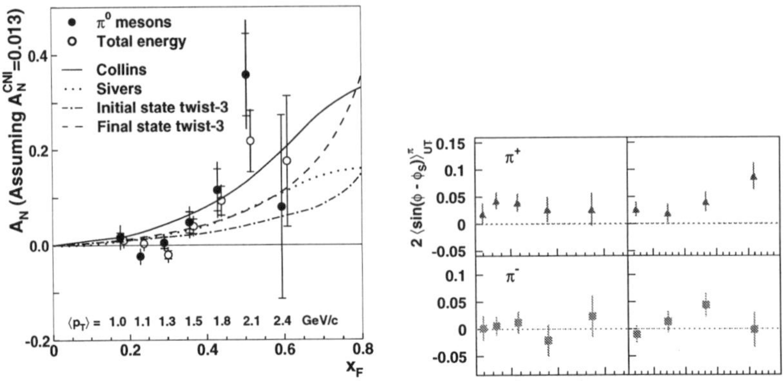

Figure 2. Transverse single spin asymmetries: A_N in $p^{\uparrow}p \to \pi^0 X$ from STAR [40] (left panel), and Sivers asymmetry in $ep^{\uparrow} \to e\pi X$ as function of x (*lhs*) and z (*rhs*) from HERMES [41] (right panel).

the factorization formula for semi-inclusive DIS is [58,59,60]

$$\sigma_{DIS} \propto \text{pdf} \otimes \text{frag} \otimes \hat{\sigma}_{part} \otimes \text{soft}\,. \tag{10}$$

For unintegrated Drell-Yan and $e^+e^- \to h_1 h_2 X$ if the two hadrons are almost back-to-back one is dealing with corresponding formulae. While time-reversal can be used to relate parton distributions in DIS which contain future-pointing Wilson lines to distributions in Drell-Yan with past-pointing lines, this is not possible for fragmentation functions. Nevertheless, by considering the analytic properties of the fragmentation correlator, it can be shown that fragmentation functions are universal [60,61]. This result, in particular, justifies to relate the Collins function in e^+e^--annihilation and semi-inclusive DIS [62]. Also for the soft factor universality between the three mentioned processes can be established [60]. Only T-odd parton distributions are non-universal in the sense that they have a reversed sign in DIS as compared to Drell-Yan, i.e.,

$$f_{1T}^{\perp}\Big|_{DY} = -f_{1T}^{\perp}\Big|_{DIS}\,. \tag{11}$$

This relation should be checked experimentally. We point out that already several publications [63,64,65,66,67] are dealing with extracting the Sivers function from the available data [41,42] in semi-inclusive DIS. There are many more interesting developments in the field of SSAs. For instance, a relation between the sign of the Sivers function and the anomalous magnetic moment of a given quark flavor was given [68]. Moreover, a sum rule relating the Sivers effect for quarks and gluons was derived [69]. It was also proposed to measure the gluon Sivers function through jet correlations in $p^{\uparrow}p$-collisions [70], and charm production ($p^{\uparrow}p \to DX$) [71].

5. Conclusions

We have briefly reviewed the status of the QCD spin structure of the nucleon. Currently, an enormous amount of activities is dealing with this vast and very interesting field.

Historically, the first subject which was studied intensely is the physics of parton helicity distributions, and today we already have a considerable knowledge about the quark helicity distribution. Uncertainties still exist in the strange quark sector and in the separation of valence and sea quark distributions, but many current activities are aiming at an improvement of this situation. In contrast to Δq, the gluon helicity distribution is still just weakly constrained. Nevertheless, a lot of new information, which is supposed to come in the near future from COMPASS and the various measurements at RHIC, will certainly increase our knowledge about Δg.

Also generalized parton distributions can provide important information in order to resolve the spin puzzle of the nucleon, because the orbital angular momentum of partons is related to these objects. Using Lattice QCD as well as phenomenological approaches people have exploited this connection to determine the orbital angular momentum of quarks. At present, the situation is not yet conclusive, but should definitely improve in the future. In particular, many new preliminay data for hard exclusive reactions on the nucleon from COMPASS, HERMES, and Jefferson Lab exist. These data will also help to clarify the role played by orbital angular momentum in the spin sum rule of the nucleon.

The discovery that time-reversal odd parton distributions in general are non-zero gave a strong boost to the interesting subject of single spin asymmetries over the past three years. Since then a lot of progress has been made on both the theoretical but also the experimental side. In this context it has been a crucial discovery that the presence of the Wilson line in transverse momentum dependent correlation functions is mandatory. Because this field in some sense is still rather young, more fundamental results are to be expected. The large amount of already existing, preliminary, and forthcoming data from lepton-nucleon and proton-proton collisions will further improve our understanding of the origin of single spin asymmetries.

References

1. R. Frisch and O. Stern, *Z. Phys.* **85**, 4 (1933).
2. V. Barone, A. Drago, and P. G. Ratcliffe, *Phys. Rep.* **359**, 1 (2002).
3. M. Anselmino, A. V. Efremov, and E. Leader, *Phys. Rep.* **261**, 1 (1995).
4. B. Lampe and E. Reya, *Phys. Rep.* **332**, 1 (2000).
5. B. W. Fillipone and X. Ji, *Adv. Nucl. Phys.* **26**, 1 (2001).
6. S. D. Bass, hep-ph/0411005.
7. HERMES Collaboration (A. Airapetian *et al.*), *Phys. Rev. Lett.* **92**, 012005 (2003); *Phys. Rev.* **D 71**, 012003 (2005).
8. A. Kotzinian, *Eur. Phys. J.* **C 44**, 211 (2005).
9. E. Christova and E. Leader, *Nucl. Phys.* **B 607**, 369 (2001).
10. A. N. Sissakian, O. Yu. Shevchenko, and O. N. Ivanov, *Phys. Rev.* **D 70**, 074032 (2004).
11. N. S. Craigie, K. Hidaka, M. Jacob, and F. M. Renard, *Phys. Rep.* **99**, 69 (1983); and references therein.
12. G. Bunce, N. Saito, J. Soffer, and W. Vogelsang, *Ann. Rev. Nucl. Part. Sci.* **50**, 525 (2000).
13. See, e.g., S. Forte, hep-ph/0501020.
14. COMPASS proposal, CERN/SPSLC 96-14, SPSLC/P297, (1996).
15. HERMES Collaboration (A. Airapetian *et al.*), *Phys. Rev. Lett.* **84**, 2584 (2000).
16. COMPASS Collaboration (E. Ageev *et al.*), hep-ex/0511028.
17. SMC Collaboration (B. Adeva *et al.*), *Phys. Rev.* **D 70**, 012002 (2004).
18. B. Jäger, A. Schäfer, M. Stratmann, and W. Vogelsang, *Phys. Rev.* **D 67**, 054005 (2003).
19. B. Jäger, S. Kretzer, M. Stratmann, and W. Vogelsang, *Phys. Rev. Lett.* **92**, 121803 (2004).
20. B. Jäger, M. Stratmann, and W. Vogelsang, *Phys. Rev.* **D 70**, 034010 (2004).
21. PHENIX Collaboration (S. S. Adler *et al.*), *Phys. Rev. Lett.* **93**, 202002 (2004).
22. X. Ji, *Phys. Rev. Lett.* **78**, 610 (1997).
23. D. Müller *et al.*, *Fortsch. Phys.* **42**, 101 (1994).
24. A. V. Radyushkin, *Phys. Rev.* **D 56**, 5524 (1997).
25. K. Goeke, M. V. Polyakov, and M. Vanderhaeghen, *Prog. Part. Nucl. Phys.* **47**, 401 (2001).
26. M. Diehl, *Phys. Rep.* **388**, 41 (2003).
27. HERMES Collaboration (A. Airapetian *et al.*), *Phys. Lett.* **B 599**, 212 (2004).
28. CLAS Collaboration (C. Hadjidakis *et al.*), *Phys. Lett.* **B 605**, 256 (2005).
29. X. Ji, *J. Phys.* **G 24**, 1181 (1998).
30. P. V. Pobylitsa, *Phys. Rev.* **D 65**, 114015 (2002).
31. M. V. Polyakov, *Phys. Lett.* **B 555**, 57 (2003).
32. M. Burkardt, *Int. J. Mod. Phys.* **A 18**, 173 (2003).
33. B. L. G. Bakker, E. Leader, and T. L. Trueman, *Phys. Rev.* **D 70**, 114001 (2004).

34. LHP Collaboration (P. Hägler *et al.*), *Phys. Rev.* **D 68**, 034505 (2003).
35. QCDSF Collaboration (M. Göckeler *et al.*), *Phys. Rev. Lett.* **92**, 042002 (2004).
36. M. Diehl, T. Feldmann, R. Jakob, and P. Kroll, *Eur. Phys. J.* **C 39**, 1 (2005).
37. M. Guidal, M. V. Polyakov, A. V. Radyushkin, and M. Vanderhaeghen, *Phys. Rev.* **D 70**, 054013 (2005).
38. J. Ossmann *et al.*, *Phys. Rev.* **D 71**, 034011 (2005).
39. E704 Collaboration, (D. L. Adams *et al.*), *Phys. Lett.* **B 264**, 462 (1991).
40. STAR Collaboration, (J. Adams *et al.*), *Phys. Rev. Lett.* **92**, 171801 (2004).
41. HERMES Collaboration, (A. Airapetian *et al.*), *Phys. Rev. Lett.* **94**, 012002 (2005).
42. COMPASS Collaboration, (V. Y. Alexakhin *et al.*), *Phys. Rev. Lett.* **94**, 202002 (2005).
43. G. L. Kane, J. Pumplin, and W. Repko, *Phys. Rev. Lett.* **41**, 1689 (1978).
44. A. V. Efremov and O. V. Teryaev, *Phys. Lett.* **B 150**, 383 (1985).
45. J. Qiu and G. Sterman, *Phys. Rev. Lett.* **67**, 2264 (1991).
46. J. P. Ralston and D. E. Soper, *Nucl. Phys.* **B 152**, 109 (1979).
47. P. J. Mulders and R. D. Tangerman, *Nucl. Phys.* **B 461**, 197 (1996).
48. D. Boer and P. J. Mulders, *Phys. Rev.* **D 57**, 5780 (1998).
49. D. W. Sivers, *Phys. Rev.* **D 41**, 83 (1990); *Phys. Rev.* **D 43**, 261 (1991).
50. J. C. Collins, *Nucl. Phys.* **B 396**, 161 (1993).
51. M. Anselmino *et al.*, *Phys. Rev.* **D 71**, 014002 (2005).
52. U. D'Alesio and F. Murgia, *Phys. Rev.* **D70**, 074009 (2004).
53. A. Bacchetta, R. Kundu, A. Metz, and P. J. Mulders, *Phys. Lett.* **B 506**, 155 (2001).
54. S. J. Brodsky, D. S. Hwang, and I. Schmidt, *Phys. Lett.* **B 530**, 99 (2002).
55. J. C. Collins, *Phys. Lett.* **B 536**, 43 (2002).
56. D. Boer, P. J. Mulders, and F. Pijlman, *Nucl. Phys.* **B 667**, 201 (2003).
57. C. J. Bomhof, P. J. Mulders, and F. Pijlman, *Phys. Lett.* **B 596**, 277 (2004).
58. J. C. Collins and D. E. Soper, *Nucl. Phys.* **B 193**, 381 (1981).
59. X. Ji, J. Ma, and F. Yuan, *Phys. Rev.* **D 71**, 034005 (2005); *Phys. Lett.* **B 597**, 299 (2004).
60. J. C. Collins and A. Metz, *Phys. Rev. Lett.* **93**, 252001 (2004).
61. A. Metz, *Phys. Lett.* **B 549**, 139 (2002).
62. A. V. Efremov, K. Goeke, and P. Schweitzer, *Phys. Lett.* **B 522**, 37 (2001).
63. A. V. Efremov *et al.*, *Phys. Lett.* **B 612**, 233 (2005).
64. M. Anselmino *et al.*, *Phys. Rev.* **D 71**, 074006 (2005).
65. M. Anselmino *et al.*, *Phys. Rev.* **D 72**, 094007 (2005).
66. W. Vogelsang and F. Yuan, *Phys. Rev.* **D 72**, 054028 (2005).
67. J. C. Collins *et al.*, hep-ph/0509076.
68. M. Burkardt, *Phys. Rev.* **D 66**, 114005 (2002).
69. M. Burkardt, *Phys. Rev.* **D 69**, 091501 (2004).
70. D. Boer and W. Vogelsang, *Phys. Rev.* **D 69**, 094025 (2004).
71. M. Anselmino *et al.*, *Phys. Rev.* **D70**, 074025 (2004).

3 Production of Hadrons and Jets

MEASUREMENTS OF α_S AND PARTON DISTRIBUTION FUNCTIONS USING HERA JET DATA

A. M. COOPER-SARKAR
Oxford University Physics Dept.,
Denys Wilkinson Building,
Keble Rd, Oxford, OX1 3RH, UK
E-mail: a.cooper-sarkar@physics.ox.ac.uk

Use of HERA jet production cross-sections can extend our knowledge of the gluon parton distribution function and yield accurate measurements of $\alpha_s(M_Z)$ in addition to illustrating the running of α_s with scale.

1. Introduction

The QCD processes that give rise to scaling violations in DIS at leading order, namely QCD Compton (QCDC) and boson–gluon fusion (BGF), also give rise to distinct jets in the final state provided that the energy and momentum transfer are large enough. These processes are illustrated in Fig. 1. Both BGF and QCDC depend on $\alpha_s(M_Z)$, but the BGF process also depends directly on the gluon parton distribution function (PDF), whereas QCDC depends on quark PDFs. The QCDC process dominates at large Q^2, where the most important quark PDFs are the well-determined u and d valence distributions (for HERA kinematics). Thus measuring jet rates

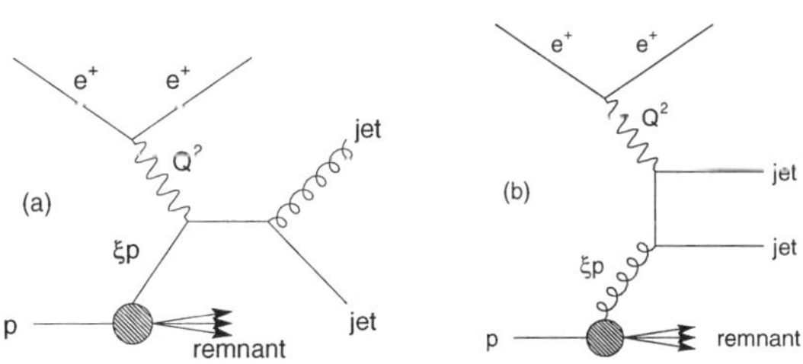

Figure 1. Leading order QCD diagrams from dijet production in DIS. (a) QCD Compton; (b) Boson-gluon fusion.

at large Q^2 allows a determination of $\alpha_s(M_Z)$, with reduced uncertainty from the less well known gluon PDF. Such measurements of $\alpha_s(M_Z)$ are described in Sec. 2.

Conversely, if $\alpha_s(M_Z)$ is well known, then the BGF process gives direct information on the gluon PDF. Historically, PDF fits have been made using NLO QCD in the DGLAP formalism to predict the inclusive cross-sections for world DIS data, but the gluon PDF affects these cross sections only indirectly through the scaling violations. Inputting jet data to such fits can improve the determinations of the gluon distribution. HERA jet data have recently been used for this purpose in the ZEUSJETS PDF fit, as described in Sec. 3. Clearly the combination of inclusive cross-section data and jet production data could lead to a simultaneous determination of $\alpha_s(M_Z)$ and the gluon PDF. An early analysis was performed by H1 [6], and this idea has recently been fully developed into a simultaneous fit of $\alpha_s(M_Z)$ and all the PDFs in the ZEUSJETS-α_s fit [7], as also described in Sec. 3.

2. Measurements of α_s from jet production data

The method used to extract α_s from jet measurements is nicely illustrated on the left hand side of Fig. 2. A programme, like DISENT or NLOJET, is used to calculate NLO QCD cross-sections for jet production, for several fixed values of $\alpha_s(M_Z)$. The parton distribution functions (PDFs) input to these calculations must be compatible with these $\alpha_s(M_Z)$ values. (In practice the CTEQ4A, MRST99 and ZEUS-S variable α_s series have been used). The predictions for each cross-section bin, i, are fitted to the form

$$\sigma_i(\alpha_s(M_Z)) = A_i\alpha_s(M_Z) + B_i\alpha_s^2(M_Z), \tag{1}$$

to determine A_i and B_i so that we have a prediction for any value of $\alpha_s(M_Z)$. We then map the measured value of σ_i and its statistical error onto this function to determine $\alpha_s(M_Z)$ and its statistical error, as illustrated in the figure. Experimental systematic uncertainties are evaluated by repeating the procedure with the measurement shifted according to each source of systematic uncertainty and theoretical uncertainties are evaluated by repeating the procedure with the predictions shifted according to different theoretical assumptions.

A summary of HERA $\alpha_s(M_Z)$ determinations, including those presented at EPS05, is given on the right hand side of Fig. 2. A HERA average has been constructed from these results [1], using the known experimental correlations within each experiment and assuming full correlation of theoretical

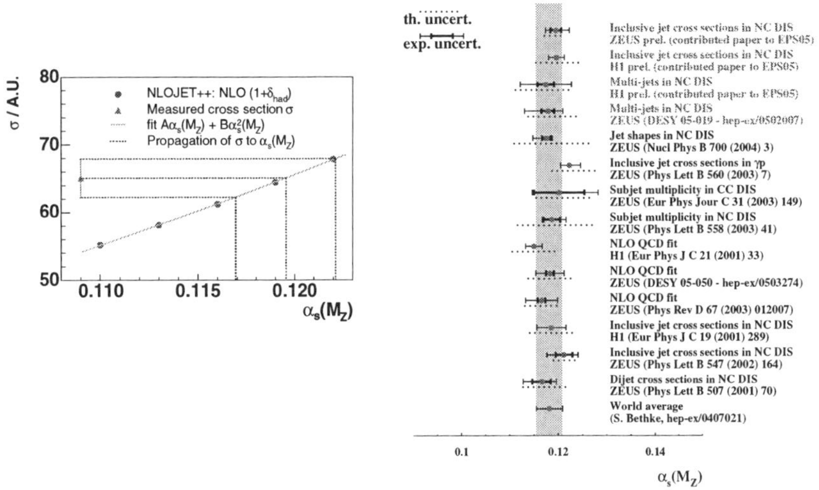

Figure 2. Left side: the method for determining $\alpha_s(M_Z)$. Right side: HERA $\alpha_s(M_Z)$ measurements, compared to the world average

scale uncertainties but no correlation between ZEUS and H1 data. The result is

$$\alpha_s(M_Z) = 0.1186 \pm 0.0011(\text{exp.}) \pm 0.0050(\text{th.})$$

The most accurate results come from inclusive jet production and tri-jet/di-jet/single jet ratios. The recent results for these processes are now discussed in more detail.

ZEUS have measured inclusive jet production in DIS using $81.7pb^{-1}$ of data from the 98-00 running period [2]. The cross-sections $d\sigma/dE_{TB}^{jet}$, where E_{TB}^{jet} is the transverse energy of the jet in the Breit frame, are compared to the predictions of DISENT, for various Q^2 regions, in Fig. 3. The method outlined above has been used to extract $\alpha_s(M_Z)$. The best result is obtained for $Q^2 > 500$ GeV^2, when the theoretical uncertainties are minimized:

$$\alpha_s(M_Z) = 0.1196 \pm 0.0011(\text{stat.}) \pm^{0.0019}_{0.0025} (\text{sys.}) \pm^{0.0029}_{0.0017} (\text{th.})$$

This can be compared to the previous ZEUS result from a similar treatment of data from the 96-97 running period:

$$\alpha_s(M_Z) = 0.1212 \pm 0.0017(\text{stat.}) \pm^{0.0023}_{0.0031} (\text{sys.}) \pm^{0.0029}_{0.0027} (\text{th.})$$

Not only do the higher statistics of the newer data reduce the statistical error, they also allow a reduced estimate of the systematic error. The biggest contribution to the experimental systematic error comes from the

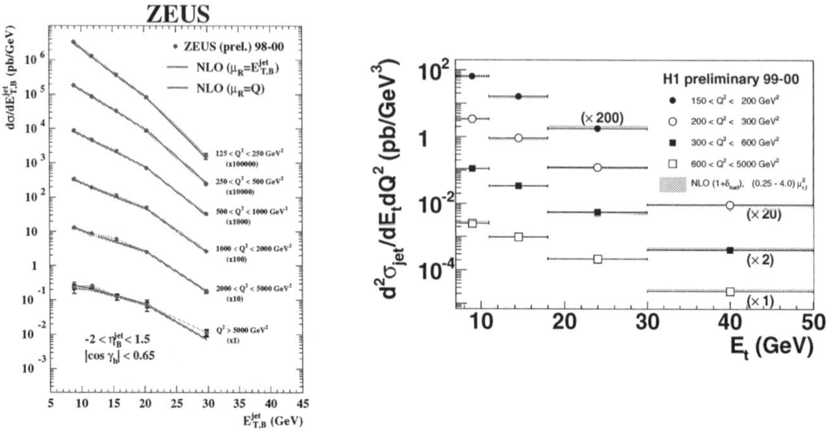

Figure 3. $d\sigma/dE_{TB}$ for inclusive jet production. Left side: ZEUS data. Right side: H1 data

absolute energy scale of the jets. For ZEUS this is $\sim 1\%$, leading to a $\sim 5\%$ error in the jet cross-sections. The theoretical uncertainties come from the usual procedure of varying the renormalisation and factorisation scales ($\mu_R = \mu_F = E_T$) by a factor of two. Fig. 3 also illustrates similar data from H1 from $61pb^{-1}$ of 99/00 running [3]. The value of $\alpha_s(M_Z)$ extracted from these data is

$$\alpha_s(M_Z) = 0.1175 \pm 0.0016(\text{exp.}) \pm^{0.0046}_{0.0048} (\text{th.})$$

The method used to extract α_s can also be used to fit for $\alpha_s(< E_T >)$ or $\alpha_s(< Q^2) >$. Fig. 4 illustrates these results as a function of Q^2, demonstrating the running of α_s with scale, from data within a single experiment.

Both experiments have used data from the same running periods to measure multijet production [4,5]. Previously the ratio of di-jet to single-jet cross-sections has been used to extract α_s. However, increased statistics now allows us to use tri-jet to di-jet cross-sections and these are more sensitive to α_s. The advantage of the ratio technique is that many correlated experimental and theoretical uncertainties cancel out in ratio. The usual method to extract α_s is applied to predictions for $R_3/R_2(\alpha_s(M_Z))$. The result for ZEUS is:

$$\alpha_s(M_Z) = 0.1179 \pm 0.0013(\text{stat.}) \pm^{0.0028}_{0.0046} (\text{sys.}) \pm^{0.0054}_{0.0046} (\text{th.})$$

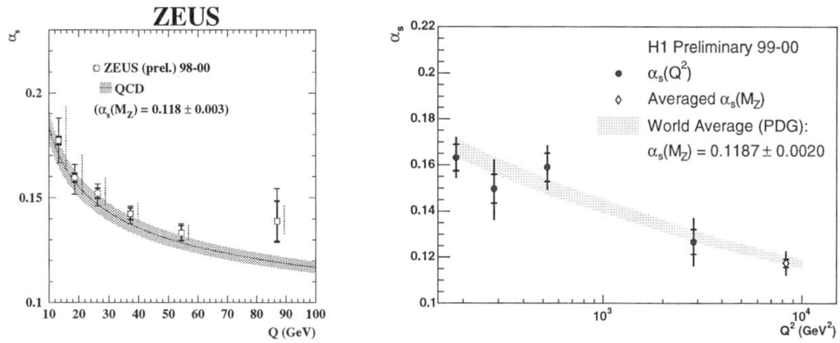

Figure 4. Left side: illustration of running α_s from ZEUS analysis of inclusive jet data. Right side: Illustration of running α_s from H1 analysis of tri-jet to di-jet cross-section ratios

and for H1:

$$\alpha_s(M_Z) = 0.1175 \pm 0.0017(\text{stat.}) \pm 0.005(\text{sys.}) \pm^{0.0054}_{0.0058} (\text{th.})$$

As usual the major contribution to the experimental systematic error is the jet energy scale uncertainty, and for the theoretical uncertainty it is the renormalisation and factorisation scale uncertainty. Fig. 4 also illustrates the running of α_s from the H1 analysis of multijet production.

3. Use of jet production data to determine the gluon and $\alpha_s(M_Z)$ in PDF fits

Both HERA collaborations have recently made PDF fits using exclusively their own data [7,8]. The advantages of a fit within one experiment are that the systematic uncertainties are well understood, and the advantages of a HERA only fit are that these data have the most extensive kinematic coverage of any DIS data, on a pure proton target, hence without the uncertainties of heavy target corrections, and without the need to make strong isospin assumptions. However, such HERA only analyses have only been possible since the 98/00 running extended HERA kinematic coverage to high Q^2, where the contributions of $W^\pm$ and Z exchange become important. This allows us to gain information on the valence d and u quarks at high x, from the e^+p and e^-p charged current cross-sections, respectively. Furthermore, the difference between the e^+p and e^-p high-Q^2 neutral current cross-sections allows us to extract the valence structure function xF_3 for all x. Valence information was the missing piece in a HERA only analysis, since HERA data have provided the major contributions to the measurement of the low-x sea and gluon distributions, through the structure function, F_2, and its scaling violations, $dF_2/dln(Q^2)$, respectively. It has been a limitation of all PDF fits, including the MRST and CTEQ global fits to world data, that information on the high-x gluon is lacking. The global fits have used Tevatron jet data to remedy this. However, the HERA jet data have the advantage of much smaller experimental uncertainties and do not suffer from possible uncertainties due to new physics.

The ZEUSJETS PDF fit [7] uses all the inclusive double differential cross-section data from HERA-I running (94-00) and adds the DIS inclusive jet production data from [9] and the di-jet data from photoproduction [10] from 96-97 running, to gain information on the gluon. As remarked earlier, the jet cross-sections depend on the gluon and quark PDFs and on α_s, and this can be used to break the strong coupling between α_s and the gluon PDF which plagues the indirect extractions from the scaling violations of inclusive data. The fit takes full account of the correlated systematic uncertainties within and between the data sets and these are propagated into the PDF uncertainties using the OFFSET method.

A remark is in order on the use of photoproduced di-jet data. It is well known that at leading order photoproduction can proceed via direct and resolved processes. In the former, the photon interacts directly as a point-like particle, whereas, in the latter, the photon has decomposed into quarks and it is one of these quarks which initiates the interaction. The resolved cross-sections will have some sensitivity to the photon PDF. We avoid this by restricting the analysis to direct enriched cross-sections by the cut

$$x_\gamma^{obs} = \Sigma_i E_T^{jet_i} exp(-\eta^{jet_i})/2yE_e > 0.75$$

Of course at NLO these distinctions cannot be made precisely, but this cut still serves to minimise the influence of the choice of photon PDF.

A new technique had to be developed to include NLO jet cross-sections in a PDF fit, since the computation of such cross-sections is very CPU intensive. Hence the NLO programmes were used to compute LO and NLO weights, $\tilde{\sigma}$, which are independent of α_s and the PDFs, and are obtained by integrating the corresponding partonic hard cross sections in bins of ξ (the proton momentum fraction carried by the incoming parton), μ_F and μ_R. The NLO QCD cross sections, for each measured bin, were then obtained

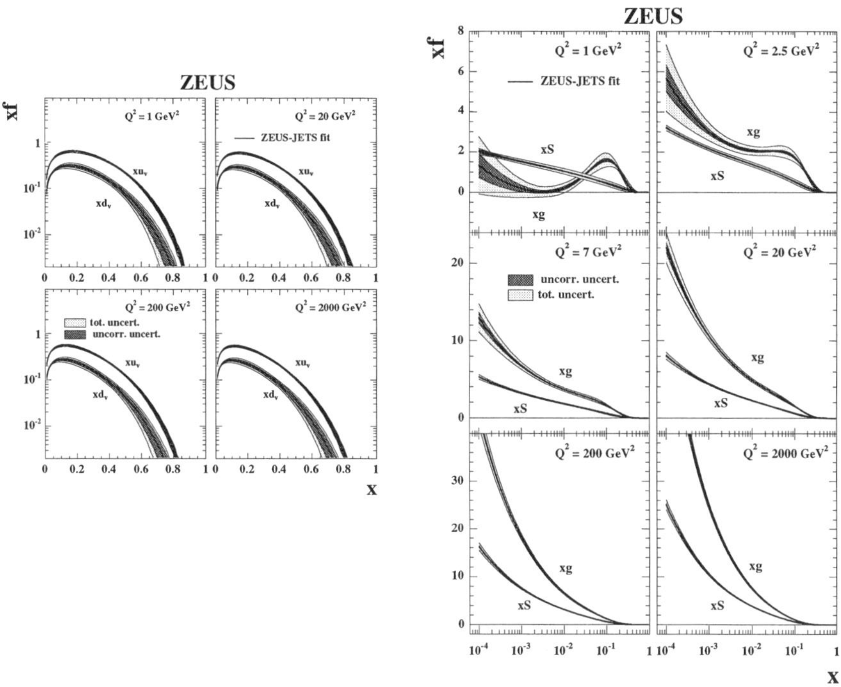

Figure 5. PDFs from the ZEUSJETS PDF fit. Left side: u and d valence PDFs. Right side: sea and gluon PDFs

by folding these weights with the PDFs and α_s according to the formula

$$\sigma = \sum_{n} \sum_{a} \sum_{i,j,k} f_a(\langle \xi \rangle_i, \langle \mu_F \rangle_j) \cdot \alpha_s^n(\langle \mu_R \rangle_k) \cdot \tilde{\sigma}^{(n)}_{a,\{i,j,k\}} \,, \tag{2}$$

where the three sums run over the order n in α_s, the flavour a of the incoming parton, and the indices (i,j,k) of the ξ, μ_F and μ_R bins, respectively. The PDF, f_a, and α_s were evaluated at the mean values $\langle \xi \rangle$, $\langle \mu_F \rangle$ and $\langle \mu_R \rangle$ of the variables ξ, μ_F and μ_R in each (i,j,k) bin. This procedure reproduces the NLO predictions to better than 0.5%.

The ZEUSJETS PDF fit gives a very good χ^2 to all of the input data sets, simultaneously describing inclusive cross-sections and jet production data, thus providing a compelling demonstration of QCD factorisation. The extracted PDFs are shown for various Q^2 in Fig. 5. The ZEUSJETS PDFs are in good agreement with world PDF extractions and are shown compared to the H1 PDFs in Fig. 6. The right hand side of this figure illustrates how the input of the jet data has improved the uncertainty on the gluon PDF, by about a factor of two, in the region $0.1 < x < 0.4$.

Most world PDF fits are performed assuming a fixed value of $\alpha_s(M_Z)$

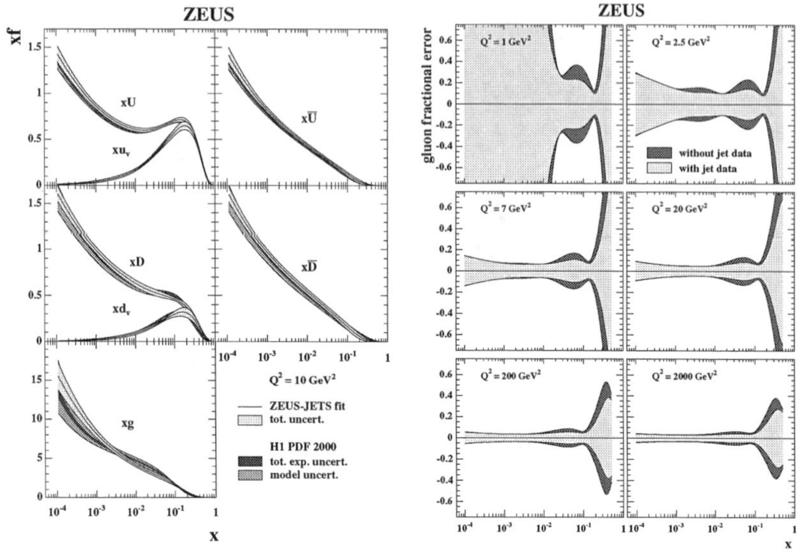

Figure 6. Left side: ZEUSJETS PDFs compared to H1 PDFs 2000. Right side: the fractional uncertainties on the gluon PDF are shown before and after the jet data are included in the ZEUSJETS PDF fit.

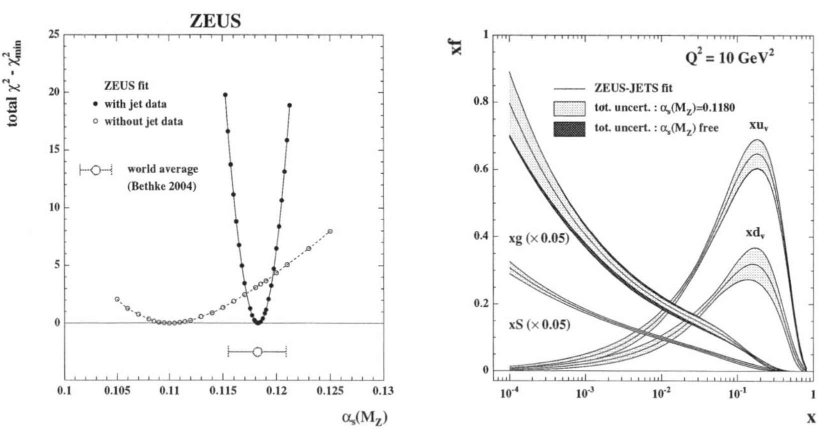

Figure 7. Left side: χ^2 profile for $\alpha_s(M_Z)$ with and without jet data. Right side: the ZEUS JETS$-\alpha_s$ PDFs with the uncertainty from variation of $\alpha_s(M_Z)$ included.

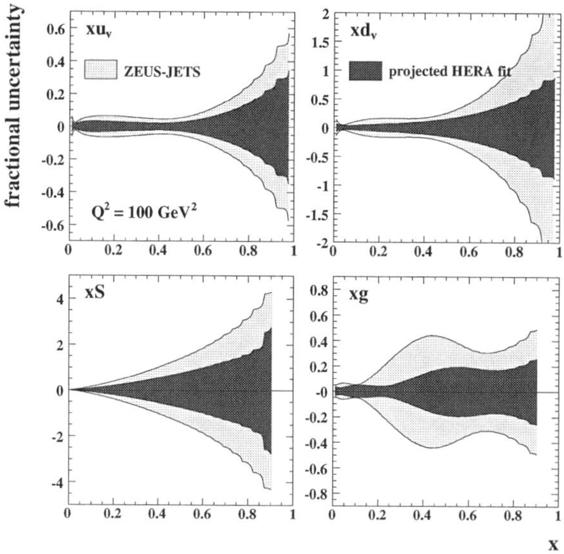

Figure 8. Projected PDF improvements after HERA-II running.

equal to the world average ($\alpha_s(M_Z) = 0.118$ for the ZEUSJETS fit described above). This is because the coupling between the gluon density and the value of $\alpha_s(M_Z)$ in the DGLAP equations does not allow a very accurate measurement of $\alpha_s(M_Z)$. However, with the added information on the gluon which comes from the jet data, we can make an accurate extraction by allowing $\alpha_s(M_Z)$ to be a free parameter simultaneously with all the PDF parameters, in the ZEUSJETS-α_s fit:

$$\alpha_s(M_Z) = 0.1183 \pm 0.0007(\text{stat.}) \pm 0.0027(\text{sys.}) \pm 0.0008(\text{model}) \pm 0.005(\text{th.})$$

where the systematic error contains all the correlated experimental systematics of the contributing data sets. The model error includes uncertainties due to changes in the choice of the proton PDF parametrisation at the starting scale, and in the choice of the photon PDF. The theoretical error comes from variation of the choice of the renormalisation and factorisation scales as usual. Fig. 7 illustrates the contribution of the jet data to the accuracy of the α_s extraction by comparing the χ^2 profile, with and without the jet data. This is the first extraction from HERA data alone because the extractions described in Sec. 2 all require input of PDFs which were extracted using other data. In the ZEUSJETS fit we are able to vary the

$\alpha_s(M_Z)$ used for predicting the jet cross-sections, simultaneously with varying the PDFs. A consequence of this is that the uncertainties on the PDFs due to the uncertainty on $\alpha_s(M_Z)$ are much reduced in the ZEUSJETS-α_s fit, as also illustrated in Fig. 7.

4. Outlook

We can look forward to improving these PDF fits with more jet data in the near future, since the jet data included in the present fit came only from the 96-97 running. Jet data from the 98-00 running are already becoming available, as described in Sec. 2, where they were used for preliminary α_s extractions [2,3]. Combination of ZEUS and H1 data will also yield greater accuracy. In the mid-term future we look forward to improving these measurements with HERA-II data. A study of possible improvements [11] has assumed the achievable goals of a luminosity of $350pb^{-1}$ for both e^+ and e^- running, and $500pb^1$ of jet data. The extra statistics of high Q^2 data will yield improved high-x valence information, and an extraction of xF_3 will extend the valence information down to $x \sim 0.01$. The extra statistics will also improve the high-x sea extraction, whereas the jet data will further improve the high-x gluon. Fig. 8 shows the expected level of improvement in PDF uncertainty.

References

1. C. Glasman, invited talk at DIS05, *hep-ex/0506035*(2005).
2. ZEUS Collaboration, S. Chekanov et al., **ZEUS-prel-05-024**.
3. H1 Collaboration, C. Adloff et al., **H1prelim-05-133**.
4. ZEUS Collaboration, S. Chekanov et al., *hep-ex/0502007* **DESY-05-019** (2005).
5. H1 Collaboration, C. Adloff et al., **H1prelim-05-033**.
6. H1 Collaboration, C. Adloff et al., *Eur. Phys. J.* **C19**, 289 (2001).
7. ZEUS Collaboration, S. Chekanov et al., *Eur. Phys. J.* **C42**, 1 (2005).
8. H1 Collaboration, C. Adloff et al., *Eur. Phys. J.* **C30**, 1 (2005).
9. ZEUS Collaboration, S. Chekanov et al., *Phys. Lett.* **B547**, 164 (2002).
10. ZEUS Collaboration, S. Chekanov et al., *Eur. Phys. J.* **C23**, 615 (2002).
11. C. Gwenlan, A. M. Cooper-Sarkar and C. Targett-Adams, *hep-ph/0509220.*

A NEW PARTON SHOWER ALGORITHM: SHOWER EVOLUTION, MATCHING AT LEADING AND NEXT-TO-LEADING ORDER LEVEL

ZOLTÁN NAGY
Institute for Theoretical Physics, University of Zürich
Winterthurerstrasse 190, CH-8057 Zürich, Switzerland
E-mail: nagyz@physik.unizh.ch

DAVISON E. SOPER
Institute of Theoretical Science, University of Oregon
5203 University of Oregon, OR 97403-5203 Eugene, USA
E-mail: soper@physics.uoregon.edu

In this paper we outline a new parton shower algorithm based on the Catani-Seymour dipole factorization. Our motivation is to have an algorithm which can naturally cooperate with the NLO calculations.

1. Introduction

One often uses perturbation theory to produce predictions for the results of particle physics experiments in which the strong interaction is involved. In order to get usable predictions one has to calculate at least at next-to-leading order to avoid large uncertainties coming from the unphysical scale dependences.

Unfortunately, standard NLO programs have significant flaws. One flaw is that the final states consist just of a few partons, while in nature final states consist of many hadrons. A worse flaw is that the weights are often very large positive numbers or very large negative numbers.

There is another class of calculational tools, the shower Monte Carlo event generators, such as HERWIG [1] and PYTHIA [2]. These have the significant advantage that the objects in the final state consist of hadrons. Furthermore, the weights are never large numbers. Finally, the programs have a lot of important structure of QCD built into them. For at least some cases, the shower Monte Carlo programs can provide a good approximation for the cross sections. The chief disadvantage of typical shower Monte Carlo event generators is that they are based on leading order perturbation theory for the basic hard process. This means that the predictions of the Monte Carlo programs usually have a large theoretical uncertainty. For this reason we cannot use the Monte Carlo prediction as theoretical result in the experimental analysis. This rises an obvious question. Is it possible to build a Monte Carlo program that is predictable?

The parton distribution function or the fragmentation function is a convolution of a perturbative and a non-perturbative part. The perturbative part is the evolution kernel which is a solution of the DGLAP equation. It is based on the collinear factorization property of QCD. Fixing the accuracy of the calculation and the factorization scheme the evolution kernel is well defined. The conventional factorization scheme is the $\overline{\text{MS}}$. The non-perturbative part cannot be calculated, fortunately it is universal and we can obtain it from a fit to the data and use it for other processes. Of course if we change the factorization scheme we have to refit the non-perturbative part.

What is the parton shower and the hadronization? We can think about the latter as a multi-hadron fragmentation function and the parton shower as its evolution kernel. The hadronization model is the non-perturbative part that we fit to the data. The perturbative part is based on the factorization theorem of QCD and is well defined up to the finite pieces. In other words, a different shower algorithm represents a different *shower scheme* but up to the leading (LL) and next-to-leading (NLL) logarithmic accuracy they must be equivalent. If we have some free parameters other than the trivial scales in the parton shower algorithm then the different values of these parameters correspond to different shower schemes and the tuning of the hadronization model must be redone.

How to define a *conventional* shower scheme? These are our criteria: i) It must be Lorentz covariant and Lorentz invariant. ii) It must be correct at LL level. It must be correct at NLL level at least in leading color approximation. iii) The soft gluon effect must be correct at least in leading color approximation. iv) In every step of the shower the phase space configuration must be the exact m-body phase space with the exact phase space weight. The infrared cutoff parameter should be the only cutoff parameter in the algorithm. v) It should "smoothly" work together with the NLO and matrix element matching schemes.

In this paper we aim to outline some new ideas for a parton shower algorithm along our criteria. In the next section we will discuss a parton shower algorithm based on the Catani-Seymour dipole factorization formulas [3] and give a general prescription for the matching of parton shower and fixed order computation at leading and next-to-leading order level. It was an obvious choice to use the dipole factorization because it fulfills our criteria and it makes the NLO level matching easier because lots of modern NLO fixed order computations are based on the dipole subtraction method, for example NLOJET++ [4] and MCFM [5].

Note that this work is still in progress and we have not published a final paper. Since this is a proceedings contribution there is no way to define every detail and prove every statement precisely.

2. Notation

2.1. *Configuration space*

In order to describe showers, we adopt a vector notation for classical statistical physics. In order not to confuse classical states with quantum states, we use the notation $|A)$ to denote a classical state A instead of using $|A\rangle$, which we reserve for a quantum state. The classical inner product is real, with $(A|B) = (B|A)$. We denote a state consisting of two incoming partons a, b and m outgoing partons with momenta and flavours

$$\{p, f\}_{\mathrm{a,b},m} \equiv \{\eta_{\mathrm{a}} p_A, a, \eta_{\mathrm{b}} p_B, b, p_1, f_1, ..., p_m, f_m\} \tag{1}$$

by $|\{p, f\}_{\mathrm{a,b},m})$. We can use these as basis states, with normalization

$$\begin{aligned}(\{p', f'\}_{\mathrm{a,b},m'}|\{p, f\}_{\mathrm{a,b},m}) = \delta_{mm'}\, \delta_{a,a'}\, \delta(\eta_{\mathrm{a}} - \eta'_{\mathrm{a}})\, \delta_{b,b'}\, \delta(\eta_{\mathrm{b}} - \eta'_{\mathrm{b}}) \\ \times \prod_{i=1}^{m} \delta_{f,f'_i}\, \delta^{(4)}(p_i - p'_i)\end{aligned} \tag{2}$$

and completeness relation

$$1 = \sum_m \int [d\{p, f\}_{\mathrm{a,b},m}]\, |\{p, f\}_{\mathrm{a,b},m})(\{p, f\}_{\mathrm{a,b},m}| \ , \tag{3}$$

where we introduced an abbrevation for the integrals and sums, that is

$$\int [d\{p, f\}_{\mathrm{a,b},m}] \equiv \prod_{i=1}^{m} \left\{ \sum_{f_i} \int d^4 p_i \right\} \sum_a \int_0^1 d\eta_{\mathrm{a}} \sum_b \int_0^1 d\eta_{\mathrm{b}} \ . \tag{4}$$

We can use the basis states to express the differential probability $A(\{p, f\}_{\mathrm{a,b},m})$ that a general state $|A)$ consists of partons $\{p, f\}_{\mathrm{a,b},m}$:

$$A(\{p, f\}_{\mathrm{a,b},m}) = (\{p, f\}_{\mathrm{a,b},m}|A) \tag{5}$$

Then

$$|A) = \sum_m \int [d\{p,f\}_{\mathrm{a,b},m}]\, |\{p,f\}_{\mathrm{a,b},m})(\{p,f\}_{\mathrm{a,b},m}|A) \ . \tag{6}$$

Since $(\{p,f\}_m|A)$ is a probability, a physical state $|A)$ must obey the normalization condition $(1|A) = 1$, where $|1)$ is the vector with $(\{p,f\}_{\mathrm{a,b},m}|1) = 1$. More explicitly the unit vector is

$$(1| = \sum_m \int [d\{p,f\}_{\mathrm{a,b},m}]\, (\{p,f\}_{\mathrm{a,b},m}| \ . \tag{7}$$

Given a measurement function F for which the value measured in a parton state $|\{p,f\}_{\mathrm{a,b},m})$ is $F(\{p,f\}_{\mathrm{a,b},m})$, it is convenient to define a vector $|F)$ by

$$(F|\{p,f\}_{\mathrm{a,b},m}) = F(\{p,f\}_{\mathrm{a,b},m}). \tag{8}$$

This is a convenient notation because then the expectation value of F in a state A is $\langle F \rangle_A = (F|A)$.

2.2. *Description of color flow*

We will need a description of the color state that is adapted to a description of shower evolution. If we use an index notation, to each parton with label l there is associated a color index A_l, which takes values $1, \dots, 3$ for a quark or antiquark and takes values $1, \dots, 8$ for a gluon. Thus the matrix element has color and spin components and we can write

$$\mathcal{M}^{A_1,\dots,A_m}_{s_1,\dots,s_m} = \sum_{\{c\}_m} V(\{c\}_m)^{A_1,\dots,A_m} \mathcal{M}(\{p,f,c\}_m)_{s_1,\dots,s_m} \ , \tag{9}$$

where the $V(\{c\}_m)$ form a basis for the color space with labels $\{c\}_m$ to be explained presently and the $\mathcal{M}(\{p,f,c\}_m)$ are expansion coefficients (which are still vectors in spin space). In a vector notation, this is

$$|\mathcal{M}\rangle_{\mathrm{c,s}} = \sum_{\{c\}_m} |V(\{c\}_m)\rangle_{\mathrm{c}} \otimes |\mathcal{M}(\{p,f,c\}_m)\rangle_{\mathrm{s}} \ . \tag{10}$$

Here $|\mathcal{M}\rangle_{\mathrm{c,s}}$ lies in the combined color-spin space while $|V(\{c\}_m)\rangle_{\mathrm{c}}$ is a vector in color space and $|\mathcal{M}(\{p,f,c\}_m)\rangle_{\mathrm{s}}$ is a vector in spin space.

The color basis vectors $|V(\{c\}_m)\rangle_{\mathrm{c}}$ are labelled by a *color string configuration* $\{c\}_m$. We define a string configuration to be a set $\{S_1, \dots, S_n\}$ of one or more *strings* S. We define strings to be of two types, *open strings* and *closed strings*. We define an open string to be an ordered set of parton indices $S = \{l_1, l_2, \dots, l_{n-1}, l_n\}$, where l_1 is the label of an antiquark,

l_n is the label of a quark, and $l_2, \ldots, l_{n-1}$ are labels of gluons. A closed string is an ordered set of parton indices $S = \{l_1, l_2, \ldots, l_{n-1}, l_n\}$, where all of the indices label gluons and where we treat sets that differ by a cyclic permutation of the indices as being the same.

We describe the color string configuration by assigning a color connection label c_i to every external parton with label $i \in \{\mathrm{a}, \mathrm{b}, 1, ..., m\}$. For a gluon, $c_i = (\tilde{\pi}(i), \pi(i))$, for a quark $c_i = (0, \pi(i))$ and for an antiquark $c_i = (\tilde{\pi}(i), 0)$. Here $\pi(i)$ is the label of the next parton on the string while $\tilde{\pi}(i)$ is the label of the previous parton. This labeling respects that every open string starts with a quark and ends with an antiquark. The allowed string configurations are those in which each parton index is an element of one of the strings.

Now we can define the basis states. We take $V(\{c\}_m)$ to be a product

$$V(\{c\}_m)^{A_1,\ldots,A_m} = V(S_1)^{\{A\}_{[1]}}\, V(S_2)^{\{A\}_{[2]}} \ldots V(S_n)^{\{A\}_{[n]}} \,. \tag{11}$$

Here we have denoted the set of color indices represented in string k by $\{A\}_{[k]} = \{A_{l_1}, \ldots, A_{l_n}\}$ if string k is $S_k = \{l_1, \ldots, l_n\}$.

We can now define the component factors $V(S)$. We first consider an open string. For notational convenience, we suppose that the partons along the string are numbered sequentially $S = \{1, 2, \ldots, n\}$. With these labels for the partons, we define

$$V(S)^{\{A\}} = \left[t^{A_2} t^{A_3} \cdots t^{A_{n-1}}\right]_{A_1 A_n} \,, \tag{12}$$

where the t^A are the SU(3) generator matrices for the fundamental representation and we take the A_1, A_n matrix element of the matrix product of the generator matrices. For a closed string with the same parton labels (now all gluons) we define

$$V(S)^{\{A\}} = \mathrm{Tr}\left[t^{A_1} t^{A_2} \cdots t^{A_n}\right] \,. \tag{13}$$

The calculation of a cross section involves a squared matrix element. Since the basis in the color space is orthogonal up to the leading color terms for the matrix element squared we have

$$|\mathcal{M}(\{p,f\}_{\mathrm{a,b},m})|^2 = \sum_{\{c\}_{\mathrm{a,b},m}} |V(\{c\}_{\mathrm{a,b},m})|^2\, |M(\{p,f,c\}_{\mathrm{a,b},m})|^2 + \cdots \,, \tag{14}$$

where the dots stand for the subleading color contributions. They are suppressed by a factor of $1/N_c^2$.

Each term in this sum represents a particular color configuration. Since the shower depends on the color configuration, from each term we generate a different shower. On the other hand, the leading color approximation is useful for nearly collinear parton configurations but it gives a poor description of the hard part of the event. For better accuracy we should begin the showering using the exact matrix element squared reweighted with the probability that the hard partons are in a certain color configuration. For a given color configuration $\{c\}_{\mathrm{a,b},m}$ this probability is

$$p(\{p,f,c\}_{\mathrm{a,b},m}) = \frac{|V(\{c\}_{\mathrm{a,b},m})|^2 \, |M(\{p,f,c\}_{\mathrm{a,b},m})|^2}{\sum\limits_{\{c'\}_{\mathrm{a,b},m}} |V(\{c\}_{\mathrm{a,b},m})|^2 \, |M(\{p,f,c\}_{\mathrm{a,b},m})|^2} \,. \tag{15}$$

Now the color configuration assigned matrix element squared is

$$|\mathcal{M}(\{p,f,c\}_{\mathrm{a,b},m})|^2 = p(\{p,f,c\}_{\mathrm{a,b},m}) \; |\mathcal{M}(\{p,f\}_{\mathrm{a,b},m})|^2 \;, \tag{16}$$

where $|\mathcal{M}(\{p,f\}_{\mathrm{a,b},m})|^2$ is the exact matrix element squared.

It is useful to extend the configuration space by adding color information. Introducing the basis vector of the color space $|\{c\}_{\mathrm{a,b},m})$ the new configuration space is defined as a tensor product

$$\big|\{p,f,c\}_{\mathrm{a,b},m}\big) \equiv \big|\{p,f\}_{\mathrm{a,b},m}\big) \otimes \big|\{c\}_{\mathrm{a,b},m}\big) \tag{17}$$

and the normalization condition reads

$$\big(\{c\}_{\mathrm{a,b},m}\big|\{c'\}_{\mathrm{a,b},m'}\big) = \delta_{mm'}\delta_{c_{\mathrm{a}}c'_{\mathrm{a}}}\delta_{c_{\mathrm{b}}c'_{\mathrm{b}}} \prod_{l=1}^{m} \delta_{c_l c'_l} \;. \tag{18}$$

The integral over the state involves a sum over all possible color configurations

$$\int [d\{p,f,c\}_{\mathrm{a,b},m}] \equiv \int [d\{p,f\}_{\mathrm{a,b},m}] \sum_{\{c\}_{\mathrm{a,b},m}} \;. \tag{19}$$

The m-parton matrix element squared as a vector in the configuration space is

$$\big|\mathcal{M}_m\big) = \int [d\{p,f,c\}_{\mathrm{a,b},m}] \, \big|\{p,f,c\}_{\mathrm{a,b},m}\big) \; |\mathcal{M}(\{p,f,c\}_{\mathrm{a,b},m})|^2 \;, \tag{20}$$

where the color assigned amplitude $|\mathcal{M}(\{p,f,c\}_{\mathrm{a,b},m})|^2$ is defined in Eq. (16).

2.3. *Phase space integrals and Born cross sections*

To define the phase space integral and include the parton distribution function it is useful to define an operator in the configuration space. Let us define

$$
\begin{aligned}
\Gamma = \sum_m \int & \left[d\{p,f,c\}_{\mathrm{a,b},m}\right] \big|\{p,f,c\}_{\mathrm{a,b},m}\big)\big(\{p,f,c\}_{\mathrm{a,b},m}\big| \\
& \times f_{a/A}(\eta_{\mathrm{a}},\mu_F^2)\, f_{b/B}(\eta_{\mathrm{b}},\mu_F^2)\, \frac{1}{2\eta_{\mathrm{a}}\eta_{\mathrm{b}}p_A\cdot p_B}\,\frac{1}{m!} \\
& \times \prod_{i=1}^{m}\left\{\frac{1}{(2\pi)^3}\delta_+(p_i^2)\right\}(2\pi)^4\delta\Big(\eta_{\mathrm{a}}p_A+\eta_{\mathrm{b}}p_B-K-\sum_{i=1}^{m}p_i\Big)\ .
\end{aligned}
\tag{21}
$$

With this definition we can easily define the phase space integral with the proper statistical $(1/m!)$ and flux $(1/2\eta_{\mathrm{a}}\eta_{\mathrm{b}}p_A\cdot p_B)$ factors. Here K is the sum of the momenta of the non-QCD particles.

Now we can define the m-parton Born level cross section as a vector in the configuration space

$$
\big|\sigma_m\big) = \Gamma\big|\mathcal{M}_m\big) \tag{22}
$$

where the vector $|\mathcal{M}_m)$ is given in Eq. (20).

With this notation we can easily define the Born level cross section of any N-jet quantity. Assuming that F is a N-jet sensitive quantity then its cross section is

$$
\sigma[F] = \big(F\big|\sigma_N\big) = \big(F\big|\Gamma\big|\mathcal{M}_m\big)\ , \tag{23}
$$

where $(F|$ is also a vector in the configuration space

$$
\big(F\big| = \sum_{m=N}^{\infty}\int \left[d\{p,f,c\}_{\mathrm{a,b},m}\right] F\big(\{p\}_{\mathrm{a,b},m}\big)\,\big(\{p,f,c\}_{\mathrm{a,b},m}\big|\ . \tag{24}
$$

3. Parton shower evolution

In this section we set up a quite general framework for describing a parton shower. We use an evolution variable t that starts at 0 and runs to ∞ and represents something like $\log(Q^2/k_\perp^2)$, where Q is some hard scale. In principle, the shower evolves to $t=\infty$ but it actually stops at an infrared cutoff t_{f}. The evolving shower is represented by a state $|A(t))$ that begins with an initial state $|A(t_0))$. The evolution is given by a linear operator $U(t',t)$, with

$$
|A(t)) = U(t,t_0)|A(t_0))\ . \tag{25}
$$

These operators have the group composition property

$$
U(t_3,t_2)\,U(t_2,t_1) = U(t_3,t_1)\ . \tag{26}
$$

The evolution operators preserve probabilities:

$$(1|U(t',t)|A) = (1|A) \ . \tag{27}$$

for any state $|A)$. Here $(1|A) = 1$ if the state is normalized to unit probability.

The class of evolution operators we will use is defined by two operators. The first is an infinitesimal generator of evolution or hamiltonian, $\mathcal{H}(t)$, which can be specified by giving its matrix elements

$$\big(\{p',f',c'\}_{\mathrm{a,b},m'}\big|\mathcal{H}(t)\big|\{p,f,c\}_{\mathrm{a,b},m}\big) \ . \tag{28}$$

The second operator is a no-splitting operator $N(t',t)$ that leaves the basis states $|\{p,f,c\}_{\mathrm{a,b},m})$ unchanged except for multiplying each of them by an eigenvalue $\Delta(\{p,f,c\}_{\mathrm{a,b},m};t',t)$:

$$N(t',t)\big|\{p,f,c\}_{\mathrm{a,b},m}\big) = \Delta(\{p,f,c\}_{\mathrm{a,b},m};t',t)\big|\{p,f,c\}_{\mathrm{a,b},m}\big) \ . \tag{29}$$

The evolution operator is expressed in terms of the hamiltonian and the no-splitting operators by

$$U(t_3,t_1) = N(t_3,t_1) + \int_{t_1}^{t_3} dt_2 \ U(t_3,t_2)\,\mathcal{H}(t_2)\,N(t_2,t_1) \ . \tag{30}$$

This equation is interpreted as saying that either the system evolves without splitting from t_1 to t_3, or else it evolves without splitting until an intermediate time t_2, splits at t_2, and then evolves (possibly with further splittings) from t_2 to t_3.

There is a relation between the eigenvalues Δ of N and the matrix elements of $\mathcal{H}$. To derive this we apply the evolution operator for a basis state and using the normalization condition one can get

$$\Delta(\{p,f,c\}_{\mathrm{a,b},m};t_2,t_1) = \exp\left(-\int_{t_1}^{t_2} dt\,\big(1\big|\mathcal{H}(t)\big|\{p,f,c\}_{\mathrm{a,b},m}\big)\right) \ . \tag{31}$$

Note that the probabilistic meaning of Eq. (31) requires that the exponent be negative definite and strictly monotonic in the variable t_2.

One of the main purposes of the parton shower programs is to resum the leading and next-to-leading logarithms which come from the soft and collinear regions. In these phase space regions the matrix elements have the factorization property that any $m+1$ parton matrix element can be written as a product of a singular factor and the corresponding m parton matrix element. It is known that the leading and next-to-leading logarithm are produced by the $1 \to 2$ type splittings. Thus it is enough to consider only the $1 \to 2$ splittings in the splitting operator $\mathcal{H}(t)$ and we have

$$\begin{aligned}(1|\mathcal{H}(t)|\{p,f,c\}_{\mathrm{a,b},m}) = \int [d\{\hat{p},\hat{f},\hat{c}\}_{\mathrm{a,b},m+1}] \\ \times \left(\{\hat{p},\hat{f},\hat{c}\}_{\mathrm{a,b},m+1}\middle|\mathcal{H}(t)\middle|\{p,f,c\}_{\mathrm{a,b},m}\right) .\end{aligned} \tag{32}$$

This is a very formal definition of the shower algorithm. Let us try to define the splitting operators.

4. Splitting operators

The formalism outlined above can be used to describe showering with a variety of splitting operators. The definition of the splitting operators that we suggest is based on the Catani-Seymour dipole factorization formulas and is given by

$$\begin{aligned}&\left(\{\hat{p},\hat{f},\hat{c}\}_{\mathrm{a,b},m+1}\middle|\mathcal{H}(t)\middle|\{p,f,c\}_{\mathrm{a,b},m}\right) \\ &= \sum_{\substack{l,k=\mathrm{a,b},1,\dots,m \\ k\neq l}} \int_0^1 \frac{dy}{y} \int_0^1 dz \int_0^{2\pi} \frac{d\phi}{2\pi}\, \delta\big(t + \log(T_{l,k}(p_l,p_k,z,y)/Q^2)\big) \\ &\times \frac{\alpha_s\big(Q^2 e^{-t}\big)}{2\pi} S_{l,k}(z,y,\hat{f}_{l,1},\hat{f}_{l,2})\, \frac{\hat{\eta}_\mathrm{a}}{\eta_\mathrm{a}} \frac{f_{\hat{a}/A}(\hat{\eta}_\mathrm{a},\mu_F^2)}{f_{a/A}(\eta_\mathrm{a},\mu_F^2)} \frac{\hat{\eta}_\mathrm{b}}{\eta_\mathrm{b}} \frac{f_{\hat{b}/B}(\hat{\eta}_\mathrm{b},\mu_F^2)}{f_{b/B}(\eta_\mathrm{b},\mu_F^2)} \\ &\times A_{f_l}\chi_{l,k}(\{c\}_{\mathrm{a,b},m}) \left(\{\hat{p},\hat{f},\hat{c}\}_{\mathrm{a,b},m+1}\middle|\mathcal{R}_{l,k}(z,y,\kappa_\perp)\middle|\{p,f,c\}_{\mathrm{a,b},m}\right) .\end{aligned} \tag{33}$$

On the right-hand side we have a sum over all the possible emitters ($l \in \{\mathrm{a,b},1,...,m\}$). In the general case when we have initial state hadrons the incoming partons are included among the allowed emitters. There is a sum also over the possible spectator partons ($k \in \{\mathrm{a,b},1,...,m\}$ with $k \neq l$). The splitting is parametrized by the splitting variables y, z and ϕ. The variable y is a virtuality-like variable, z is a momentum fraction variable and ϕ is an azimuthal angle that parametrizes the space-like unit vector $\kappa_\perp$ ($\kappa_\perp^2 = -1$) which is perpendicular to the emitter and spectator momenta.

The dimensionless evolution variable t is given by the function $T_{l,k}$ and depends on the $\{p,f\}_{\mathrm{a,b},m}$ configuration and on the splitting variables y, and z. This function could be any fully infrared sensitive variable. We prefer to use the transverse momentum of the emitted partons as defined in the centre-of-mass frame of the emitter and spectator. The transverse momentum is perpendicular to the momenta of both emitter and spectator.

With this definition, it is boost invariant when we go back to the original frame. The strong coupling is also calculated at the evolution scale.

The splitting kernel is given by the the function $S_{l,k}(z, y, \hat{f}_{l,1}, \hat{f}_{l,2})$. It is derived from the soft and collinear behavior of the tree level matrix elements and related to the Altarelli-Parisi splitting functions. In the case when we have initial state splitting then according to the backward evolution the flavor $\hat{f}_{\mathrm{a},1} \equiv \hat{a}$ is the flavor of the new initial state parton. The spectator can be either a final state or an initial state parton and of course the functional form of the splitting kernel can depend on the type of the spectator parton. Since the functional form of the splitting function depends on the emitter and spectator, one can distinguish four cases. They are

$$S_{l,k}(z, y, \hat{f}_{l,1}, \hat{f}_{l,2}) = \begin{cases} S_{\mathrm{fin}}(z, y, \hat{f}_{l,1}, \hat{f}_{l,2}) & \text{if } l, k \in \{1, ..., m\} \\ \tilde{S}_{\mathrm{fin}}(z, y, \hat{f}_{l,1}, \hat{f}_{l,2}) & \text{if } l \in \{1, ..., m\} \;\; \& \;\; k \in \{\mathrm{a}, \mathrm{b}\} \\ S_{\mathrm{ini}}(z, y, \hat{f}_{l,1}, \hat{f}_{l,2}) & \text{if } l \in \{\mathrm{a}, \mathrm{b}\} \;\; \& \;\; k \in \{1, ..., m\} \\ \tilde{S}_{\mathrm{ini}}(z, y, \hat{f}_{l,1}, \hat{f}_{l,2}) & \text{if } l, k \in \{\mathrm{a}, \mathrm{b}\} \end{cases} \tag{34}$$

When the emitter or the spectator parton is one of the initial state partons, the momentum fraction of the incoming parton changes after the splitting. This effect is reflected by the the ratios of the parton distribution functions in Eq. (33). The parton distribution functions depend on the arbitrary factorization scale μ_F^2. In a shower program μ_F^2 is usually set to the evolution variable. However, the exact choice affects only subleading logarithms.

The operator $\mathcal{R}_{l,k}$ in Eq. (33) describes the $\{p, f, c\}_{\mathrm{a,b},m} \rightarrow \{\hat{p}, \hat{f}, \hat{c}\}_{\mathrm{a,b},m+1}$ transformation and provides the correspondent Jacobians.

In the dipole formalism, a splitting always involves two partons. One is the emitter (labeled by l in Eq. (33)) that splits and produces two daughter partons. The other is the spectator (labeled by k in Eq. (33)) which absorbs the recoiled momentum and plays an important role in the color dynamics. With this construction, one ensures that the momenta are on-shell and that momentum conservation is maintained exactly. Furthermore, the splitting probability includes the exact phase space weight.

How should one choose the spectator parton in a shower picture? In keeping with the role of the spectator in the color dynamics for a soft emission in the Catani-Seymour scheme, the spectator parton should be color connected to the emitter parton. In order to have a good approximation in the soft limit we should include the color connection between the emitter and spectator $(-\boldsymbol{T}_l \cdot \boldsymbol{T}_k/\boldsymbol{T}_l^2)$. Expanding this operator in $1/N_c$ we have

$$\frac{\boldsymbol{T}_l \cdot \boldsymbol{T}_k}{-\boldsymbol{T}_l^2} = A_{f_l}\, \chi_{l,k}(\{c\}_{\mathrm{a,b},m}) + \mathcal{O}\left(\frac{1}{N_c^2}\right) \ , \tag{35}$$

where factor $A_{f_l=g} = 1/2$ for a gluon emitter and for quark or antiquark emitter it is $A_{f_l=q,\bar{q}} = 1$. The function $\chi_{l,k}$ is

$$\chi_{l,k}(\{c\}_{\mathrm{a,b},m}) = \begin{cases} 1 & \text{if } c_l = (k, \pi(l)) \text{ or } c_l = (\pi'(l), k) \\ 0 & \text{otherwise} \end{cases} \ . \tag{36}$$

Remember, our color basis is based on open and closed color strings. The spectator parton is one of the neighbor partons of the emitter on the color string. When the gluon is the emitter there are two possible spectators because the gluon always has two neighbors on the color string while for quarks and antiquarks there is always one possible spectator since they are always at the end of the color string.

4.1. *Final state splitting with final state spectator*

Let us see an example for the splitting functions and for the transformations. In the case when both the emitter and the spectator are final state partons ($l, k \in \{1, ..., m\}$) the splitting function is given by

$$\begin{aligned} S_{\mathrm{fin}}(z, y, f_1, f_2) = & \sum_{r=u,\bar{u},d,\bar{d},\ldots} \Big[\delta_{rf_1}\delta_{\bar{r}f_2}\, S_{q\bar{q}}(z,y) + \delta_{rf_1}\delta_{gf_2}\, S_{qg}(z,y) \\ & \qquad + \delta_{gf_1}\delta_{rf_2}\, S_{qg}(1-z,y) \Big] \\ & + \delta_{gf_1}\delta_{gf_2}\, S_{gg}(z,y) \ . \end{aligned} \tag{37}$$

Then for the splitting of a quark or antiquark into the same flavor quark or antiquark plus a gluon the splitting function is

$$S_{qg}(z,y) = C_{\mathrm{F}} \left[\frac{2}{1 - z(1-y)} - (1+z) \right] \ . \tag{38}$$

For the splitting of a gluon into a quark and an antiquark, we define

$$S_{q\bar{q}}(p, z, y) = T_{\mathrm{R}} \left[1 - 2z(1-z) \right] \ . \tag{39}$$

Finally for the splitting of a gluon into two gluons the splitting function is

$$S_{gg}(z,y) = 2C_{\mathrm{A}} \left[\frac{1}{1 - z(1-y)} + \frac{1}{1 - (1-z)(1-y)} - 2 + z(1-z) \right] \ . \tag{40}$$

The momentum and flavor transfomation is given by the matrix element of the $\mathcal{R}_{l,k_l}$ operator and it is

$$
\begin{aligned}
&\big(\{\hat{p},\hat{f}\}_{\mathrm{a,b},m+1}\big|\mathcal{R}_{l,k}(z,y,\kappa_\perp)\big|\{p,f\}_{\mathrm{a,b},m}\big) \\
&\quad = \frac{1}{2}(1-y)\,\delta^{f_l}_{\hat{f}_{l,1}+\hat{f}_{l,2}}\,\delta^{a}_{\hat{a}}\,\delta(\hat{\eta}_\mathrm{a}-\eta_\mathrm{a})\,\delta^{b}_{\hat{b}}\,\delta(\hat{\eta}_\mathrm{b}-\eta_\mathrm{b})\prod_{\substack{i=1\\ i\neq l}}^{m}\delta^{f_i}_{\hat{f}_i} \\
&\qquad \times \delta^{(4)}(\hat{p}_k-(1-y)p_k)\prod_{\substack{i=1\\ i\neq l,k}}^{m}\delta^{(4)}(\hat{p}_i-p_i) \\
&\qquad \times \delta^{(4)}\big(\hat{p}_{l,1}-zp_l-y(1-z)p_k-[2p_l\cdot p_k yz(1-z)]^{1/2}\kappa_\perp\big) \\
&\qquad \times \delta^{(4)}\big(\hat{p}_{l,2}-(1-z)p_l-yzp_k+[2p_l\cdot p_k yz(1-z)]^{1/2}\kappa_\perp\big)\ .
\end{aligned}
\tag{41}
$$

The evolution variable is the magnitude of the transverse momentum, that is

$$
T_{l,k}(p_l,p_k,z,y)=2p_l\cdot p_k\,y\,z(1-z)\ . \tag{42}
$$

4.2. *Including heavy quark contributions and SUSY-QCD*

The masses of the top quark and the strongly interacting SUSY particles are very big, thus the probability of their production is low. In this work we do not want to deal with these cases because the phase space requires a more complicated treatment. In the future we want to extend our algorithm to heavy and SUSY particles.

5. Shower cross sections

In order to calculate the shower cross sections in shower approximation we have to define the physical state at the starting (hard) scale t_0. The hardest part of the event is always the simplest process which is kinematically possible. For example in e^+e^- annihilation the simplest process is $e^+e^-\to q\bar{q}$, or for Higgs production in proton-proton collision it is $pp\to H$.

For instance, if we want to study jet production in proton antiproton collision then the simplest process is the $2\to 2$ QCD processes and $m=2$. The t_0 is given by the hard part which could be the the invariant mass of the outgoing hard partons. The physical state at the starting scale is the simplest state that is kinematically possible

$$
\big|\sigma(t_0)\big)=\big|\sigma_2\big)\ , \tag{43}
$$

where $|\sigma_2)$ is given in Eq. (22). The shower cross section is just the evolution of the initial state

$$\big|\sigma(t_{\mathrm{f}})\big) = U(t_{\mathrm{f}}, t_0)\big|\sigma(t_0)\big) \ , \tag{44}$$

where the t_{f} scale is the infrared cutoff scale. And the cross section of any jet quantity given by F is

$$\sigma[F] = \big(F\big|D(t_{\mathrm{f}})\big|\sigma(t_{\mathrm{f}})\big) = \big(F\big|D(t_{\mathrm{f}})U(t_{\mathrm{f}}, t_0)\big|\sigma(t_0)\big) \ , \tag{45}$$

where the operator $D(t_f)$ represents the hadronization.

Hadronization is a long distance effect that we cannot calculate from QCD but we have some QCD motivated model for it. In principle the model is universal (as the hadronization). Once the model has been fit to a set of data (*e.g.* LEP data) we do not have to retune the model for another type of process. Although the hadronization model is universal, it is strongly coupled with the shower evolution. That means if we change the shower evolution we have to retune the hadronization model.

6. Adjoint splitting operator

The whole parton shower idea is based on the factorization properties of the matrix elements in the soft and collinear regions. In these regions an $m+1$ parton matrix element can be written as the product of a m-parton matrix element and an universal singular factor. Using this approximation, one can start from a very simple configuration and generate large multiplicity multiparton states.

In our formalism it is essential that the phase space is the exact m-body phase space with the exact weight in every step of the shower. Furthermore the shower is Lorentz invariant. We saw that applying the operator $\mathcal{H}(t)$ to an m-parton state, one obtains an $m+1$ parton state. The probability for the emission is the $m+1$ parton phase space weight times a universal singular factor.

Now we can ask if it is possible to do this the other way around. Suppose that we have generated an $m+1$ parton phase space point. Can we define an operator that acts on this state and finds all the possible m-parton states with corresponding weights such that these would be the weights of an emission if the operator $\mathcal{H}$ was applied on these m-parton states? We claim it is possible to define this operator since the factorization works exactly this way.

Let us define the adjoint splitting operator $\mathcal{H}^\dagger(t)$ that obeys the following condition:

$$\big(F\big|\mathcal{H}(t)\Gamma\big|A\big) = \big(A\big|\mathcal{H}^\dagger(t)\Gamma\big|F\big) \ , \tag{46}$$

where the states $|A)$ and $|F)$ are arbitrary and the operator Γ makes the integrals phase space integrals as defined in Eq. (21). We can see immediately that the operator $\mathcal{H}^\dagger(t)$ always decreases the number of partons because $\mathcal{H}(t)$ always increases it. Since in $\mathcal{H}(t)$ we consider only $1 \to 2$ splittings in $\mathcal{H}^\dagger(t)$ we have only $2 \to 1$ merging. Thus there is only one non-zero matrix element

$$\begin{aligned}&\big(1\big|\mathcal{H}^\dagger(t)\big|\{p,f,c\}_{\mathrm{a,b},m}\big)\\ &\quad = \int \big[d\{\tilde{p},\tilde{f},\tilde{c}\}_{\mathrm{a,b},m-1}\big]\big(\{\tilde{p},\tilde{f},\tilde{c}\}_{\mathrm{a,b},m-1}\big|\mathcal{H}^\dagger(t)\big|\{p,f,c\}_{\mathrm{a,b},m}\big) \ .\end{aligned} \tag{47}$$

From Eq. (46) we can immediately see that for several emissions we can write

$$\big(F\big|\mathcal{H}(t_m)\mathcal{H}(t_{m-1})\cdots\mathcal{H}(t_3)\Gamma\big|A\big) = \big(A\big|\mathcal{H}^\dagger(t_3)\cdots\mathcal{H}^\dagger(t_{m-1})\mathcal{H}^\dagger(t_m)\Gamma\big|F\big) \ . \tag{48}$$

Using Eq. (33) and from the Catani-Seymour dipole subtraction terms [3] for the non-zero matrix elements of operator $\mathcal{H}^\dagger(t)$ we have

$$\begin{aligned}&\big(\{\tilde{p},\tilde{f},\tilde{c}\}_{\mathrm{a,b},m}\big|\mathcal{H}^\dagger(t)\big|\{p,f,c\}_{\mathrm{a,b},m+1}\big)\\ &\quad = \sum_{\substack{i,j\\ \text{pairs}}}\sum_{k\neq i,j}\frac{1}{2p_i\cdot p_j}\frac{\eta_{\mathrm{a}}}{\tilde{\eta}_{\mathrm{a}}}\frac{\eta_{\mathrm{b}}}{\tilde{\eta}_{\mathrm{b}}}\,\langle \boldsymbol{V}_{ij,k}(p_i,p_j,p_k,f_i,f_j)\rangle\\ &\qquad\times A_{f_{ij}}\chi_{ij,k}(\{\tilde{c}\}_{\mathrm{a,b},m})\,\delta\big(t+\log(T_{ij,k}(p_i,p_j,p_k)/Q^2)\big)\\ &\qquad\times \big(\{\tilde{p},\tilde{f},\tilde{c}\}_{\mathrm{a,b},m}\big|\mathcal{Q}_{ij,k}\big|\{p,f,c\}_{\mathrm{a,b},m+1}\big) \ .\end{aligned} \tag{49}$$

Here we sum first over all the possible pairs $\{i,j\}$ of partons that could be the daughter partons from the splitting. The daughter partons must be color connected. Then we sum over possible choices k of the spectator parton. The Dirac delta function maps out the dimensionless evolution variable t, which is the logarithm of the transverse momentum $T_{ij,k}(p_i,p_j,p_k)$ in the splitting. We have the spin averaged splitting function $\langle \boldsymbol{V}_{ij,k}\rangle$ that depends on the momenta of partons $i,j,k \in \{\mathrm{a},\mathrm{b},1,..m+1\}$ and the flavors of partons i and j. It contains a factor of the strong coupling $\alpha_s(k_\perp^2)$. The color connection between the emitter and spectator $(-\boldsymbol{T}_{ij}\cdot\boldsymbol{T}_k/\boldsymbol{T}_{ij}^2)$ is considered only at leading color level, that is $A_{f_{ij}}\chi_{ij,k}$. The momentum, flavor and color transformation prescription is specified by the operator $\mathcal{Q}_{ij,k}$.

7. Matching parton shower and Born level matrix elements

In Section 5 we defined the shower cross section as the shower evolution of the kinematically simplest hard process, for example the $2 \to 2$ process in hadron-hadron collisions. It is obvious that this approximation is very poor if we are interested in cross sections other than the 2-jet cross section. In order to have a better approximation one should consider higher multiplicity exact matrix elements.

Expanding the evolution operator up to the n-th step we get the shower cross section

$$\begin{aligned}
\big|\sigma(t_{\mathrm{f}})\big) &= N(t_{\mathrm{f}}, t_2)\big|\sigma_2\big) \\
&+ \sum_{m=3}^{n-1} \int_{t_2}^{t_{\mathrm{f}}} dt_3 \int_{t_3}^{t_{\mathrm{f}}} dt_4 \cdots \int_{t_{m-1}}^{t_{\mathrm{f}}} dt_m\, N(t_{\mathrm{f}}, t_m)\mathcal{H}(t_m) \\
&\qquad \times N(t_m, t_{m-1})\mathcal{H}(t_{m-1}) \cdots N(t_4, t_3)\mathcal{H}(t_3)\, N(t_3, t_2)\big|\sigma_2\big) \\
&+ \int_{t_2}^{t_{\mathrm{f}}} dt_3 \int_{t_3}^{t_{\mathrm{f}}} dt_4 \cdots \int_{t_{n-1}}^{t_{\mathrm{f}}} dt_n\, U(t_{\mathrm{f}}, t_n)\mathcal{H}(t_n) \\
&\qquad \times N(t_n, t_{n-1})\mathcal{H}(t_{n-1}) \cdots N(t_4, t_3)\mathcal{H}(t_3)\, N(t_3, t_2)\big|\sigma_2\big) \ .
\end{aligned} \tag{50}$$

In every term we have a string of operator products $H(t)N(t, t')$, which provide the probability of non-emission between "times" t' and t followed by a splitting at t. This corresponds to the approximate matrix element squared times the proper Sudakov factors. We want to replace the approximate matrix element with the exact m-parton tree level matrix element while keeping the proper Sudakov weights.

We use the approximation of the matrix elements in the soft and collinear regions when emissions are ordered. One can show that if the state $|\{p, f, c\}_{\mathrm{a,b},m})$ was generated according to the shower procedure then the following is a good approximation

$$\begin{aligned}
\big(\mathcal{M}_m\big|\{p, f, c\}_{\mathrm{a,b},m}\big) &\approx \big(\mathcal{A}_m(t_{\mathrm{f}}, t_2)\big|\{p, f, c\}_{\mathrm{a,b},m}\big) \\
&\equiv \int_{t_2}^{t_{\mathrm{f}}} dt_3 \int_{t_3}^{t_{\mathrm{f}}} dt_4 \cdots \int_{t_{m-1}}^{t_{\mathrm{f}}} dl_m \prod_{k=3}^{m} \frac{\alpha_s(\mu_R^2)}{\alpha_s(Q^2 e^{-t_k})} \\
&\quad \times \big(\mathcal{M}_2\big|\mathcal{H}^\dagger(t_3)\mathcal{H}^\dagger(t_4) \cdots \mathcal{H}^\dagger(t_m)\big|\{p, f, c\}_{\mathrm{a,b},m}\big) \ .
\end{aligned} \tag{51}$$

It is clear that knowing the state $|\{p, f, c\}_{\mathrm{a,b},m})$, the approximate matrix element $(\mathcal{A}_m(t_{\mathrm{f}}, t_2)|\{p, f, c\}_{\mathrm{a,b},m})$ is calculable according to the definition of the operator $\mathcal{H}^\dagger(t)$. However this could be a very big sum with lots of terms. One can immediately see that this approximation is valid only in those regions where the emissions are ordered and the phase space is dominated by soft and collinear radiation.

In the next step we reweight every term in Eq. (50) by a matrix element factor. Whenever $(\mathcal{M}_m|$ is known and the state $|\{p,f,c\}_{\mathrm{a,b},m})$ is generated by the shower procedure, the following ratio is well defined:

$$w_M(\{p,f,c\}_{\mathrm{a,b},m},t_\mathrm{f},t_2) = \frac{(\mathcal{M}_m\big|\{p,f,c\}_{\mathrm{a,b},m})}{(\mathcal{A}_m(t_\mathrm{f},t_2)\big|\{p,f,c\}_{\mathrm{a,b},m})} \; . \tag{52}$$

When $|\mathcal{M}_m)$ is unknown, we can simply set $w_M(\{p,f,c\}_{\mathrm{a,b},m},t_\mathrm{f},t_2) = 1$. Let us define the matrix element reweighting operator

$$\begin{aligned} W_M(t_\mathrm{f},t_2) = \sum_m \int \big[d\{p,f,c\}_{\mathrm{a,b},m}\big] \, \big|\{p,f,c\}_{\mathrm{a,b},m}\big)\big(\{p,f,c\}_{\mathrm{a,b},m}\big| \\ \times w_M(\{p,f,c\}_{\mathrm{a,b},m},t_\mathrm{f},t_2) \; . \end{aligned} \tag{53}$$

Acting on an m-parton state generated by the shower algorithm, this operator replaces the approximate squared matrix element by the exact one (if m is small enough that the squared matrix element is known). Acting on a state with larger m it acts as the unit operator. Using this reweighting operator, the improved shower cross section is

$$\begin{aligned} \big|\sigma_M(t_\mathrm{f})\big) = {} & N(t_\mathrm{f},t_2)\big|\sigma_2\big) \\ & + \sum_{m=3}^{n-1} \int_{t_2}^{t_\mathrm{f}} dt_3 \int_{t_3}^{t_\mathrm{f}} dt_4 \cdots \int_{t_{m-1}}^{t_\mathrm{f}} dt_m \, N(t_\mathrm{f},t_m) W_M(t_\mathrm{f},t_2) \\ & \qquad \times \mathcal{H}(t_m) N(t_m,t_{m-1}) \cdots \mathcal{H}(t_3)\, N(t_3,t_2)\big|\sigma_2\big) \\ & + \int_{t_2}^{t_\mathrm{f}} dt_3 \int_{t_3}^{t_\mathrm{f}} dt_4 \cdots \int_{t_{n-1}}^{t_\mathrm{f}} dt_n \, U(t_\mathrm{f},t_n) W_M(t_\mathrm{f},t_2) \\ & \qquad \times \mathcal{H}(t_n) N(t_n,t_{n-1}) \cdots \mathcal{H}(t_3)\, N(t_3,t_2)\big|\sigma_2\big) \; . \end{aligned} \tag{54}$$

Here it is assumed that the matrix elements $|\mathcal{M}_2),\ldots,|\mathcal{M}_n)$ are known. The nice feature of this formula is that the event is generated according to a shower algorithm starting from the trivial $2 \to 2$ kinematics and whenever the exact m-parton matrix element is known the event is reweighted by the ratio of the tree level amplitude and the approximate shower amplitude. This form is suitable for numerical implementation.

With the exact matrix element correction we do not change the leading and next-to-leading logarithms of the cross section. This is easy to see if we rewrite Eq. (54) in an equivalent form by adding and subtracting the same terms. Thus we have

$$\begin{aligned}\big|\sigma_M(t_{\rm f})\big) &= U(t_{\rm f},t_2)\big|\sigma_2\big) \\ &+ \sum_{m=3}^{n}\int_{t_2}^{t_{\rm f}} dt_3 \int_{t_3}^{t_{\rm f}} dt_4 \cdots \int_{t_{m-1}}^{t_{\rm f}} dt_m\, U(t_{\rm f},t_m)\big[W_M(t_{\rm f},t_2),\, \mathcal{H}(t_m)\big] \\ &\qquad \times N(t_m,t_{m-1})\mathcal{H}(t_{m-1})\cdots\mathcal{H}(t_3)\,N(t_3,t_2)\big|\sigma_2\big)\end{aligned} \tag{55}$$

and the commutator is defined in the usual way, that is

$$\big[W_M(t_{\rm f},t_2),\, \mathcal{H}(t_m)\big] = W_M(t_{\rm f},t_2)\mathcal{H}(t_m) - \mathcal{H}(t_m)W_M(t_{\rm f},t_2)\ . \tag{56}$$

In Eq. (55) in the terms with higher multiplicity matrix element correction the leading (LL) and next-to-leading logarithms (NLL) are removed by the commutator leaving only subleading logarithms and finite pieces. All the LL and NLL logarithms are accumulated in the first term which is actually the standard shower.

7.1. *Matching with Sudakov reweighting*

In this subsection we rewrite our matching formula in an equivalent form. This form is less efficient from the point of numerical implementation but it is very useful for further studies.

After some algebraic manipulation one can find that Eq. (54) can be written in the following form:

$$\begin{aligned}\big|\sigma_M(t_{\rm f})\big) &= \big|\sigma_\Delta(t_{\rm f})\big) \\ &\equiv N(t_{\rm f},t_2)\big|\sigma_2\big) + \sum_{m=3}^{n-1}\int_{t_2}^{t_{\rm f}} dt_m\, N(t_{\rm f},t_m)W_\Delta(t_{\rm f},t_m,t_2)\big|\sigma_m\big) \\ &+ \int_{t_2}^{t_{\rm f}} dt_n\, U(t_{\rm f},t_n)W_\Delta(t_{\rm f},t_n,t_2)\big|\sigma_n\big)\ ,\end{aligned} \tag{57}$$

where the state $|\sigma_m)$ is defined in Eq. (22) and the Sudakov reweighting operator is

$$\begin{aligned}W_\Delta(t_{\rm f},t,t_2) &= \sum_m \int \big[d\{p,f,c\}_{\rm a,b,m}\big]\,\big|\{p,f,c\}_{\rm a,b,m}\big)\big(\{p,f,c\}_{\rm a,b,m}\big| \\ &\times \int_{t_2}^{t} dt_{m-1}\int_{t_2}^{t_{m-1}} dt_{m-2}\cdots\int_{t_2}^{t_4} dt_3 \\ &\times \frac{\big(\mathcal{M}_2\big|N(t_3,t_2)\mathcal{H}^\dagger(t_3)\cdots N(t,t_{m-1})\mathcal{H}^\dagger(t)\big|\{p,f,c\}_{\rm a,b,m}\big)}{\big(\mathcal{A}_m(t_{\rm f},t_2)\big|\{p,f,c\}_{\rm a,b,m}\big)}\ .\end{aligned} \tag{58}$$

This operator provides all the possible emission history with the proper Sudakov factor.

7.2. *Slicing method*

The slicing method was originally defined by Catani, Krauss, Kuhn and Webber [6] for e^+e^- annihilation and was later generalized for processes with initial state partons by Krauss [7] and Mrenna and Richardson [8]. Nowadays this method is very popular and one can find several implementations in the literature [9]. It is usually called the CKKW method. The basic idea is to divide the evolution region into two parts, a hard part and a soft part. In the hard region one uses the exact matrix element supplemented by an appropriate Sudakov factor. The soft part is dominated by soft and collinear radiation, so the shower cross section provides a good approximation.

Defining the matching scale $t_{\mathrm{f}} > t_{\mathrm{ini}} > t_0$ and using the group decomposition property of the evolution operator we then have

$$\big|\sigma(t_{\mathrm{f}})\big) = U(t_{\mathrm{f}}, t_{\mathrm{ini}})U(t_{\mathrm{ini}}, t_2)\big|\sigma_2(t_2)\big) \ . \tag{59}$$

Now we can apply the matching formula for $U(t_{\mathrm{ini}}, t_2)|\sigma_2(t_2))$ according to Eq. (57). The CKKW method uses a further approximation for the Sudakov reweighting operator. Instead of using (57), the emission history and the Sudakov factors are defined according to the $k_\perp$ jet finding algorithm. With this method, a unique emission history is associated with each given n-parton configuration. For the CKKW cross section we have

$$\begin{aligned}\big|\sigma_{\mathrm{CKKW}}(t_{\mathrm{f}})\big) &= U(t_{\mathrm{f}}, t_{\mathrm{ini}})N(t_{\mathrm{ini}}, t_2)\big|\sigma_2\big) \\ &+ \sum_{m=3}^{n-1}\int_{t_2}^{t_{\mathrm{ini}}} dt_m\, U(t_{\mathrm{f}}, t_{\mathrm{ini}})N(t_{\mathrm{ini}}, t_m)W_{\mathrm{CKKW}}(t_{\mathrm{ini}}, t_m, t_2)\big|\sigma_m\big) \\ &+ \int_{t_2}^{t_{\mathrm{ini}}} dt_n\, U(t_{\mathrm{ini}}, t_n)W_{\mathrm{CKKW}}(t_{\mathrm{ini}}, t_n, t_2)\big|\sigma_n\big) \ .\end{aligned} \tag{60}$$

The CKKW method has the advantage that the Sudakov reweighting is very simple. Thus the method is easy to implement. However, we have to pay the price that the result depends on the arbitrary matching scale. Although the t_{ini} dependence is canceled at next-to-leading logarithmic level, this dependence could still be a source of uncertainty [8].

8. Matching parton shower and matrix elements at next-to-leading order level

Matching the parton shower and the NLO computation is a long outstanding problem. Recently some very important progress has been made. One example is the program of Frixione, Nason, and Webber [10], which so far has been applied to cases with massless incoming partons but not to cases with massless final state partons. The other example is that of [11], which concerns three-jet observables in electron-positron annihilation.

In the following we review a general algorithm for matching fixed order NLO computations and parton shower.

8.1. *NLO cross section*

Before we discuss the matching at NLO level it is useful to have a brief review of the structure of the NLO fixed order calculation. According to the dipole method in the general case the cross section is

$$\begin{aligned}\sigma_{\text{NLO}} = &\int_N d\sigma^B + \int_{N+1} \left[d\sigma^R - d\sigma^A\right] \\ &+ \int_N \left[d\sigma^B \otimes \boldsymbol{I}(\epsilon) + d\sigma^V\right]_{\epsilon=0} + \int_N d\sigma^B \otimes \left[\boldsymbol{K} + \boldsymbol{P}(\mu_F)\right] \ .\end{aligned} \tag{61}$$

The first term $d\sigma^B$ is the Born contribution. The second term is an $N+1$ parton phase space integral which gives the contribution of the real radiations $d\sigma^R$. $d\sigma^R$ is singular in some region of the phase space. These singularities are regularized by the dipole subtraction terms $d\sigma^A$. The dipole contributions are based on the dipole factorization formulas [3].

The next term is the contribution of the virtual graphs $d\sigma^V$ with N outgoing external partons. They have also infrared divergences in $d = 4$ dimensions in terms of $1/\epsilon$ (where epsilon is the parameter of the dimensional regularization) but these divergences are cancelled by $d\sigma^B \otimes \boldsymbol{I}$. Furthermore we have some finite N-parton contributions $d\sigma^B \otimes [\boldsymbol{K} + \boldsymbol{P}(\mu_F)]$.

8.2. *Cross section with shower*

Let us calculate the cross section of an infrared safe N-jet quantity. Assuming we know $|\sigma_N)$ and $|\sigma_{N+1})$ the matrix element improved cross section from Eq. (57) is

$$
\begin{aligned}
\big(F_N\big|D(t_{\rm f})\big|\sigma_\Delta(t_{\rm f})\big) = \int_{t_2}^{t_{\rm f}} dt_N \,\big(F_N\big|D(t_{\rm f})N(t_{\rm f},t_N)W_\Delta(t_{\rm f},t_N,t_2)\big|\sigma_N\big) \\
+ \int_{t_2}^{t_{\rm f}} dt_{N+1} \,\big(F_N\big|D(t_{\rm f})U(t_{\rm f},t_{N+1})W_\Delta(t_{\rm f},t_{N+1},t_2)\big|\sigma_{N+1}\big) \ ,
\end{aligned}
\tag{62}
$$

where the operator $D(t_{\rm f})$ represents the hadronization and the function F_N defines the N-jet observable. Since F_N is an infrared safe quantity it is safe to expand this formula in α_s and drop the power corrections which are provided by the hadronization. After the expansion one can show that

$$
\begin{aligned}
\big(F_N\big|D(t_{\rm f})\big|\sigma_\Delta\big) = \int_N d\sigma^B \left(1 + C + \frac{\alpha_s}{2\pi} W_\Delta^{(1)}\right) + \int_{N+1} \left[d\sigma^R - d\sigma^A\right] \\
+ \mathcal{O}(\alpha_s^2) + \mathcal{O}(1\,\text{GeV}/\sqrt{s}) \ .
\end{aligned}
\tag{63}
$$

On the right-hand side we have two integrals. The second term is the $N+1$ parton integral of the real radiation like in Eq. (61). The first term consists of three contributions. We got the Born term $d\sigma^B$. The NLO contribution $d\sigma^B W_\Delta^{(1)}$ comes from the expansion of the Sudakov reweighting operator W_Δ.

The contribution $d\sigma^B C$ is the *error term* of the leading color approximation. Since the splitting operator $\mathcal{H}(t)$ is valid only at leading color level in the soft gluon limit this error term generally is not zero. It is a leading order contribution but it is only a subleading color contribution

$$
C = C^{(0)} + \frac{\alpha_s}{2\pi} C^{(1)} = \mathcal{O}\left(\frac{1}{N_c^2}\right) \ . \tag{64}
$$

On the other hand only next-to-leading logarithms contributes to this term. One can neglect this term but with minor redefinition of Eq. (62) it is possible to make this term vanish [12]. In this proceeding we simply neglect this term.

One can immediately see that the expression on the right-hand side of Eq. (63) is almost the NLO cross section. We got the Born term and the real contributions right but we missed the virtual and some finite NLO contributions. In order to match the shower and the NLO fixed order computation we have to modify Eq. (62). Thus we have

$$
\begin{aligned}
\big(F_N\big|D(t_\mathrm{f})\big|\sigma_\mathrm{NLO}(t_\mathrm{f})\big) &= \int_{t_2}^{t_\mathrm{f}} dt_N\,\big(F_N\big|D(t_\mathrm{f})N(t_\mathrm{f},t_N)W_\Delta(t_\mathrm{f},t_N,t_2)\big|\sigma_N\big) \\
&+ \int_{t_2}^{t_\mathrm{f}} dt_{N+1}\,\big(F_N\big|D(t_\mathrm{f})U(t_\mathrm{f},t_{N+1})W_\Delta(t_\mathrm{f},t_{N+1},t_2)\big|\sigma_{N+1}\big) \\
&+ \int_{t_2}^{t_\mathrm{f}} dt_N\,\big(F_N\big|D(t_\mathrm{f})U(t_\mathrm{f},t_N)W_\Delta(t_\mathrm{f},t_N,t_2)\big|\sigma_N^{NLO}\big)\ .
\end{aligned}
\tag{65}
$$

Here we have an extra term that is basically the shower evolution of the state $|\sigma_N^{NLO})$. This state provides all the missing terms for the NLO matching

$$
\begin{aligned}
\int [d\{c\}_{\mathrm{a,b},N}]\big(\sigma_N^{NLO}\big|\{p,f,c\}_{\mathrm{a,b},N}\big) &= -\frac{\alpha_s}{2\pi}W_\Delta^{(1)}\left|\mathcal{M}_N\right|^2 \\
+\left|\mathcal{M}_N\right|^2 \otimes \big(\boldsymbol{K}+\boldsymbol{P}(\mu_F)\big) &+ \Big[\left|\mathcal{M}_N\right|^2_{1-loop} + \left|\mathcal{M}_N\right|^2 \otimes \boldsymbol{I}(\epsilon)\Big]_{\epsilon=0}\ ,
\end{aligned}
\tag{66}
$$

where $|\mathcal{M}_N|^2$ and $|\mathcal{M}_N|^2_{1-loop}$ are the tree and 1-loop level squared matrix elements, respectively, and the insertion operators $\boldsymbol{I}(\epsilon)$, $\boldsymbol{K}$ and $\boldsymbol{P}(\mu_F)$ are defined according to Eq. (61).

However this matching formula is very *simple* and we still have some caveats. Even if the problem with the error term of the leading color approximation is solved the matching is exact only for the $2 \to 2$ like processes. That is because in the shower the emissions are ordered and we use only the leading-order splitting kernels. This ordering cuts out a certain region of the available phase space which is usually considered in the fixed order NLO calculations. On the other hand this error is negligible because in this region the NLO computation is not reliable.

8.3. *Slicing method at NLO level*

In order to have a simpler Sudakov reweighting procedure one can do the NLO matching with the slicing method. This algorithm is completely worked out for processes in e^+e^- annihilation and the details can be found in Ref. [12].

9. Conclusions

In this paper we outlined a new parton shower algorithm based on the Catani-Seymour dipole factorization formulas. We also introduced a new formalism for describing the shower and we find it very powerful. The main advantages of this new algorithm are the following: i) It is a Lorentz invariant and covariant formalism. ii) The evolution is a transverse momentum ordered algorithm and it is managed that the partons are always onshell, the momentum conservation is maintained and the phase space weight is the exact weight in steps of the evolution. iii) There are no *ambiguous technical* parameters. The only parameter is the infrared cutoff scale parameter. iv) Improved soft gluon treatment. Although the soft gluon limit is still considered in large N_c limit we could make some improvements by taking into account not only the final-final but also the initial-final and initial-initial state color connections.

We derived a more general formula for matching Born level matrix elements and the parton shower. Our matching algorithm is free from any arbitrary matching parameter but we show that the CKKW method is a specialization of our matching procedure.

Since the shower evolution is based on the dipole factorization, the matching of the shower and NLO fixed order computation (based on the Catani-Seymour subtraction method) is rather straightforward.

Acknowledgments

I am grateful to the organizers of the Ringberg workshop for their invitation as well as for providing a pleasant atmosphere during the meeting. This work was supported in part by the Swiss National Science Foundation (SNF) under contract number 200020-109162 and by the Hungarian Scientific Research Fund grant OTKA T-038240.

References

1. G. Marchesini, B. R. Webber, G. Abbiendi, I. G. Knowles, M. H. Seymour and L. Stanco, Comput. Phys. Commun. **67**, 465 (1992); G. Corcella *et al.*, JHEP **0101**, 010 (2001).
2. T. Sjöstrand, Comput. Phys. Commun. **39**, 347 (1986); T. Sjöstrand, P. Eden, C. Friberg, L. Lönnblad, G. Miu, S. Mrenna and E. Norrbin, Comput. Phys. Commun. **135**, 238 (2001); T. Sjöstrand and P. Z. Skands, Eur. Phys. J. C **39**, 129 (2005).
3. S. Catani and M. H. Seymour, Nucl. Phys. B **485**, 291 (1997) [Erratum-ibid. B **510**, 503 (1997)] [arXiv:hep-ph/9605323].
4. Z. Nagy and Z. Trócsányi, Phys. Rev. D **59**, 014020 (1999) [Erratum-ibid. D **62**, 099902 (2000)] Phys. Rev. Lett. **87**, 082001 (2001); Z. Nagy, Phys. Rev. Lett. **88**, 122003 (2002); Phys. Rev. D **68**, 094002 (2003).

5. J. M. Campbell and R. K. Ellis, Phys. Rev. D **60**, 113006 (1999); Phys. Rev. D **62**, 114012 (2000); Phys. Rev. D **65**, 113007 (2002).
6. S. Catani, F. Krauss, R. Kuhn and B. R. Webber, JHEP **0111**, 063 (2001) [arXiv:hep-ph/0109231].
7. F. Krauss, JHEP **0208**, 015 (2002) [arXiv:hep-ph/0205283].
8. S. Mrenna and P. Richardson, JHEP **0405**, 040 (2004) [arXiv:hep-ph/0312274].
9. A. Schalicke and F. Krauss, JHEP **0507**, 018 (2005) [arXiv:hep-ph/0503281]; N. Lavesson and L. Lonnblad, JHEP **0507**, 054 (2005) [arXiv:hep-ph/0503293].
10. S. Frixione and B. R. Webber, JHEP **0206**, 029 (2002); S. Frixione, P. Nason and B. R. Webber, JHEP **0308**, 007 (2003).
11. M. Krämer and D. E. Soper, Phys. Rev. D **69**, 054019 (2004); D. E. Soper, Phys. Rev. D **69**, 054020 (2004).
12. Z. Nagy and D. E. Soper, JHEP **0510**, 024 (2005) [arXiv:hep-ph/0503053].

JET PRODUCTION AT HERA

D. TRAYNOR
Department of Physics
Queen Mary, University of London,
Mile End Road, London, E1, UK
E-mail: d.traynor@qmul.ac.uk

1. Introduction

Both H1 and ZEUS have recently presented new results on jet production using the high statistics and well understood data set from the 98/2000 running period of HERA at a centre-of-mass energy of $\sqrt{s} = 319$ GeV. Of particular interest are: inclusive jet production in high Q DIS [1,2], multi-jet production in high Q DIS [3,4] and high E_t dijet production in photoproduction [5]. The large scales involved in these processes (Q or jet E_t) provide a safe region of phase space for a comparison to DGLAP based NLO QCD predictions [6,7,8], the extraction of the strong coupling constant α_s and for inclusion into fits to extract the proton structure[a]. This paper will concentrate on a discussion of the uncertainties contributing to the experimental and theoretical errors on the jet cross sections.

The Breit frame is the preferred frame to measure jet production in DIS. In this frame contributions from the Born level and jets induced by the beam remnant are suppressed. In photoproduction the laboratory frame is used. Jets are reconstructed using the $k_\perp$ cluster algorithm in the longitudinally invariant inclusive mode.

2. Common Experimental Issues

The experimental data are corrected for limited detector acceptance and resolution and, in DIS, for QED radiative effects. Monte Carlo simulations are used to calculate these correction factors. In DIS the programs

[a]For a discussion on the last two issues see the contribution in these proceedings by Amanda Cooper-Sarkar: Measurements of α_s and Parton Distribution Functions using HERA Jet Data.

DJANGO[9] and RAPGAP[10] are used, implementing either the colour-dipole model or the parton shower model for the parton cascade. In photoproduction PYTHIA[11] and HERWIG[12] are used, implementing respectively the Lund string or cluster hadronisation models. The uncertainty in the correction factors arising from the model implementation is taken from the differences in the results obtained from the two models.

The uncertainty in the hadronic energy scale is quoted for ZEUS[13] as $\pm 1\%$ for jets with $E_t^{jet} > 10$ GeV, else $\pm 3\%$. For H1 the hadronic energy scale uncertainty is $\pm 2\%$ in high Q DIS and $\pm 1.5\%$ in high E_t^{jet} dijet photoproduction. The resultant systematic errors are highly correlated from bin to bin.

3. Common Theoretical Issues

The jet production cross section in perturbative QCD is given by the convolution of the proton PDF with the hard subprocess cross section. This provides a parton level prediction for the jet cross section,

$$\sigma_{jet} = \sum_{i=q,\bar{q},g} \int dx f_i(x, \mu_F, \alpha_s) \hat{\sigma}_{QCD}(x, \mu_F, \mu_R, \alpha_s(\mu_R)).(1 + \delta_{had}),$$

where x is the fraction of the proton's momentum taken by the interacting parton, f_i is from the proton PDF, μ_F is the factorisation scale, $\hat{\sigma}_{QCD}$ is the subprocess cross section and μ_R is the renormalisation scale. In order to compare with the measured cross section an additional correction factor is required, δ_{had} , to take into account the nonperturbative effect of hadronisation. Each component of the theoretical calculation has an associated uncertainty.

The proton parton density function has to be extracted from data and the global analyses presently used rely on a variety of data to make fits of the proton structure. In recent years significant progress has been made in the estimation of the uncertainties of these PDFs. Most analyses presented here include a contribution to the theoretical uncertainty calculated from the 40 eigenvectors of the CTEQ6 PDF analysis [14].

The choice of factorisation and renormalisation scales is to some extent arbitrary although sensible choices are: Q the hard scale in DIS, E_t the transverse energy of the jet (the only choice in photoproduction), or a function of the two. The theoretical uncertainty due to terms beyond NLO is obtained by varying the choice of the scale for μ_R and μ_F by a factor of four.

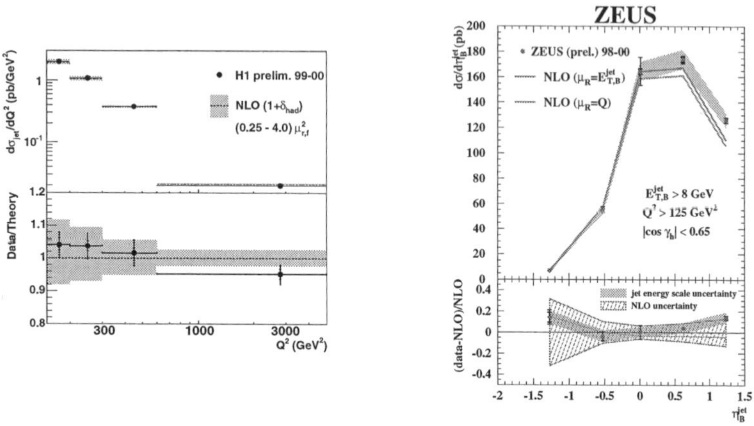

Figure 1. Left: $d\sigma_jet/dQ^2$ measured by H1. Right: $d\sigma/d\eta^{jet}_{Breit}$ measured by ZEUS. Both are compared with NLO QCD predictions corrected for hadronisation effects.

The hadronisation correction factor, δ_{had}, is calculated using the Monte Carlo models. An assumption is made that the description of the final state by the Monte Carlo simulation after the parton cascade is equivalent to that of the NLO calculation. Then the effect of the hadronisation on the Monte Carlo jet cross sections can directly be applied to the NLO prediction. The uncertainty on δ_{had} is calculated from the difference between two Monte Carlo models.

4. Inclusive Jet Production in DIS

The inclusive jet cross section has been measured by H1[1] as a function of jet E_t in four Q^2 bins, as a function of Q^2 integrated over E_t (figure 1 left) and of E_t integrated over Q^2. H1 have used a luminosity of about 61 pb^{-1} and find about 20,000 events in the phase space defined by:

$$150 < Q^2 < 5000\ \text{GeV}^2,\ 0.2 < y < 0.6,\ E^{\text{jet}}_{\text{t,Breit}} > 7\ \text{GeV},\\ -1.0 < \eta^{\text{jet}}_{\text{lab}} < 2.5.$$

ZEUS[2] have used a luminosity of about 82 pb^{-1} and also about 20,000 events in the phase space defined by:

$$Q^2 > 125\ \text{GeV}^2,\ |\cos\gamma_h| < 0.65,\ E^{\text{jet}}_{\text{t,Breit}} > 8\ \text{GeV},\ -2.0 < \eta^{\text{jet}}_{\text{Breit}} < 1.5,.$$

where γ_h corresponds to the angle of the scattered quark in the quark parton model and is calculated from the hadronic final state. In addition ZEUS measure the inclusive jet cross section as a function of jet pseudorapidity in the Breit frame (figure 1 right).

From figure 1 it can be seen that the NLO QCD predictions provide a good description of the data within the quoted uncertainties. Typical errors on the measured cross section are $\pm5\%$ from the hadronic energy scale and $\pm7\%$ from the model uncertainty. Typical errors on the predicted cross sections are $\pm5\%$ from the scale uncertainty, $\pm3\%$ from the PDF uncertainty and $\pm3\%$ from the hadronisation uncertainty. It can also be seen that the scale uncertainty is smaller at higher Q and that the choice of renormalisation scale (Q or E_t) has little effect on the predicted cross section.

5. Multi Jet Production in DIS

An analysis of multi jet production has recently been published by ZEUS[3] presenting dijet and trijet cross sections as a function of $E^{jet}_{t,Breit}$, η^{jet}_{lab} and Q^2 (figure 2). The ratio of dijet to trijet events as a function of Q^2 is used to extract α_s. ZEUS have used a luminosity of about 82 pb^{-1} and find about 37,000 dijets and 13,500 trijets in the phase space defined by:

$$10 < Q^2 < 5000\ \mathrm{GeV}^2,\ 0.04 < y < 0.6,\ E^{\mathrm{jet}}_{\mathrm{t,Breit}} > 5\ \mathrm{GeV},$$
$$-1.0 < \eta^{\mathrm{jet}}_{\mathrm{lab}} < 2.5,\ M_{2jets(3jets)} > 25\ \mathrm{GeV}.$$

H1[4] have used a luminosity of about 65 pb^{-1} and find about 5,500 dijets and 1,800 trijets in the phase space defined by:

$$150 < Q^2 < 15000\ \mathrm{GeV}^2,\ 0.2 < y < 0.6,\ E^{\mathrm{jet}}_{\mathrm{t,Breit}} > 5\ \mathrm{GeV},$$
$$-1.0 < \eta^{\mathrm{jet}}_{\mathrm{lab}} < 2.5,\ M_{2jets(3jets)} > 25\ \mathrm{GeV}.$$

In figure 2 the dijet and trijet cross sections are shown as a function of Q^2. The NLO QCD predictions give a good description of the cross section over about four orders of magnitude. At high Q^2 the measurement is statistically limited. At low Q^2 the scale uncertainty is large ($\pm 20\%$). Typical errors on the measured dijet cross section are $\pm6\%$ from the hadronic energy scale and $\pm2\%$ from the model uncertainty. Typical errors on the predicted dijet cross sections are $\pm10\%$ from the scale uncertainty, $\pm2\%$ from the PDF uncertainty and $\pm6\%$ from the hadronisation uncertainty. For the trijet cross section most of the uncertainties are larger and the statistics are smaller. In the ratio of dijet to trijet cross sections (figure 3) many of the uncertainties cancel and the extraction of α_s from this ratio provides a competitive result.

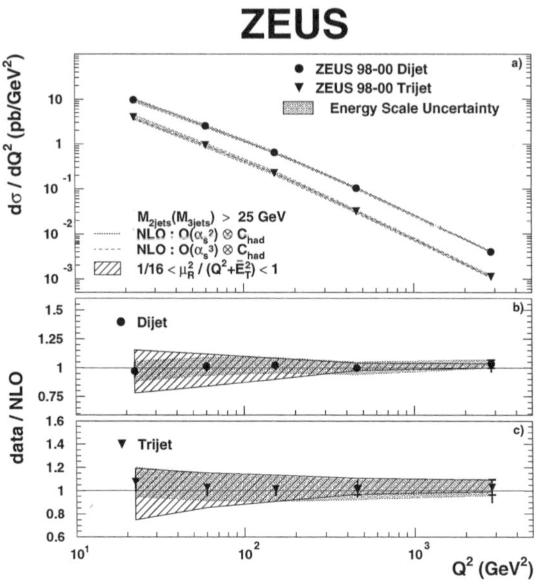

Figure 2. (a) The inclusive dijet and trijet cross sections (ZEUS) as functions of Q^2. The predictions of pQCD in NLO, corrected for hadronisation effects, are compared to the data. (b) and (c) show the ratio of the data to the predictions.

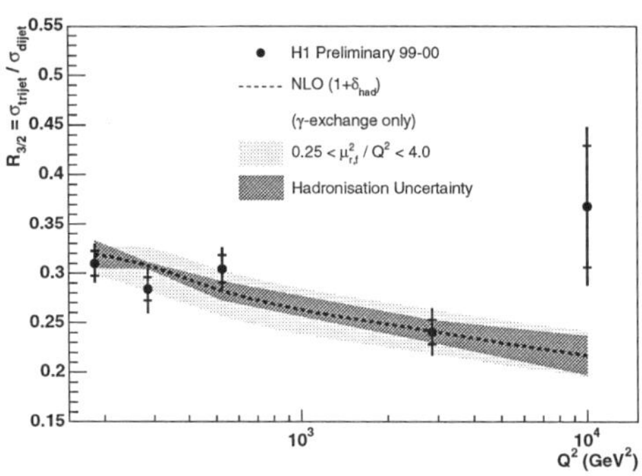

Figure 3. The ratio of dijet to trijet cross sections (H1) as a function of Q^2 compared with a NLO pQCD prediction, with hadronisation corrections. The light shaded band shows the scale uncertainty. The dark shaded band shows the hadronisation correction uncertainty.

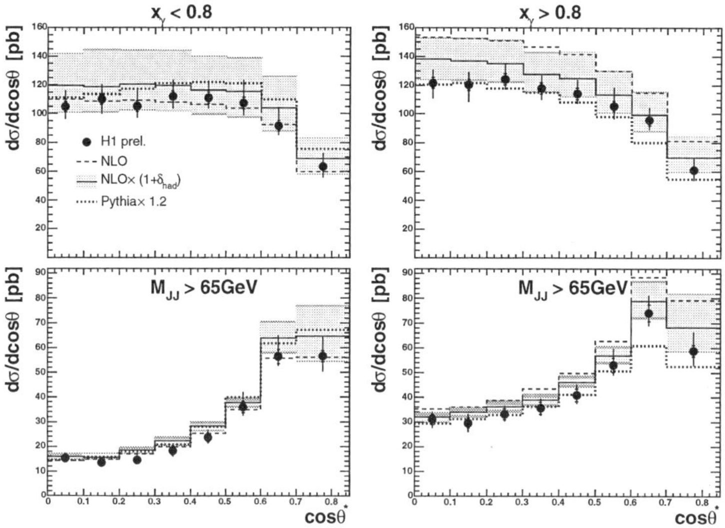

Figure 4. Cross section vs. $|\cos\theta^*|$ for data (points), NLO with (solid line) and without (dashed) hadronisation corrections δ_{had} and for PYTHIA (dotted) scaled by a factor of 1.2. The inner band of the NLO $(1+\delta_{had})$ result reflects the scale uncertainty, the outer band is the total uncertainty which includes also the one from PDFs and hadronisation.

6. High E_t Dijet Production in Photoproduction

A new measurement of high E_t dijet photoproduction has recently been made by H1[5]. Results for the dijet cross section as a function of $|\cos\theta^*|$ (figure 4), x_γ, x_p, and E_t^{jet} have been obtained. The measurements are studied in resolved photon enhanced ($x_\gamma < 0.8$) and direct photon enhanced ($x_\gamma > 0.8$) samples and with different jet topologies. H1 have used a luminosity of about 67 pb^{-1} and find about 14,000 dijet events in the phase space defined by:

$$Q^2 < 1 \text{ GeV}^2,\ 0.1 < y < 0.9,\ E_{\text{t,lab}}^{\text{jet},1} > 25 \text{ GeV},\ E_{\text{t,lab}}^{\text{jet},2} > 15 \text{ GeV},$$
$$-0.5 < \eta_{\text{lab}}^{\text{jet}} < 2.75.$$

Both the NLO QCD prediction and the PYTHIA Monte Carlo simulation generally provide a good description of the data. There is a significant reduction in the theoretical scale uncertainty for $M_{JJ} > 65$ GeV. In this region the cross section is sensitive to the dynamics of the hard interaction.

Typical errors on the measured dijet cross section are ±10% to 20% from the hadronic energy scale and ±6% from the model uncertainty. Typical errors on the predicted dijet cross sections are ±3% to 30% from the scale uncertainty, ±4% to 20% from the PDF uncertainty and ±5% from the hadronisation uncertainty. High E_t dijet photoproduction has been shown to be sensitive to the gluon density in the proton at medium and high x[15].

7. Future Improvements to Experimental Measurements

The HERA II running period should provide about seven times more data useful for analysis compared to the HERA I measurements presented here. Most obviously this is important for jet production at the highest Q and E_t as well as for trijet (and four-jet) production, which are all presently statistically limited. In addition the increased statistics allow for the use of a higher E_t jet selection where the smaller uncertainty on the hadronic energy scale might provide for a reduced total error on the cross section.

With increasing data and time also comes an improved understanding of our detectors. It is expected that the hadronic energy scale uncertainty should improve as $1/\sqrt{N_{\text{event}}}$. If this is true then as an example for HERA II H1 should be able to quote an uncertainty of $< 1\%$ for $E_t^{jet} > 7$ GeV resulting in a 2.5% error on the measured inclusive jet cross section.

The model uncertainty calculated using Monte Carlo programs, tuned to e^+e^- data, is becoming a large source of error. In order to understand and reduce this error a greater variety of models need to be used and possibly they need to be tuned to HERA data directly.

8. Future Improvements to Theoretical Predictions

The uncertainty due to the scale dependence is often one of the largest contributing errors to a jet analysis. The uncertainty is smaller at high jet E_t and large M_{JJ} but this is in part due to the fact that the NLO contribution to the cross section is small (i.e. the phase space for additional jet production is reduced) and NLO QCD is no longer being tested. The uncertainty is also smaller at high Q but here the steeply falling cross section will require the full exploitation of the HERA II data set to allow for an improved analysis of jet production.

The scale uncertainty is there to take into account the beyond next to leading order contributions to the predicted cross section. One way to reduce this uncertainty is to calculate these higher orders. Unfortunately due to the complexity of the calculation we can expect only inclusive and dijet predictions to be available in any reasonable time scale[b].

It has been suggested by Brodsky[16] that the renormalisation scale ambiguities can be eliminated. In his procedure it is n_f, the number of light fermion flavours, that sets the renormalisation scale in NLO QCD (although the ambiguity due to the choice of factorisation scale remains). This procedure has been demonstrated to work in QED but not yet for QCD.

A large error from the PDF uncertainty indicates that the predicted cross section is sensitive to the parton distributions in the proton and could be used to constrain these distributions[15]. Careful choice of cross section measurements and different event and jet selections can be used to enhance (or reduce) the sensitivity to the proton PDF. Studies of these effects can be done to maximise impact of future jet analyses in PDF fits.

The calculation of the hadronisation corrections applied to the NLO QCD parton level predictions rely on the fact that the Monte Carlo parton level simulation matches the NLO QCD predictions. Different Monte Carlo models provide different predictions of the parton level and the difference is used as the uncertainty. It could be said that this double counts the model uncertainty since a similar error exists for the measured cross sections. It is possible to correct the data to the parton level and compare directly to the parton level NLO QCD predictions. Then there is only one model uncertainty and in certain cases this is smaller than when both data and theory are corrected to the hadron level. This procedure implicitly assumes local parton hadron duality.

In the future the Monte Carlo programs such as MC@NLO[17] which implement NLO matrix elements matched with parton showers will be available for DIS. This could provide NLO QCD predictions at the hadron level with less uncertainty than with the present (leading order) Monte Carlo programs.

9. Conclusions

Several new results on jet production at high Q and high E_t in DIS and photoproduction have recently been made by H1 and ZEUS. They improve on previous measurements by using higher statistics and an improved understanding of the detector systematics. In general there is good agreement with NLO QCD predictions.

---

[b]See the contribution in these proceedings by Zoltan Trocsanyi: Multi-jet production in lepton-proton scattering at next-to-leading order accuracy

For future HERA II data analyses a significant increase in the statistics and an expected improved understanding of the detectors should result in further improvements to jet production measurements. These improvements will need to be matched by improvements in the uncertainties of the theoretical predictions for the full benefit of the HERA II data to be realised.

References

1. H1 Collaboration, C. Adloff et al., **H1prelim-05-133**.
2. ZEUS Collaboration, S. Chekanov et al., **ZEUS-prel-05-024**.
3. ZEUS Collaboration, S. Chekanov et al., *hep-ex/0502007* **DESY-05-019**.
4. H1 Collaboration, C. Adloff et al., **H1prelim-05-033**.
5. H1 Collaboration, C. Adloff et al., **H1prelim-05-134**.
6. S. Catani and M.H. Seymour, *Nucl. Phys. J.* **B 485**, 291 (1997); Erratum in *Nucl. Phys.* **B 510**, 503 (1998).
7. Z. Nagy and Z. Trocsanyi, *Phys. Rev. Lett.* **87**, 082001 (2001).
8. S. Frixione, *Nucl. Phys.* **B 507**, 295 (1997); S. Frixione and G. Ridolfi, *Nucl. Phys.* **B 507**, 315 (1997).
9. K. Charchula, G. Schuler and H. Spiesberger, *Comp. Phys. Comm.* **81**, 381 (1994).
10. H. Jung, *Comp. Phys. Comm.* **86**, 147 (1995).
11. T. Sjöstrand, *Comp. Phys. Comm.* **82**, 74 (1994). T. Sjöstrand et al., *Comp. Phys. Comm.* **135**, 238 (2001).
12. G. Marchesini et al., *Comp. Phys. Comm.* **67**, 465 (1992); G. Corcella et al., *hep-ph/0201201.*
13. ZEUS Collaboration, S. Chekanov et al., *Phys. Lett.* **B 547**, 164 (2002).
14. J. Pumplin et al., *JHEP* **0207**, 12 (2002).
15. ZEUS Collaboration, S. Chekanov et al., *hep-ex/0503274* **DESY-05-050**.
16. S. Brodsky, G. P. Lepage and P.B. Mackenzie, *Phys. Rev.* **D28**, 228 (1983).
17. S. Frixione and B. R. Webber, *JHEP* **0206**, 029 (2002).

MULTI-JET PRODUCTION IN LEPTON-PROTON SCATTERING WITH NEXT-TO-LEADING ORDER ACCURACY

Z. TRÓCSÁNYI

University of Debrecen and Institute of Nuclear Research of the Hungarian Academy of Science,
H-4001 Debrecen P.O.Box 51, Hungary,
E-mail: Z.Trocsanyi@atomki.hu

I summarize the theoretical and experimental status of multijet production in DIS. I present the state-of-the-art theoretical predictions and compare those to the corresponding experimental results obtained by analysing the data collected by the H1 and ZEUS collaborations at HERA. I also show new predictions for three-jet event-shape distributions at the NLO accuracy.

1. Introduction

Deep inelastic lepton-hadron scattering (DIS) has played a decisive role in our understanding of the deep structure of matter. The latest version of the experiments performed with colliding electrons or positrons and protons at HERA yields increasingly precise data so that not only fully inclusive measurements can be used to study the physics of hadronic final states. In fact, the study of jet-rates and event shapes has become an important project at HERA which yields results with continuously increasing accuracy [1]. Thus HERA is considered a machine for performing precision measurements for understanding Quantum Chromodynamics (QCD), the theory of strong interactions.

In order to perform precision measurements one needs precision tools for analysing the data. In the case of studying hadronic final states in high-energy particle collisions such tools have been developed in the framework of perturbative QCD. In order to make precision quantitative predictions in perturbative QCD, it is essential to perform the computations (at least) at the next-to-leading order (NLO) accuracy. Such computations, however, yield reliable predictions only in a limited part of the phase space, where the statistics of the data are relatively small. In order to increase the predictive power of the theory, the fixed-order predictions must be improved by matching those to predictions obtained by resumming large logarithmic contributions to all orders.

In the case of DIS, fixed-order and resummation computations have so far been completed for one-jet inclusive, 2- or 3(+1 beam[a])-jet cross sections and means or distributions of event shapes. Since the main goal of the experimental analyses is to compare data to precision theoretical predictions (beyond the LO accuracy), this implies that the experimental analysis of multi-jet events is generally constrained to considering three-jet events. Therefore, in this talk "multi" will mean three. This is in contrast to hadron collider studies, where the multi-jet events are backgrounds to various new-particle signatures, therefore, even the predictions at LO are considered valuable information and the construction of parton-level event generators is an important research topic [2]. One of the main lines of this research is the construction of public computer programs that could be used for the automated production of multi-parton events and thus for computing multi-jet cross sections. These computer programs could also be used to study high-multiplicity final states in DIS. Note, however, that the studies that can be made at HERA are not directly applicable at the LHC because the events at HERA have jets of typical energy in the order of 10 GeV while jets at the LHC will be triggered at the order of 100 GeV.

The automation of computing cross sections at the NLO accuracy has also been considered, but has not yet yielded mature results. Thus one has to recourse to programs for specific processes. The state of the art in the fixed-order computations of cross sections in DIS is represented by the NLOJET++ program that can be used for computing two- and three-jet observables [3,4]. Other related programs for leptoproduction of two jets are DISENT, DISASTER++ and JETVIP [5,6,7]. The predictions obtained by the NLOJET++ and DISASTER++ codes for two-jet cross sections agree within statistical accuracy of the numerical integrations.[b] The DISENT code is known to have a small bug [8] leading to slightly different predictions (the cross sections agree within 1–2 % [6]). At the time of the comparison of DISENT and JETVIP the latter code was not able to produce reliable predictions over the whole phase space [9], which was due to a bug in the binning routine that has been corrected since [10].

The state of the art in the resummed predictions is represented by the

[a]In the following, I omit the reference to the beam-jet, i.e., I do not count the beam-jet.
[b]The essential difference is that NLOJET++ is significantly faster.

recent analytic computations of the distribution of the multijet event shape $K_{\rm out}$ [11], the di-jet rates with symmetric $E_{\rm t}$ cuts [12], as well as by the CAESAR program that can be used for computing cross sections of two- and three-jet event shapes in a semi-automatic way [13]. In my talk I shall present NLO predictions of three-jet event-shape distributions for which resummed predictions already exist, but the fixed-order radiative corrections have not been computed before.

2. Fixed-order predictions

There are several process-independent ways to compute QCD radiative corrections. In computing the NLO corrections to multijet cross sections, the dipole subtraction scheme of Catani and Seymour [14] is a convenient formalism. It is used both in the DISENT and the NLOJET++ programs. The comparison of the two-jet predictions of the three programs has been performed in Ref. [3] and complete agreement was found apart from the slight difference in the DISENT predictions mentioned above. These programs use matrix elements that take into account only virtual photon exchange. Neglecting the exchange of the Z^0 boson means that the predictions are not reliable for large Q^2 values around $(90\,\text{GeV})^2$ and above.

The subtraction scheme applied in the NLOJET++ program is modified slightly as compared to the original one in [14] in order to have a better control on the numerical computation. The main idea is to cut the phase space of the dipole subtraction terms as introduced in Ref. [15]. The details of the computations are given in Ref. [16].

Once the phase space integrations are carried out, one can write the NLO jet cross section in the following form:

$$\sigma^{(J)}(p,q) = \tag{1}$$
$$\sum_a \int_0^1 \mathrm{d}\eta\, f_{a/P}(\eta,\mu_F^2)\, \sigma^{(J)}_{a,\mathrm{NLO}}\left(\eta p, q, \alpha_{\rm s}(\mu_R^2), \mu_R^2/Q_{HS}^2, \mu_F^2/Q_{HS}^2\right),$$

where p^μ and q^μ are the four-momenta of the incoming proton and the exchanged virtual photon, respectively. The function $f_{a/P}(\eta,\mu_F^2)$ is the density of the parton of type a in the incoming proton at momentum fraction η and factorization scale μ_F. The partonic cross section $\sigma^{(J)}_{a,\mathrm{NLO}}$ represents the sum of the LO and NLO contributions, given explicitly in Ref. [16], with jet function J. In addition to the parton momenta and possible parameters of the jet function, it also depends explicitly on the renormalized strong coupling $\alpha_{\rm s}(\mu_R^2)$, the renormalization and factorization scales $\mu_R = x_R Q_{\rm H.S.}$

and $\mu_F = x_F Q_{\rm H.S.}$, where $Q_{\rm H.S.}$ is the hard scale that characterizes the parton scattering, set event by event. Furthermore, the cross section also depends on the electromagnetic coupling, for which the NLOJET++ code uses $\overline{\rm MS}$ running $\alpha_{\rm EM}(Q^2)$ at the scale of the virtual photon momentum squared $Q^2 = -q^2$.

The publicly available version of the NLOJET++ program [4] is based on the tree-level and one-loop matrix elements given in Refs. [15,17], crossed into the photon-parton channel. It uses a C/C++ implementation of the LHAPDF library [18] with CTEQ6M [19] parton distribution functions and with the corresponding α_s expression for the renormalized coupling which is included in this library. The CTEQ6M set was fitted using the two-loop running coupling with $\alpha_s(M_{Z^0}) = 0.118$.

3. Comparison of fixed-order predictions to data

During the last few years, the experimental groups at HERA have performed extensive studies of multijet cross sections and compared their results to NLO predictions. The H1 collaboration already presented their results at this workshop four years ago [20]. The analysis was carried out parallel to our theoretical work with Z. Nagy that led to the NLOJET++ code, but without knowing about each other. When we finished testing our program and started to think of what to compute, we learnt about the H1 analysis accidentally. At that time preliminary H1 results showed rather large differences between data and LO predictions as seen in Fig. 1, even in the shapes of distributions, not only the absolute normalization.[c] We decided to make predictions of the same distributions at NLO accuracy.

H1 defined the jets using the inclusive $k_\perp$ algorithm implemented in the Breit frame (the precise definition can be found in Ref. [22]), selected three-jet events and plotted differential distributions of the DIS kinematical variables Q^2, $x_{\rm B}$ and the invariant three-jet mass $M_{\rm 3jet}$. We used the same jet algorithm. Furthermore, in our computations we chose the same kinematical region as H1 did [23], namely, for the basic DIS kinematic variables Q^2, $x_{\rm B}$ and $y = Q^2/(s\, x_{\rm B})$ we required $5\,{\rm GeV}^2 < Q^2 < 5000\,{\rm GeV}^2$, $0 < x_{\rm Bj} < 1$, $0.2 < y < 0.6$. Following the H1 analysis, we also restricted the (pseudo)rapidity-range in the laboratory frame and the minimum transverse energy of the jets in the Breit frame as $-1 < \eta^{\rm jet}_{\rm lab} < 2.5$, $E^{\rm jet}_{\rm T,B} > 5\,{\rm GeV}$. For the hard

[c]For obtaining the predictions at LO accuracy we used the CTEQ5L parton distribution functions [21] and the running coupling at one-loop with $\alpha_s(M_Z) = 0.127$.

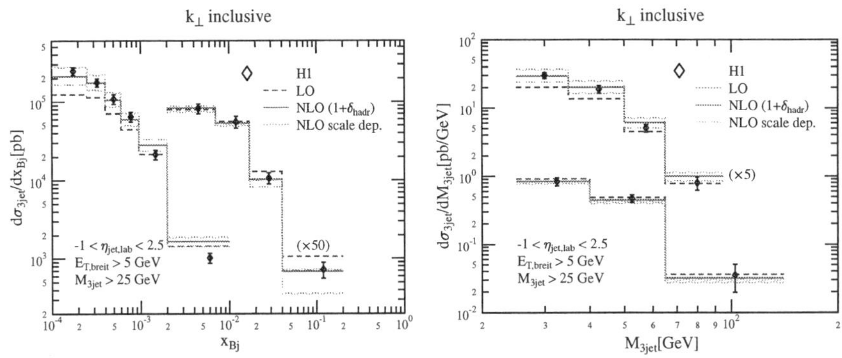

Figure 1. The differential distributions of the Bjorken variable x_{B} and the three-jet invariant mass $M_{3\mathrm{jet}}$. The histograms with larger values of the cross section correspond to the range $5\,\mathrm{GeV}^2 < Q^2 < 100\,\mathrm{GeV}^2$ and those with lower values to $150\,\mathrm{GeV}^2 < Q^2 < 5000\,\mathrm{GeV}^2$

scattering scale we chose the average transverse momentum of the jets,

$$Q_{\mathrm{H.S.}} = \frac{1}{3}\sum_j E_{\mathrm{T,B}}^{\mathrm{jet},\,j}\,. \tag{2}$$

We also studied the other usual choice, when $Q^2_{\mathrm{H.S.}} = Q^2$, but have not found significant differences. Finally, in order to compare our parton-level prediction to the hadron level data, we asked for the bin-wise correction factors of hadronization as estimated by H1 (the correction factors were between 1.2–1.3). With the inclusion of the NLO corrections the improvement in the theoretical description was spectacular, see Fig. 1.

Recently, the ZEUS collaboration has also performed an analysis of the three-jet events [24]. They published measurements of the inclusive three-jet cross section as a function of Q^2, the jet transverse energy in the Breit frame, $E^{\mathrm{jet}}_{\mathrm{T,B}}$ and the jet pseudorapidity in the laboratory frame $\eta^{\mathrm{jet}}_{\mathrm{lab}}$ compared to the NLO predictions obtained with the NLOJET++ code, corrected for hadronization (the correction factors C_{had} were in the range of 1.15–1.35.) The NLO QCD predictions were found to describe both the shapes of the predictions as well as the absolute normalization of the measured cross sections. The two competing most significant sources of uncertainty are the energy scale uncertainty from the experimental side and renormalization scale uncertainty from the theoretical side.

Such an agreement between data and theory promised a precise measurement of the strong coupling and its running by fitting the cross section ratio $R_{3/2}$ of the three-jet cross section to the two-jet one as a function of Q^2. The correlated systematic and renormalization-scale uncertainties mostly cancel in the ratio. According to the studies made by ZEUS [24], the total experimental and theoretical uncertainties are about 5 % and 7 %, respectively. The reduction in the errors is very large. For instance, at low Q^2 (below $100\,\text{GeV}^2$), the theoretical uncertainties in the ratio are a fourth of those in the three-jet distributions. The cited value of $\alpha_s(M_Z)$ as determined from the measurements of $R_{3/2}$ is

$$\alpha_s(M_Z) = 0.1179 \pm 0.0013\,(\text{stat.})\,^{+0.0028}_{-0.0046}\,(\text{exp.})\,^{+0.0064}_{-0.0046}\,(\text{theo.})\,.$$

The dominant source of uncertainty is still the theoretical one which calls for further efforts in improving the predictions by computing even higher order corrections.

4. Recent developments: predictions for multi-jet event shapes

There are two directions in computing higher-order corrections. One is the exact fixed-order computations that I discussed previously by considering the NLO corrections. Going beyond the NLO accuracy is very difficult and so far has only been achieved for totally inclusive quantities such as structure functions. For jet cross sections the first step in order to make advances in this direction is the computation of inclusive jet and dijet cross sections at the next-to-next-to-leading order (NNLO) accuracy. The recent advances in computing the NNLO corrections in the crossed channel of electron-positron annihilation into three jets [25] raise hopes that for jet cross sections in DIS the NNLO prediction will also be available in the not too distant future.[d] In order to compute NNLO corrections to the multi-jet cross sections, a major bottleneck is the computation of the necessary virtual corrections, and I do not expect quick progress in this direction.

The other possibility to improve the predictions is to resum the most important logarithmic corrections, due to collinear and soft radiation, to all orders, which leads to predictions at the next-to-leading logarithmic (NLL) accuracy. Such computations are not available for jet rates. However, much progress has been achieved recently in resumming the LL and NLL contributions to multi-jet event-shape distributions [27]. These works lead to much deeper insight about the structure of QCD cross sections. In

[d] Note that the naive iterative extension of the dipole subtraction scheme to NNLO is not possible [26].

particular, prior to these studies it was believed that distributions of DIS observables, measured in the current hemisphere in the Breit frame, were trivially related to their well studied counterparts in electron-positron annihilation, where the resummed logarithms are due to soft radiation over the whole phase space (hence they are called global observables). However, it was found that there were also important single-logarithmic non-global effects due to radiation into one hemisphere [28]. In this talk I want only to collect the currently available theoretical information on multi-jet event shapes without going into the details of the theoretical studies.

There are two multi-jet event shapes computed to NLL accuracy so far. One quantifies the out-of-plane QCD radiation (the sum of the momentum components perpendicular to the event plane), called $K_{\rm out}$, defined precisely in Ref. [11]. The other is the y_3 observable that is defined to be the largest value of the jet resolution variable $y_{\rm cut}$ such that the event is clustered into three jets with $k_\perp$-clustering [29]. The $K_{\rm out}$ distribution has been studied experimentally in Ref. [30], however, with different definition of the observable as done in the resummation computation, therefore, conclusions cannot be drawn from the results.

One may ask why the computation of the NLO corrections is necessary if resummed predictions are known. The reason is that the NLL and NLO predictions are valid in rather distinct parts of the phase space, which can be clearly seen on the left panel of Fig. 2, where the differential distributions in $K_{\rm out}/Q$ at fixed values of $Q^2 = (35{\rm GeV})^2$ and $x_{\rm B} = 0.02$, normalized to the Born cross section, are presented. The dotted line is the LO prediction, the dashed is the NLL one. Expanding the NLL prediction in $\alpha_{\rm s}$ and changing the leading term to the exact LO one, we obtain the matched prediction shown with the dash-dotted line. We see that in the $K_{\rm out}$-region where the best precision experimental data can be collected, neither the fixed-order nor the resummed values are reliable, but one should use the matched prediction. On the right panel, I show the effect of including the power corrections both to the NLL and the matched predictions. The importance of matching is clear also in this case.

In analysing the multihadron data collected in electron-positron annihilation, a very accurate theoretical description of event-shape distributions was found with matched resummed and fixed-order predictions improved with hadronisation corrections (see e.g. [31]). I expect it will also be interesting to compare the HERA results for multi-jet event-shape distributions to predictions of the same level. Conclusions of such studies could also be important for analyses at the LHC, where the presence of the incoming

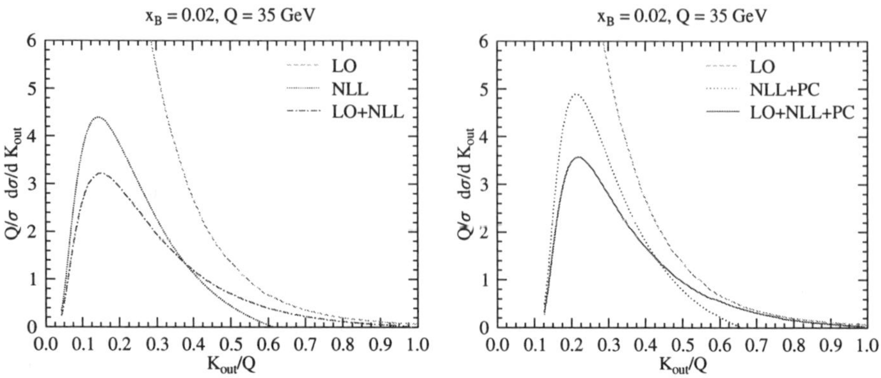

Figure 2. The differential distribution of K_{out}. Left panel: perturbative computations. Right panel: perturbative predictions with power corrections.

two hard partons in the event means that even the dijet event shapes need at least four hard partons which is a multi-jet event shape configuration in DIS. Thus dijet event shapes at hadron colliders represent kinematical situations where NLL resummations and power corrections are as yet untested.

The semi-automatic computation of NLL predictions with the program CAESAR is currently being interfaced to the output of the NLOJET++ code. With this interface matched fixed-order and resummed predictions improved with power corrections will be obtained in a semi-automatic way soon [32]. Here I would like to present NLO predictions for the distribution of the event-shape observable K_{out}, computed recently [16]. I used the same definitions of the observables and performed the computations at fixed values of the DIS kinematic variables $Q^2 = (35\text{GeV})^2$, $x_{\text{B}} = 0.02$ as in the resummed computations [33]. Figure 3 shows the LO and NLO predictions. The shaded bands in the left panel correspond to the range of scales $1/2 \leq x_R = x_F \leq 2$. We find that the radiative corrections are in general large, thus the scale dependence reduces only relatively to the cross sections. They also increase with increasing value of K_{out} because the phase space for events with large out-of-plane radiation with three partons in the final state (at LO) is much smaller than that with four partons in the final state (real corrections). The boundary of the phase space in K_{out} is about 20 % larger for the NLO computation than at LO. The cross sections decrease rapidly with increasing K_{out}. The small cross section for medium or large values of K_{out} leaves the small K_{out}-region for experimental analysis.

In the small K_{out}-region, the logarithmic contributions of the type

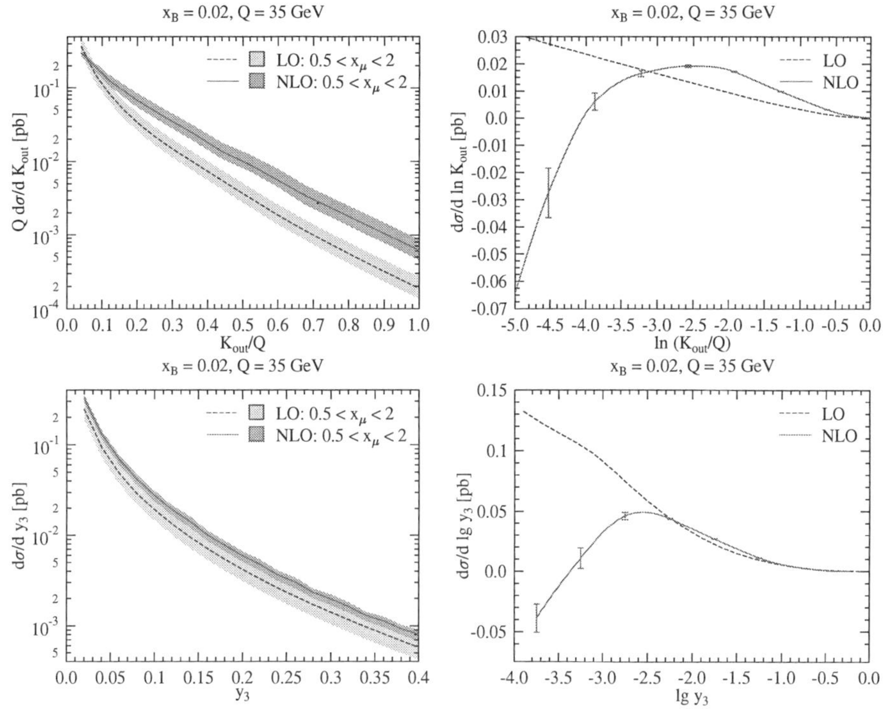

Figure 3. The differential distributions of the K_{out} and y_3 observables at fixed value of x_{B} and Q^2. The left panel shows the distributions as a function of K_{out}/Q and y_3, the right panel shows the distributions as a function of $\ln K_{\mathrm{out}}/Q$ and $\lg y_3$ in order to exhibit the logarithmic behaviour for small values of the observable. Both the LO predictions (dashed line) and the NLO predictions (solid line) were obtained with the CTEQ6M parton distribution functions. The error bars indicate the uncertainty of the numerical integration that is negligible for the computation at LO accuracy.

$\ln K_{\mathrm{out}}/Q$ are dominant as can be seen on the plot in the right panel. At LO, the logarithmic dominance starts at about $\ln K_{\mathrm{out}}/Q = -2$, at NLO, it starts at about $\ln K_{\mathrm{out}}/Q = -4$. Below these values the cross section is a linear function of $\ln K_{\mathrm{out}}/Q$ and the fixed-order predictions diverge with $K_{\mathrm{out}} \to 0$ with alternating signs, which makes the resummation of these large logarithmic contributions mandatory. Reliable theoretical predictions can be obtained by matching the cross sections valid at the NLO and NLL accuracy. This matching is obtained by expanding the NLL prediction in α_{s} and changing the first two terms in that expansion with the exact values of the NLO computation. Qualitatively similar conclusions can be drawn from the distributions for the observable y_3 [16], but the corrections are smaller.

5. Conclusions and Outlook

In this talk I discussed the present status of predicting distributions of multi-jet cross sections in lepton-proton scattering. The only existing program for computing the three-jet observables at NLO accuracy in DIS is the NLOJET++ code. This program is well-tested, but has the slight disadvantage that the Z-boson exchange diagrams are not included. The predictions for three-jet rates agree well with the data collected at HERA, although the main source of uncertainty remains the theoretical one, which calls for taking into account the higher order corrections.

The other option of taking into account higher orders is the matching with resummed predictions valid at the NLL accuracy. Recent years yielded a lot of progress in this area of research. The CAESAR program can be used for computing NLL predictions to multi-jet distributions in a semi-automatic way. I showed the importance of matching the NLO and NLL predictions. Both are available for certain three-jet event-shape observables, like the out-of-plane momentum $K_{\rm out}$and the y_3 variable. The matching of the NLO and NLL predictions is expected to be available soon [32].

Acknowledgments

I am greatful to the organizers of the Ringberg workshop for their invitation as well as for providing a pleasant atmosphere during the meeting. This work was supported by the Hungarian Scientific Research Fund grant OTKA T-038240.

References

1. C. Adloff *et al.* [H1 Collaboration], *Eur. Phys. J.* C **6**, 575 (1999) [hep-ex/9807019]; C. Adloff *et al.* [H1 Collaboration], *Eur. Phys. J.* C **13**, 415 (2000) [hep-ex/9806029]; C. Adloff *et al.* [H1 Collaboration], *Eur. Phys. J.* C **14**, 255 (2000) [Erratum-ibid. C **18**, 417 (2000)] [hep-ex/9912052]; C. Adloff *et al.* [H1 Collaboration], *Phys. Lett.* B **515**, 17 (2001) [hep-ex/0106078]; M. Derrick *et al.* [ZEUS Collaboration], *Phys. Lett.* B **363**, 201 (1995) [hep-ex/9510001]; J. Breitweg *et al.* [ZEUS Collaboration], *Phys. Lett.* B **479**, 37 (2000) [hep-ex/0002010]; S. Chekanov *et al.* [ZEUS Collaboration], *Eur. Phys. J.* C **27**, 531 (2003) [hep-ex/0211040];
2. M. L. Mangano, eConf **C030614**, 015 (2003) [hep-ph/0312117].
3. Z. Nagy and Z. Trócsányi, *Phys. Rev. Lett.* **87**, 082001 (2001) [hep-ph/0104315].
4. Z. Nagy, NLOJET++: www.cpt.dur.ac.uk/ nagyz/nlo++/.
5. M. H. Seymour, DISENT 1.1: hepwww.rl.ac.uk/theory/seymour/nlo.
6. D. Graudenz, "DISASTER++ version 1.0" hep-ph/9710244.

7. B. Pötter, *Comput. Phys. Commun.* **133**, 105 (2000) [hep-ph/9911221]; JETVIP homepage: www.desy.de/~poetter/jetvip.html.
8. M. Dasgupta and G. P. Salam, *JHEP* **0208**, 032 (2002) [hep-ph/0208073].
9. C. Duprel, T. Hadig, N. Kauer and M. Wobisch, hep-ph/9910448.
10. M. Klasen, private communication.
11. A. Banfi, G. Marchesini, G. Smye and G. Zanderighi, *JHEP* **0111**, 066 (2001) [hep-ph/0111157].
12. A. Banfi and M. Dasgupta, *JHEP* **0401**, 027 (2004) [hep-ph/0312108].
13. A. Banfi, G. P. Salam and G. Zanderighi, *JHEP* **0503**, 073 (2005) [hep-ph/0407286].
14. S. Catani and M. H. Seymour, *Nucl. Phys.* B **485**, 291 (1997) [Erratum-ibid. B **510**, 291 (1997)] [hep-ph/9605323].
15. Z. Nagy and Z. Trócsányi, *Phys. Rev.* D **59**, 014020 (1999) [Erratum-ibid. D **62**, 014020 (1999)] [hep-ph/9806317].
16. Z. Nagy and Z. Trócsányi, hep-ph/0511328.
17. Z. Bern, L. Dixon, D. A. Kosower and S. Weinzierl, *Nucl. Phys.* B **489**, 3 (1997) [hep-ph/9610370]; Z. Bern, L. Dixon and D. A. Kosower, *Nucl. Phys.* B **513**, 3 (1998) [hep-ph/9708239].
18. W. T. Giele, S. A. Keller and D. A. Kosower, hep-ph/0104052.
19. J. Pumplin, D. R. Stump, J. Huston, H. L. Lai, P. Nadolsky and W. K. Tung, *JHEP* **0207**, 012 (2002) [hep-ph/0201195].
20. M. Wing, *J. Phys.* G **28**, 857 (2002) [hep-ex/0109039].
21. H. L. Lai *et al.* [CTEQ Collaboration], *Eur. Phys. J.* C **12**, 375 (2000) [hep-ph/9903282].
22. C. Adloff *et al.* [H1 Collaboration], *Nucl. Phys.* B **545**, 3 (1999) [hep-ex/9901010].
23. C. Adloff *et al.* [H1 Collaboration], *Phys. Lett.* B **515**, 17 (2001) [hep-ex/0106078].
24. S. Chekanov *et al.* [ZEUS Collaboration], hep-ex/0502007.
25. A. Gehrmann-De Ridder, T. Gehrmann and E. W. N. Glover, *JHEP* **0509**, 056 (2005) [hep-ph/0505111].
26. G. Somogyi, Z. Trócsányi and V. Del Duca, *JHEP* **0506**, 024 (2005) [hep-ph/0502226].
27. M. Dasgupta and G. P. Salam, *J. Phys.* G **30**, R143 (2004) [hep-ph/0312283].
28. M. Dasgupta and G. P. Salam, *Phys. Lett.* B **512**, 323 (2001) [hep-ph/0104277].
29. S. Catani, Y. L. Dokshitzer and B. R. Webber, *Phys. Lett.* B **285**, 291 (1992).
30. A. Everett, "Event shapes in deep inelastic $ep \to eX$ scattering at HERA", Proceedings of the XIII International Workshop on Deep Inelastic Scattering.
31. Z. Nagy and Z. Trócsányi, *Nucl. Phys. Proc. Suppl.* **74**, 44 (1999) [hep-ph/9808364].
32. A. Banfi and G. Zanderighi, private communication.
33. A. Banfi, G. Zanderighi and G. Salam, CAESAR homepage: qcd-caesar.org.

DIJET RATES WITH SYMMETRIC E_T CUTS

A. BANFI

Dipartimento di Fisica G. Occhialini
Università degli Studi di Milano-Bicocca
Piazza della Scienza, 3, 20126 Milano, Italy

University of Cambridge, Cavendish Laboratory
Madingley Road, CB3 0HE Cambridge, UK
E-mail: abanfi@hep.phy.cam.ac.uk

DAMTP, Centre for Mathematical Sciences
Wilberforce Road, CB3 0WA Cambridge, UK

We discuss the physics underlying an all-order resummation of logarithmic enhanced contributions to dijet cross sections, and present preliminary results for the distribution in the dijet transverse energy difference in DIS.

1. Introduction

Clustering hadrons into jets is a very useful tool to make QCD predictions. Measuring jet cross section removes the theoretical uncertainty due to fragmentation functions, and makes it possible to directly compare data with perturbative (PT) QCD predictions, whose degrees of freedom are partons and not hadrons. This comparison works extremely well when considering high transverse energy (E_T) jets. In this kinematical situation one is expected to probe quarks and gluons at small distances. This joint theoretical and experimental effort has led to the measurement of the QCD coupling α_s from jet inclusive E_T spectra and to better constrain the gluon density in the proton[1].

There are actual limitations in the use of perturbation theory. First of all any perturbative series diverges. This is related to the fact that the observed degrees of freedom are hadrons and not partons, and results in an ambiguity in PT predictions that fortunately is suppressed by inverse powers of the hard scale of the process. The second limitation has to do with the fact that the coefficients in the PT expansion can be logarithmically enhanced due to incomplete real-virtual cancellations in the infrared (IR). This has the consequence that in particular phase space regions one observes a breakdown of the PT expansion, which can be cured only by performing an all-order resummation of such logarithms.

It is therefore mandatory, in order to have predictions valid in the whole of the phase space, to complement any fixed order calculation with the all-order resummation of logarithmic enhanced contributions, and, if possible, to remove the ambiguity in the PT expansion with a power-suppressed non-perturbative (NP) correction, which, unless it can be computed from the lattice, has to be taken as a phenomenological input[2].

2. The observable

The observable we consider is the dijet rate, i.e. the fraction of events with at least two jets. To select hard jets we put a cut on the transverse energy of the two highest E_T jets, requiring $E_{T1} > E_{T2} > E_m$. This particular choice is referred to as "symmetric E_T cuts". It was noted by Klasen and Kramer[3] that symmetric E_T cuts produce IR instabilities in next-to-leading order (NLO) QCD predictions. It was later Frixione and Ridolfi[4] who proposed to perform an asymmetry study by considering

$$\sigma(\Delta) \equiv \sigma(E_{t1} > E_m + \Delta; E_{t2} > E_m)\,. \tag{1}$$

They computed $\sigma(\Delta)$ in photoproduction at NLO, and obtained that while $\sigma(0)$, the total dijet rate with symmetric E_T cuts, is finite, the slope of the curve $\sigma'(\Delta) \equiv d\sigma/d\Delta$ diverges for $\Delta = 0$. This behaviour is not present at all in the data, as one can see from the asymmetry study performed by ZEUS in DIS[5], and reported in fig. 1. There the data is plotted against NLO QCD predictions obtained from the numerical program DISENT[6]. Again, the slope of the NLO curve diverges for $\Delta = 0$, while the data decreases smoothly with increasing Δ. This can be easily understood by noting that $\sigma'(\Delta)$ is just (minus) the differential distribution of the highest E_T jet:

$$\sigma'(\Delta) = -\left.\frac{d\sigma}{dE_{T1}}\right|_{E_{T1}=E_m+\Delta}\,. \tag{2}$$

Since the latter quantity is a physical cross section, the slope $\sigma'(\Delta)$ has to be negative. This corresponds to what is seen in the data, and also implies that any turnover in the NLO curves is unphysical.

To better understand the origin of the divergence in $\sigma'(\Delta)$, we consider the value of the jet transverse energy difference $E_{T1} - E_{T2}$. In the specific case of dijet production in DIS, with an incoming quark p, at the Born level,

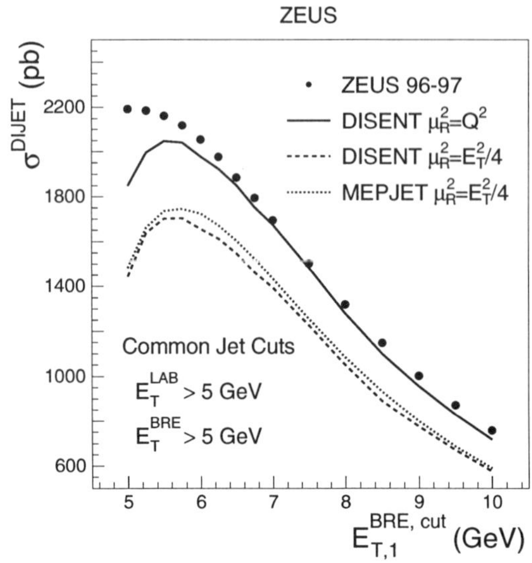

Figure 1. The dijet rate $\sigma(\Delta)$ as a function of E_{T1} in DIS. The dots are the data, while the lines are NLO predictions.

a dijet event consists only of two outgoing hard partons p_1 and p_2, in this case a quark and a gluon. Their transverse momenta $\vec{p}_{t1}$ and $\vec{p}_{t2}$ are back-to-back, so that, defining $E_{Ti} = |\vec{p}_{ti}|$, we have $E_{T1} = E_{T2}$. After emission of a soft gluon k not clustered with any of the two outgoing partons, we obtain from transverse momentum conservation $E_{T1} - E_{T2} \simeq |k_x|$, where k_x is the component of the gluon $\vec{k}_t$ parallel to $\vec{p}_{t1}$. When considering $\sigma'(\Delta)$, E_{T1} is forced to lie on the line $E_{T1} = E_m + \Delta$, but from kinematics $E_{T1} = E_{t2} + |k_x|$. These two lines should then intersect somewhere in the allowed phase space $E_{T1} > E_{T2} > E_m$, and this is possible only for $|k_x| < \Delta$, as can be seen from fig. 2. Squeezing soft radiation is known to give rise to large logarithms in fixed order calculations. Indeed, if we consider the emission of a soft gluon collinear to the incoming quark together with the corresponding virtual correction, we obtain

$$\sigma'(\Delta) = \sigma'_0(\Delta) \times \left(1 - \frac{\alpha_s C_F}{\pi} \ln^2 \frac{Q}{\Delta}\right) , \tag{3}$$

so that the slope $\sigma'(\Delta)$ diverges for $\Delta \to 0$. A quick solution of the problem can be achieved by choosing kinematical cuts such that the Born slope $\sigma'_0(\Delta) \to 0$ for $\Delta \to 0$. This can be achieved for instance by imposing asymmetric E_T cuts[7], or choosing a suitable range for the dijet invariant

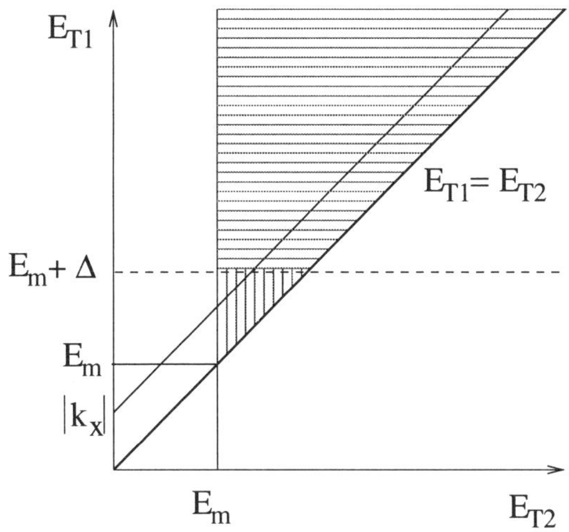

Figure 2. The phase space for the dijet rate in the $E_{T1} - E_{T2}$ plane.

mass and rapidities[1]. However, the actual solution relies on the all-order resummation of soft gluon effects, as will be explained in the next section.

3. Resummation

Considering an arbitrary number of partons one finds $E_{T1} - E_{T2} \simeq |\sum_i' k_{ix}|$, where the primed sum runs on all partons not clustered with the jets. Introducing $S(\vec{k}_t)$, the probability that $\sum_i' \vec{k}_{ti} = \vec{k}_t$, we have $\sigma'(\Delta) = \sigma_0'(\Delta)W(\Delta)$, where the K-factor $W(\Delta)$ is in general given by

$$W(\Delta) = \int d^2\vec{k}_t \, S(\vec{k}_t) \, \Theta(\Delta - |k_x|) = \frac{2}{\pi} \int \frac{db}{b} \sin(b\Delta)\Sigma(b) \, . \qquad (4)$$

For emissions soft and collinear to the incoming quark, no secondary partons are clustered with p_1 or p_2, so that an all-order resummation of such contributions yields $\Sigma(b) = \exp[-C_F \frac{\alpha_s}{\pi} \ln^2 b]$. From eq. (4) we have then $W(\Delta) \sim \Delta$ for small Δ, which roughly corresponds to the behaviour seen in the data.

In general, $\Sigma(b) = \exp[Lg_1(\alpha_s L) + g_2(\alpha_s L) + \alpha_s g_3(\alpha_s L) \ldots]$, with $L = \ln b$, and we aim at next-to-leading logarithmic (NLL) accuracy, i.e. the knowledge of g_1 and g_2. In order to perform an all-order resummation, one needs to reorganise the PT series and collect all contributions to $\Sigma(b)$ up to a given accuracy. First of all one can see that at NLL accuracy, $\Sigma(b)$ can be interpreted as the probability that $|\sum_i' k_{ix}| < e^{-\gamma_E}/b$. Furthermore, given a variable V, a generic function of all final state momenta, multi-parton matrix elements and phase space constraints can be approximated by using the following general properties:

(1) all leading logarithms (LL) originate from configurations of soft and collinear gluons for which the hardest emission k_1 dominates, that is $V(k_1, \ldots, k_n) \simeq V(k_1)$;
(2) hard-collinear and soft-large-angle emissions give a NLL contribution, and again one can approximate $V(k_1, \ldots, k_n) \simeq V(k_1)$;
(3) the fact that $V(k_1, \ldots, k_n) \neq V(k_1)$ needs only to be taken into account for soft and collinear emissions; these "multiple emission effects" are NLL, and can be treated either analytically with a suitable integral transform, or numerically with Monte Carlo (MC) techniques;
(4) secondary splittings are accounted for by taking the QCD coupling in the physical CMW scheme, at a scale of the order of the transverse momentum of the parent parton, i.e. $\alpha_s \to \alpha_s(k_t)$.

The strongest implication of these four statements is that at NLL accuracy soft and/or collinear (SC) partons can be considered as emitted independently from the hard parton antenna, in spite of the fact that QCD is a non-abelian gauge theory. This crucial simplification is known as the "independent emission" approximation, and leads to exponentiation of leading logarithms, and factorisation of NL logarithms. If we assume the validity of independent emission, we obtain

$$\Sigma(b) = \frac{f(1/b)}{f(Q)} \otimes e^{-R(b)} , \qquad R(b) = \int [dk]\, w(k)\, \Theta\left(|k_x| - e^{-\gamma_E}/b\right) , \quad (5)$$

where the 'radiator' $R(b)$ is (minus) the contribution to $\Sigma(b)$ from a single SC gluon k emitted with probability $w(k)$. The ratio of parton densities $f(1/b)/f(Q)$ is a general feature of resummations involving incoming partons, and is due to the fact that the scale of the incoming parton density is set by the upper limit of observed parton transverse momenta, which in this case is of order $1/b$.

Unfortunately, independent emission approximation does not hold for all variables, but only for those who are recursively infrared and collinear (rIRC) safe and (continuously) global[8]. The lack of either of these two properties causes one (or more) of the above statements not to be true.

For instance, for the three-jet rate in the JADE algorithm, which is global but not rIRC safe, statement 1 is false, since there are LL contributions originated by multiple emissions, with the consequence that leading logarithms do not exponentiate[9].

Globalness means simply that V is sensitive to emissions in the whole of the phase space. This is not the case for $E_{T1} - E_{T2}$, since only partons that are not clustered with the two hard jets contribute to the dijet transverse energy difference. The first consequence of non-globalness is that secondary splittings cannot be accounted for by simply setting the proper scale for α_s, because new NLL contributions, called "non-global logs"[10], arise when a cascade of energy ordered partons outside the measure region emits a softer gluon inside. In this case phase space boundaries have to be taken into account exactly for all emissions, so that one has to rely on MC methods. Moreover, due to the complicated colour structure of multi-gluon matrix elements, non-global logs at the moment can be resummed only in the large N_c limit.

In our case, if particles are clustered into jets with a cone algorithm, after having specified a procedure to deal with overlapping cones, one has that to NLL accuracy a jet consists of all particles that flow into a cone around p_1 or p_2, and equation (5) gets modified as follows[11]:

$$\Sigma(b) = \frac{f(1/b)}{f(Q)} \otimes e^{-R(b)} \times S[t(b)]\,, \qquad t(b) = \frac{1}{2\pi}\int_{1/b}^{Q} \frac{dk_t}{k_t}\alpha_s(k_t)\,, \tag{6}$$

where $S(t)$ represents the contribution of non-global logs, and can be computed with a MC procedure as a function of the evolution variable t.

If jets are clustered with the k_t algorithm[12], further complications arise. In this case the fact that a particle belongs to a jet depends on all other emitted particles. This has interesting implications for non-global logs. Dasgupta and Salam[13] have shown that the dominant contribution to non-global logs arises when the observed gluon is close to the boundary of the measure region. Appleby and Seymour[14] noted then that the k_t algorithm requirement forces such gluons to be clustered with gluons outside the measure region, thus reducing the magnitude of non-global logs.

One can also ask whether for the independent emission contribution to $\Sigma(b)$, the exponential form of equation (5) is still valid. This is true for cones (the "unclustered" case) since the phase space constraint is factorised, but should be checked for the k_t algorithm (the "clustered" case). In order to answer this question we considered a much simpler variable, the transverse energy flow away from the jets. Given a pair of hard jets (taken back-to-back for simplicity), one defines a region Ω away from the jet axis (which in e^+e^- annihilation roughly coincides with the thrust axis), and the away-from-jet E_T flow

$$E_{T,\Omega} = \sum_{i\in\Omega} k_{ti} \,, \tag{7}$$

with k_{ti} the transverse momentum of the i-th *jet* with respect to the jet axis[13,14]. The quantity $\Sigma(Q, Q_\Omega)$, the probability that $E_{T,\Omega} < Q_\Omega$, in the region $Q_\Omega \ll Q$ is sensitive only to soft gluons at large angles. One then wishes to resum LL contributions to $\Sigma(Q, Q_\Omega)$, whose order is $\alpha_s^n \ln^n(Q/Q_\Omega)$.

At LL accuracy $\Sigma(Q, Q_\Omega)$ is given by

$$\Sigma(Q, Q_\Omega) = \Sigma_{\Omega,P}(t) \cdot S(t) \,, \tag{8}$$

where $t \equiv t(Q_\Omega)$ is the evolution variable defined in equation (6). Gluons emitted directly from the two hard partons give rise to $\Sigma_{\Omega,P}(t)$, while non-global logs are embodied in $S(t)$. Appleby and Seymour[14] assumed that also for the clustered case $\Sigma_{\Omega,P}(t) = e^{-R(t)}$, that is the single gluon contribution to $\Sigma(Q, Q_\Omega)$ exponentiates. This naive expectation can be motivated by the fact that if emissions are assumed to be independent, multiple emission effects are usually relevant only for soft and collinear gluons (statement 3). However, since $E_{T,\Omega}$ is manifestly non-global, one cannot exclude multiple emission effects coming from soft gluons at large angles. Consider for instance two gluons k_1 and k_2, with $\omega_1 \gg \omega_2$, k_2 inside Ω and k_1 outside. It is possible that the jet algorithm clusters k_2 with k_1, thus spoiling the exponentiation of the single gluon result[15]. This is indeed seen when comparing the resummed expression for the differential distribution $\sigma^{-1}d\sigma/dL$, with $L = \ln(Q_\Omega/Q)$, with the NLO program EVENT2[6]. If LL are correct, the difference of the two distributions should go to a constant for large (negative) L. This happens for the coefficient of $C_F C_A \alpha_s^2$, indicating that both in the unclustered and the clustered case non-global logarithms are correctly taken into account. If one however uses for the clustered case the same expression for $\Sigma_{\Omega,P}$ as for the unclustered case, one finds a discrepancy in the coefficient of $C_F^2 \alpha_s^2$, the plot of fig. 3. In the clustered case $\Sigma_{\Omega,P}$ must be corrected by taking into account the fact that the softest gluon, in spite of the fact that it is emitted in Ω, can be nevertheless clustered with the hard jets, and therefore does not contribute to $\Sigma_{\Omega,P}$. Even for primary emissions this constitutes a LL contribution. The resummation of these multiple emission effects can be performed with the same MC used for non-global logs, and is the last bit that is needed to resum the dijet rate with the k_t algorithm at NLL accuracy.

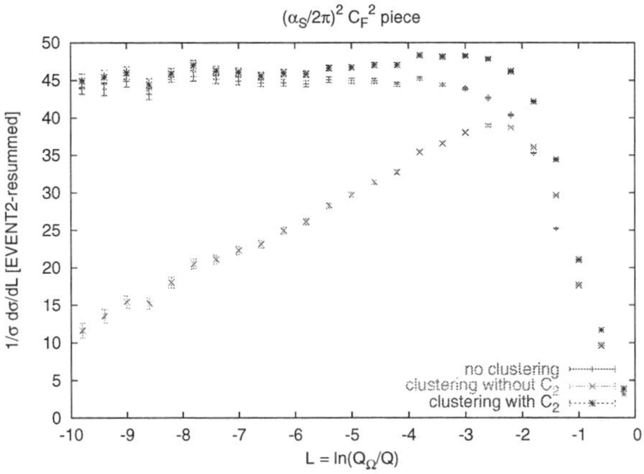

Figure 3. The coefficient of $(\alpha_s/2\pi)^2 C_F^2$ for the difference between resummed and NLO predictions for $\sigma^{-1} d\sigma/dL$, as explained in the text. Jets are clustered with the k_t algorithm with $R = 1$, and Ω is a rapidity slice with $\Delta\eta = 1$. C_2 is the correction that needs to be applied at order α_s^2 to take into account multiple emission effects.

4. Phenomenology of dijet observables

Instead of going through the details of the resummed calculation for the dijet rate in DIS, we discuss what information on QCD dynamics we can gain from allowing symmetric E_T cuts. Since in the region $\Delta \to 0$ we are almost in a three-jet configuration, this measurement is complementary to three-jet event shapes, and allows one to investigate the coherence properties of QCD emission from a three parton antenna. This observable has also the advantage that NP corrections are smaller than in event shapes distributions. This is mainly due to the fact that setting $\Delta \to 0$ does not put a direct veto on emitted parton transverse momenta, but rather on their vector sum. We have then emissions with large transverse momenta contributing also in the extreme $\Delta \to 0$ region. This fact also affects the magnitude of non-global logs, which here[11] give a correction of order 10%, while for event shapes[2] their contribution can be as large as 30%.

One can consider other observables that have the same resummation as the dijet rate and can be more easily handled theoretically or experimentally. Among these are the distributions in the transverse energy difference $\Delta = E_{T1} - E_{T2}$ and in the azimuthal angle $\Delta\phi$ between the jets. For any of these variables V, one studies

$$\frac{1}{\sigma}\frac{d\sigma}{dV} = \frac{d}{dV}\Sigma(V)\,. \tag{9}$$

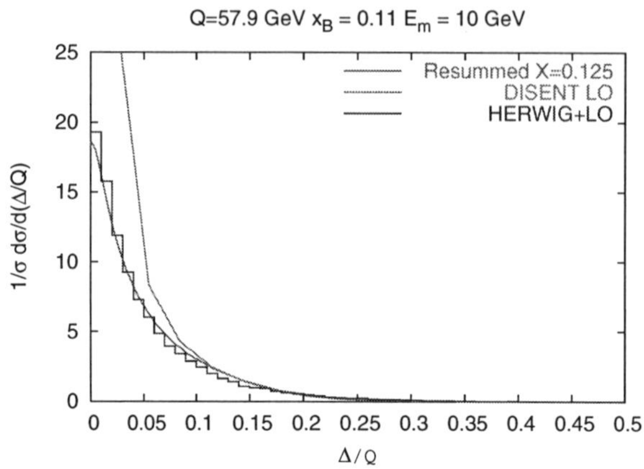

Figure 4. Preliminary plots for the transverse energy differential distribution in DIS. The parameter X is used to estimate NNLL effects.

As already stated, no further effort is needed to resum these distributions, since $\Sigma(V) = \sigma_0(0)W(V)$, where W is the same as in eq. (4). Since from the previous analysis we have seen that $\Sigma(V) \sim V$ for small V, one expects any of these differential distributions to approach a constant for $V \to 0$. This is what is already seen in the data for the $\Delta\phi$ distribution in hadronic dijet production at the Tevatron[16]. This observable is particularly interesting because it gives access to the QCD *fifth form factor*, which is present when there are at least four hard emitting partons[17].

A plateau for small Δ is also seen in the differential distribution in the dijet transverse energy difference in DIS computed with HERWIG[18]. Fig. 4 shows the comparison between the resummed result and that obtained from HERWIG. In this case jets are identified with the k_t algorithm, and in the resummation non-global logs have been approximated with their value for the cone algorithm, which has been taken as an upper bound. The pretty good agreement between the two predictions, together with the fact that HERWIG does not fully contain the matrix elements giving rise to non-global logs, suggests that the observable is dominated by soft and collinear emissions.

Another interesting possibility could be that of making $E_{T1} - E_{T2}$ global by modifying the particle recombination scheme. Actually, in a massless E_0 scheme, one finds that emissions clustered with the jets do contribute to $E_{T1} - E_{T2}$, so that one can use the program CAESAR[8] to obtain automatically the resummed expression for its distribution.

The concluding remark of this overview is that pushing measurements in phase space regions where fixed order predictions are not expected to give an accurate description of the data opens up the possibility to investigate properties of QCD otherwise unaccessible. In particular, cross sections for jets with almost equal E_T's represent a yet poorly explored field, which could be complementary to the traditional event-shape measurements.

Acknowledgements

This work would not have been possible without the collaboration of Gennaro Corcella and Mrinal Dasgupta. I am grateful to the organisers for the invitation and for the possibility to enjoy the beauty of Ringberg castle and Tegernsee together with my wife and my baby daughter.

References

1. S. Chekanov *et al.* [ZEUS Collaboration], hep-ex/0502007.
2. M. Dasgupta and G. P. Salam, *J. Phys.* **G30**, R143 (2004) [hep-ph/0312283], and references therein.
3. M. Klasen and G. Kramer, *Phys. Lett.* **B366**, 385 (1996) [hep-ph/9508337].
4. S. Frixione and G. Ridolfi, *Nucl. Phys.* **B507**, 315 (1997) [hep-ph/9707345].
5. S. Chekanov *et al.* [ZEUS Collaboration], *Eur. Phys. J.* **C23**, 13 (2002) [hep-ex/0109029].
6. S. Catani and M. H. Seymour, *Phys. Lett.* **B378**, 287 (1996) [hep-ph/9602277]; *Nucl. Phys.* **B485**, 291 (1997) [*Erratum-ibid.* **B510**, 503 (1997)] [hep-ph/9605323].
7. A. Aktas *et al.* [H1 Collaboration], *Eur. Phys. J.* **C33**, 477 (2004) [hep-ex/0310019].
8. A. Banfi, G. P. Salam and G. Zanderighi, *JHEP* **0503**, 073 (2005) [hep-ph/0407286].
9. N. Brown and W. J. Stirling, *Phys. Lett.* **B252**, 657 (1990).
10. M. Dasgupta and G. P. Salam, *Phys. Lett.* **B512**, 323 (2001) [hep-ph/0104277].
11. A. Banfi and M. Dasgupta, *JHEP* **0401**, 027 (2004) [hep-ph/0312108].
12. S. Catani, Y. L. Dokshitzer, M. H. Seymour and B. R. Webber, *Nucl. Phys.* **B406**, 187 (1993); S. D. Ellis and D. E. Soper, *Phys. Rev.* **D48**, 3160 (1993) [hep-ph/9305266].
13. M. Dasgupta and G. P. Salam, *JHEP* **0203**, 017 (2002) [hep-ph/0203009].
14. R. B. Appleby and M. H. Seymour, *JHEP* **0212**, 063 (2002) [hep-ph/0211426].
15. A. Banfi and M. Dasgupta, *Phys. Lett.* **B628**, 49 (2005) [hep-ph/0508159].
16. V. M. Abazov *et al.* [D0 Collaboration], *Phys. Rev. Lett.* **94**, 221801 (2005) [hep-ex/0409040].
17. Y. L. Dokshitzer and G. Marchesini, hep-ph/0508130.
18. G. Corcella *et al.*, *JHEP* **0101**, 010 (2001) [hep-ph/0011363].

QCD DYNAMICS FROM FORWARD HADRON AND JET MEASUREMENTS

L. GOERLICH *

Institute of Nuclear Physics
Polish Academy of Sciences
Radzikowskiego 152, 31-342 Kraków, Poland
E-mail: lidia.goerlich@ifj.edu.pl

Recent measurements of deep inelastic events containing jets and π^0 mesons, produced close to the proton remnant in the laboratory frame, are discussed. The data are used to discriminate between QCD models with different parton evolution approximations.

1. Introduction

The HERA collider has extended the available kinematic region for deep-inelastic scattering (DIS) to regions of small Bjorken-x ($x_{B_j} \approx 10^{-5}$) at moderate four-momentum transfers Q^2 of a few GeV^2. The large $\gamma^* p$ centre of mass energy available at small x_{B_j} gives rise to a large phase space for gluon cascades exchanged between the proton and the photon. The fast rise of the gluon density with decreasing x_{B_j} suggests that at some point the gluon density will be so large that non-linear recombination effects leading to saturation phenomena may become important.

In perturbative QCD multiparton emissions are described only in approximation and in different regions of Q^2 and x_{B_j} different QCD-based prescriptions are expected to describe the radiation of partons. At large Q^2 the initial state radiation is described by standard DGLAP evolution equations[1] which at leading order resum $(\alpha_S \ln Q^2)^n$ terms. In this approach the struck quark originates from a parton cascade ordered in virtuality what is reflected by a strong ordering in transverse momenta k_T of subsequent gluon emissions. It is believed that at asymptotically high energies the theoretically correct description in the region of very small x_{B_J}

*Representing the H1 and ZEUS collaborations

is given by the BFKL equation[2]. In this approximation the evolution is dominated by large leading $\alpha_S \ln(1/x)$ terms which are resummed to all orders. Here the cascade is ordered strongly in fractional longitudinal momenta, while there is no ordering in transverse momentum of the partons along the ladder. The DGLAP and BFKL approaches are examples of two "extreme" choices for the transverse momentum ordering of the radiated gluons. The CCFM evolution[3] is an attempt to unify these two approaches. In the limit of asymptotic energies the CCFM evolution equation is almost equivalent to BFKL and for large x and high Q^2 it behaves like the DGLAP approximation. This behaviour follows from the fact that the CCFM evolution resums in addition to large logarithms of $(\alpha_s \ln(1/x))^n$ also terms $(\alpha_s \ln(1/(1-x))^n$. Furthermore, this approach introduces angular ordering of emissions to correctly treat gluon coherence effects.

A partially unordered parton chain can be also realized within DGLAP evolution scheme by allowing the exchanged virtual photon to have a partonic structure. In this approach two k_T-ordered DGLAP cascades are initiated: one from the proton and one from the photon. Resolved photon processes may become important in the region in which the k_T^2 of the parton at the photon end of the parton cascade is greater than the photon virtuality Q^2.

One of the important questions in high energy collisions at HERA is whether deviations from DGLAP type predictions can be observed and whether there is a need for small-x effects. At low x_{B_j} there is a lot of gluon radiation and to make a reliable comparison of measurements with theoretical predictions it is essential to calculate higher order terms in the relevant perturbative expansion.

Different measurements of DIS at low x_{B_j} have been performed to study the QCD parton dynamics in this region. Measurements of the proton structure function $F_2(x, Q^2)$ in a wide range of x and Q^2, and in particular the strong rise of F_2 with decreasing x_{B_j}, are well described by NLO DGLAP calculation. However, in this calculation there is an interplay of perturbative (NLO DGLAP evolution) and non-perturbative effects, coming from the input form of the parton distribution functions of the proton, which may obscure the underlying dynamics effects at low x_{B_j}. The observable F_2 is too inclusive to distinguish between the different alternatives. The hadronic final state observables which reflect the kinematic structure of gluon emissions are expected to be able to discriminate between different evolution approximations.

Studies of DIS events accompanied by an energetic jet of high transverse momentum close to the proton remnant are considered to be especially sensitive to the QCD dynamics at low x. These measurements, inspired by the proposal of Mueller[4], are restricted to the phase space region where the transverse momentum of the jet is approximately equal to the photon virtuality, $k_{T,jet}^2 \approx Q^2$, to suppress DGLAP evolution in Q^2. Sufficiently large values of transverse momentum also minimize the k_T diffusion into the infrared region[5]. In addition, the ratio x_{jet}/x_{B_j}, where $x_{jet} = E_{jet}/E_p$, should be large to enhance the phase space for BFKL effects[a]. The selected jets are close to the proton remnant direction in the laboratory frame, the so-called forward region.

Studies of single forward particles are complementary to measurements of forward jet production. An advantage of studying single particles as opposed to jets is the easier experimental identification without the dependence on the jet-finding algorithm and the potential to reach angles closer to the proton remnant direction. A disadvantage is that the cross section for the process is suppressed in comparison to forward jet production, and that the fragmentation effects are more significant.

In this analysis we present recent results of the H1 and ZEUS Collaborations on high tranverse momentum forward jet and π°-mesons produced in DIS at low x_{B_j}. The data are confronted with various QCD-based models.

2. Monte Carlo Programs and Theoretical Calculations

The presented measurements will be compared either with predictions of Monte Carlo (MC) programs which model higher order terms by leading logarithm parton showers, or with parton level fixed order QCD calculations. Different MC event generators which adopt various QCD approaches to modelling the parton cascade are used.

LEPTO[6] and RAPGAP[7] match LO QCD matrix elements for direct photon processes to DGLAP based parton showers with strong k_T ordering. In addition to the direct photon processes, RAPGAP simulates resolved photon interactions. The RAPGAP description of the hadronic final state is very similar to that of LEPTO when only direct photon interactions are considered.

ARIADNE[8] is an implementation of the Color Dipole Model (CDM)[9] in which all radiation is assumed to come from the dipole formed by the struck quark and the proton remnant. Subsequent parton emissions come

[a] $E_{jet}(E_P)$ denotes the energy of the forward jet(proton) in the laboratory system.

from a chain of independently radiated dipoles formed by emitted gluons. In consequence, transverse momenta of emitted gluons perform a random walk and in this sense ARIADNE is a BFKL-like program.

CASCADE[10] uses off-shell QCD matrix elements, supplemented with parton emissions based on the CCFM equation. An unintegrated gluon density, obtained using CCFM evolution and fitted to describe the inclusive DIS cross section is used as input to this model. Two different sets of unintegrated gluon density are used: one with only singular terms of the gluon splitting function and another one including also non-singular terms.

The studied data samples with selected forward jets are multijet events and are compared to fixed order QCD predictions of the following programs:

DISENT[11], which provides calculations at LO(α_S) and NLO(α_S^2) accuracy for dijet production in DIS via direct photon interactions;

NLOJET++[12], which calculates cross sections for three-jet production in DIS at NLO(α_S^3) accuracy. This program is the only NLO generator for three-jet configuration and is used in the description of the class of events with the forward jet accompanied by two additional jets.

Recently, three independent NLO calculations of the cross section for the production of high transverse momentum hadrons in DIS interactions have been made[13–15], which describe measurements of forward π° production. A large part of the cross section is generated by higher order contributions which correspond to lowest order BFKL and resolved photon processes. In the analysis of Aurenche et al.[13] also higher order corrections to the resolved photon processes are included. NLO calculations of high p_T hadrons are discussed in more detail in Ref. 16.

Analytical calculations based on LO BFKL effects with a consistency constraint which effectively includes part of higher order BFKL effects have been discussed in Ref. 17 for DIS events containing a forward jet or π°. These predictions are very sensitive to the choice of the scale for α_S and the infra-red cut-off.

3. Measurements of Forward Jets

Recently the ZEUS Collaboration has published results on inclusive forward jet measurements in the region $Q^2 > 25$ GeV^2 and $y > 0.04$ [18] based on a data sample corresponding to an integrated luminosity of 38.7 pb^{-1}. Jets are reconstructed using the inclusive k_T algorithm in the laboratory frame and then special conditions are applied to define the BFKL phase space region. The hadronic angle corresponding to the polar angle of the massless struck quark in the Quark-Parton Model (QPM) is required to be in the backward direction ($cos\gamma_h < 0$) and at least one jet with $p_{T,jet} >$ 6 GeV is found in the forward region of the detector ($0 < \eta_{jet} < 3$). These conditions suppress QPM-type events with single jets and enhance the kinematic region for multijet production. In addition, the requirement $0.5 < E_T^2/Q^2 < 2$ restricts the jet kinematics to the region where the DGLAP effects are suppressed and the BFKL effects are expected to be large.

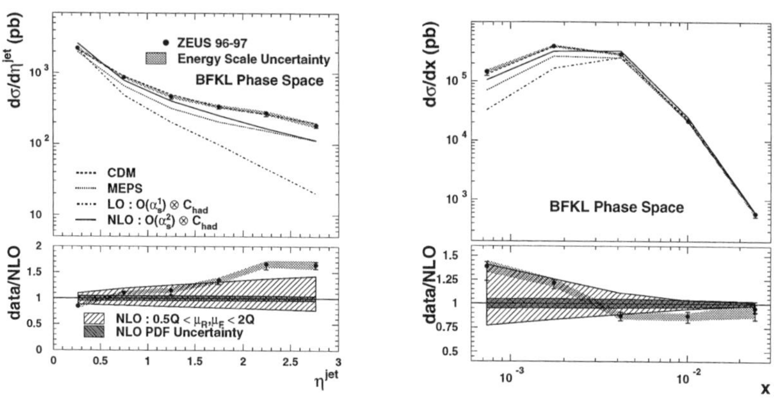

Figure 1. The forward jet cross section measurements of ZEUS as a function of η_{jet} and x_{B_j} compared with the predictions of CDM, LEPTO(MEPS) and DISENT.

In Fig. 1 the inclusive forward jet cross section is shown as a function of η_{jet} and x_{B_j}. The data are compared with LO(α_S^1) and NLO(α_S^2) predictions provided by DISENT with the renormalisation scale set to Q. The lower part of each plot shows the ratio of data to the QCD calculations and the theoretical uncertainties. The theoretical uncertainty is estimated in the conventional way by varying renormalisation or factorisation scales by a factor of two up and down. NLO calculations give a reasonable description of the data except for the forward region ($\eta_{jet} > 1$) and for low x_{B_j}. In these regions NLO corrections are large and the ratio NLO/LO reaches values as large as five for $\eta_{jet} \approx 3$. The large NLO/LO factors at low x_{B_j} and large η_{jet} are associated with the contribution from t-channel gluon exchange diagrams. Large scale variation uncertainties, related mainly with the renormalisation scale, suggest that higher order terms are needed to improve the description of the data. Predictions of the BFKL-like model CDM are also shown in Fig. 1 and describe the data well.

Further new measurements of forward jet production are available from both H1[19] and ZEUS[20] Collaborations. In both analyses jets are identified using the inclusive k_T-algorithm in the Breit frame and therefore QPM type events with single jets are kinematically forbidden. The kinematical cuts and jet selection criteria are summarized in Table 1.

Table 1. Selection cuts of forward jets by H1 and ZEUS.

$exp.$	$Q^2[GeV^2]$	x	E_T^2/Q^2	η_{jet}	$p_{T,jet}[GeV]$	x_{jet}
$H1$ 13.7 pb^{-1}	5-85	0.0001-0.004	0.5-5.0	1.7-2.8	> 3.5	> 0.035
$ZEUS$ 82 pb^{-1}	20-100	0.0004-0.005	0.5-2.0	2.0-3.5	> 5.0	> 0.036

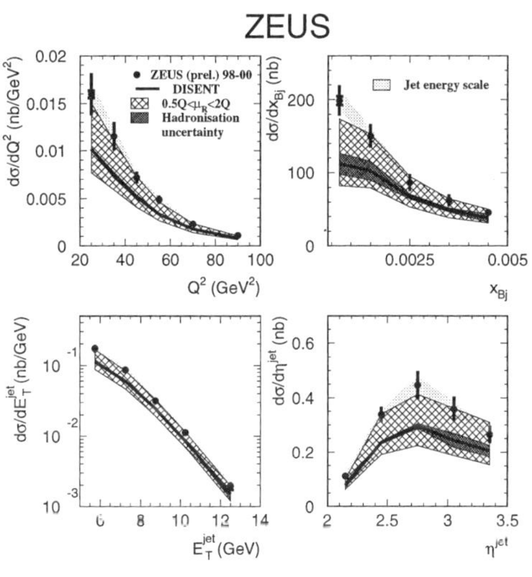

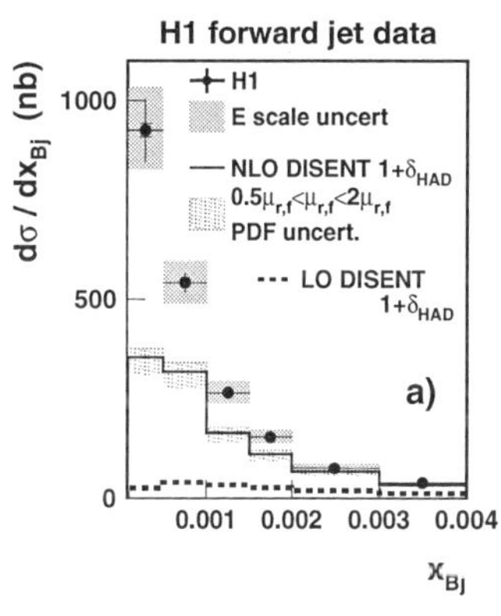

Figure 2. The forward jet cross section measurements of ZEUS (left) and H1 (right) compared with LO and NLO predictions of DISENT.

The H1 analysis reaches lower x_{B_j} values as it covers the phase space region at lower Q^2. The requirement on E_T^2/Q^2 is more restrictive in the ZEUS measurement. ZEUS also extends the measurements to the more forward region up to $\eta_{jet} = 3.5$. Fig. 2 shows the forward jet cross sections as a function of Q^2, x_{B_j}, $E_{T,jet}$ and η_{jet} obtained by ZEUS. The results are compared with NLO(α_S^2) predictions of DISENT with the choice of the renormalisation scale $\mu_R = Q$. The NLO QCD predictions are below the data but within theoretical uncertainties except for the region of low x_{B_j}. The large renormalisation scale uncertainty indicates the importance of higher order terms in this analysis.

The new H1 forward jet measurements are also shown in Fig. 2. The forward jet cross section as a function of x_{B_j} is compared with LO(α_S^1) and NLO(α_S^2) calculations from DISENT with the renormalisation scale defined by the average p_T^2 of the dijets from the matrix element $\mu_R^2 = < p_{T,dijets}^2 >$. The scale uncertainty in this analysis is estimated by simultaneously changing the renormalisation and factorisation scales. The NLO calculation is closer to the data than the LO contribution and at low x_{B_j} there is one order of magnitude difference between LO and NLO predictions. This reflects the fact that at LO the forward jet production is suppressed by kinematics. However, the NLO predictions are still a factor two below the data at low x_{B_j}.

Conclusions from measurements of the forward jet cross section by H1 and ZEUS seem to be contradictory. However, the main difference in the NLO(α_S^2) description comes from different choice of the renormalisation scale which leads to a significantly different estimate of the theoretical uncertainty.

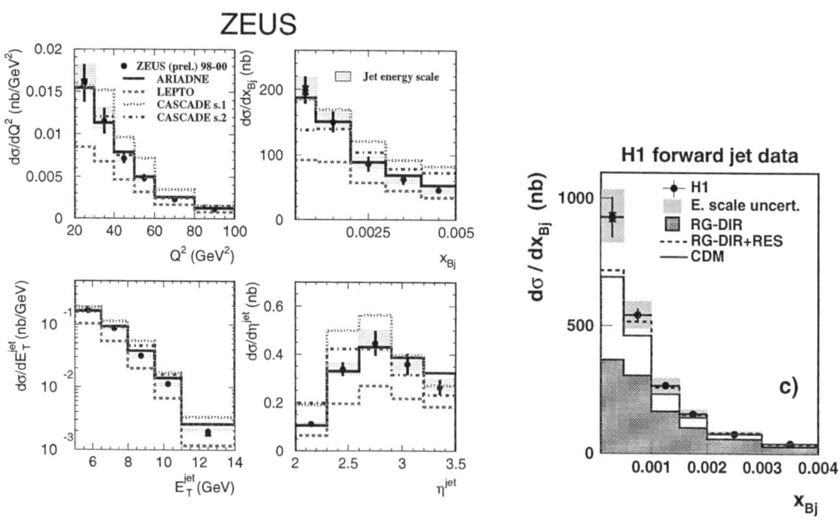

Figure 3. The forward jet cross section measurements of ZEUS (left) and H1 (right) compared with different Monte Carlo models.

The experimental results presented in Fig. 2 are again shown in Fig. 3, now in a comparison with different MC models. Predictions of DGLAP-based models with only the direct photon processes included, marked LEPTO or RG-DIR, are well below the data. The addition of the resolved photon contribution, as shown for the H1 data (RG-DIR+RES), significantly improves the description. However, there is still a discrepancy in the lowest bin of x_{B_j}. A similar good description is given by the CDM model with emissions non-ordered in transverse momentum. The ZEUS results are best described by the CDM model as given by ARIADNE. Predictions of the CCFM model as implemented in CASCADE with two different parametrizations of the unintegrated gluon density fail to describe the shape of the distributions in x_{B_j} and η_{jet}. This might be caused by the parametrization of the unintegrated gluon density and/or missing contributions from quark induced cascades and gluon splittings into quark pairs.

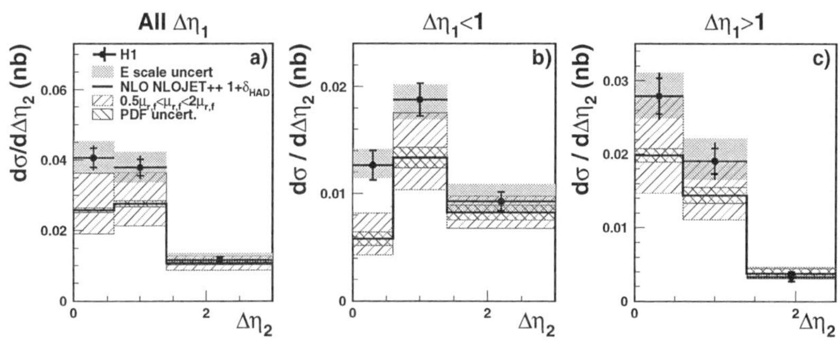

Figure 4. The cross section for events with "forward jet + 2 additional jets" as a function of the rapidity separation $\Delta_{\eta 2}$ (see text) compared with NLO predictions.

The H1 Collaboration has also performed a more exclusive analysis of events with the forward jet accompanied by two additional jets. All jets have transverse momenta larger than 6 GeV and are ordered in rapidity according to $\eta_{fwdjet} > \eta_{jet2} > \eta_{jet1} > \eta_e$, where η_e is the rapidity of the scattered electron. The cross section is measured in two bins of $\Delta\eta_1$ as a function of $\Delta\eta_2$, where $\Delta\eta_1 = \eta_{jet2} - \eta_{jet1}$ is defined as a difference in rapidity between two additional jets and $\Delta\eta_2 = \eta_{fwdjet} - \eta_{jet2}$ is a separation in rapidity between the forward jet and the closer additional jet. The direction of the forward jet is restricted in space as being close to the proton axis and the directions of additional jets are selected through $\Delta\eta$

requirements. In general, by choosing different $\Delta\eta$ values one can select different event topologies and test the breaking of the k_t ordering at different location along the ladder between the virtual photon and the forward jet. Fig. 4 shows a comparison of the data with NLO(α_S^3) predictions of the NLOJET++ program. Good agreement with the data is observed if additional jets are produced in the central region (large $\Delta\eta_2$). NLO QCD calculations deviate from the data if both values $\Delta\eta_1$ and $\Delta\eta_2$ are small i.e. when all three jets are in the forward region. This configuration is suppressed in NLOJET++.

In Fig. 5 we show that the more exclusive measurement of "forward jet + 2 additional jets" is able to differentiate different QCD models. The data are best described by the CDM model. The DGLAP-based model with only direct as well as with resolved photon processes is below the data. The conclusion is that a different type of the k_T-order breaking is needed than that provided by the resolved photon model.

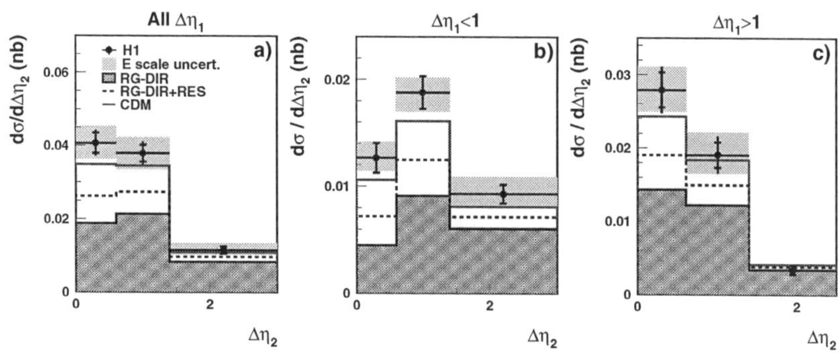

Figure 5. The cross section for events with "forward jet + 2 additional jets" compared with predictions of CDM and RAPGAP with direct and resolved processes .

4. Inclusive Forward π° Cross Sections and Transverse Energy Flow

Measurements of forward π° mesons[21] in the H1 detector are restricted in the kinematic range $2 < Q^2 < 70$ GeV^2, $0.1 < y < 0.6$, $5^\circ < \theta_\pi < 25^\circ$, $x_\pi = E_\pi/E_p > 0.01$ and $p^*_{T,\pi} > 2.5$ or 3.5 GeV, where $E_\pi(E_P)$ denotes the energy of the forward π(proton) in the LAB and $p^*_{T,\pi}$ is defined in the photon-proton centre-of-mass system (CMS). The resulting range in Bjorken-x extends down to $x \approx 4 \cdot 10^{-5}$.

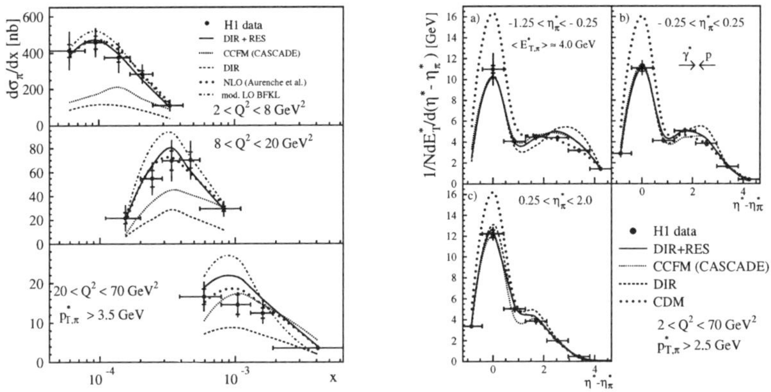

Figure 6. Inclusive forward π° cross section as a function of x_{B_j} in three regions of Q^2 for $p^*_{t,\pi} > 3.5$ GeV (left) and transverse energy flow relative to the π° in various ranges of η^*_π (right).

Inclusive forward π° cross sections as a function of x_{B_j} for $p^*_{T,\pi} > 2.5$ GeV in comparison with MC calculations which implement various QCD models have been discussed in Ref. 22. The conclusions are similar to those for the forward jet analysis. In Fig. 6 of this paper we show results for $p^*_{T,\pi} > 3.5$ GeV which are also compared with recent NLO DGLAP calculations[13]. The NLO prediction describes the data well, however large NLO/LO factors and large theoretical uncertainties indicate that a NNLO analysis would be required for a reliable prediction.

Studies of transverse energy flow provide a complementary means of investigating QCD processes. The transverse energy flow in the CMS, $\frac{1}{N}dE^*_T/d(\eta^* - \eta^*_\pi)$, in events containing at least one forward π° is shown in Fig. 6. The energy flow, which includes the contribution from the forward π°, is plotted as a function of the distance in pseudorapidity $\Delta\eta^* = (\eta^* - \eta^*_\pi)$ from the forward π°. The spectra are presented in three intervals of the π° pseudorapidity ranging from close (Fig. 6a) to far (Fig. 6c) from the proton direction. The QCD-based models except CDM all describe the transverse energy flow around the π° but give different predictions in the current region. The differences between the models can be qualitatively understood as a consequence of the ordering or otherwise of the transverse momentum in the parton cascades implemented within MC models. The CDM predicts too much transverse energy in the vicinity of the π°, which is a consequence of the too hard π° transverse momentum distribution predicted by this model.

5. Future Studies

At HERA II one expects about 700 pb^{-1} of ep data until the end of data taking in 2007, which would exceed the luminosity of HERA I by a factor of 5. With increased statistics further improvements in the precision of the forward region measurements can be gained by selecting forward jets with higher transverse momentum and higher jet energy. The investigation of the dependence on additional variables like p_T^2/Q^2, x_{B_j}/x_{jet}, $cos\gamma_h$ may help to differentiate various QCD models.

The interpretation of the forward production data indicates the importance of resolved photon processes. Therefore, in addition to the primary hard scattering, additional interactions between the remnants of the proton and the photon could contribute in the reaction. A better understanding of multiple interactions could help in precision studies of the forward jet and particle production.

Continuation of the studies of multijet production with different forward jet configurations, the jet shape analysis of forward jets and measurements of azimuthal correlation between the scattered electron and the forward jet will provide additional information about the QCD dynamics at low x_{B_J}.

Another possibility is to perform a measurement of forward photon production in DIS[23]. In this process there is no dependence on the hadronization mechanism, however the rate for this reaction is low and background from neutral pions may turn out to be large.

6. Conclusions

A new kinematic regime of low values of x_{B_j} provided by HERA for DIS measurements allows for tests of the QCD dynamics in this region. The production of forward jets and hadrons is especially sensitive to the dynamics of parton evolution and has been studied by the H1 and ZEUS Collaborations for the last ten years.

The measured cross sections show deviations from the LO DGLAP predictions. The inclusion of resolved photon processes improves the overall description. In some regions of phase space NLO DGLAP predictions give reasonable agreement with the data, however higher order calculations would be needed to improve the agreement and to reduce theoretical uncertainties. The CCFM evolution as implemented in the MC program CASCADE does not describe the data. The BFKL-like CDM model describes the data in most of the phase space region. Summarizing, existing theoretical approaches do not explain all features of forward jet and particle production. Recently important progress in measurements and theoretical calculations has been achieved, however the correct interpretation of the low x_{Bj} data at HERA is still a challenge.

Acknowledgments

I would like to thank the organisers for the invitation to this stimulating Workshop with a pleasant atmosphere certainly enhanced by so beautiful surroundings.

References

1. V. Gribov and L. Lipatov, *Sov. J. Phys.* **15**, 438 and 675 (1972).
   L. Lipatov, *Sov. J. Phys.* **20**, 94 (1975).
   G. Altarelli and G. Parisi, *Nucl. Phys.* **B126**, 298 (1977).
   Y. Dokshitzer *Sov. Phys. JETP* **46**, 641 (1977).
2. E. Kuraev, L. Lipatov and V. Fadin, *Sov. Phys. JETP* **44**, 443 (1976).
   E. Kuraev, L. Lipatov and V. Fadin, *Sov. Phys. JETP* **45**, 199 (1977).
   Y. Balitsky and L. Lipatov, *Sov. J. Nucl. Phys.* **28**, 822 (1978).
3. M. Ciafaloni, *Nucl. Phys.* **296**, 49 (1988).
   S. Catani, F. Fiorani and G. Marchesini, *Phys. Lett.* **B234**, 339 (1990).
   S. Catani, F. Fiorani and G. Marchesini, *Nucl. Phys.* **B336**, 18 (1990).
4. A. H. Mueller, *Nucl. Phys.* **B** *(Proc. Suppl.)* **18C**, 125 (1990).
5. A. Martin, Lect. XXI Int. Meet. Fund. Phys., Madrid (1993) DTP/93/66.
6. G. Ingelman, A. Edin and J. Rathsman, *Comp. Phys. Comm.* **101**, 108 (1997).
7. H. Jung, *Comp. Phys. Comm.* **86**, 147 (1995).
8. L. Lönnblad, *Comp. Phys. Comm.* **71**, 15 (1992).
9. B. Andersson, G. Gustafson and L. Lönnblad, *Nucl. Phys.* **B339**, 393 (1990).
10. H. Jung, *Comp. Phys. Comm.* **143**, 100 (2002).
11. S. Catani and M. H. Seymour, *Nucl. Phys.* **B485**, 291 (1997).
    Erratum in *Nucl. Phys.* **B510**, 503 (1998).
12. Z. Nagy and Z. Trocsanyi, *Phys. Rev. Lett.* **87**, 082001 (2001).
13. P. Aurenche, R. Basu, M. Fontannaz and R. M. Godbole, *Eur. Phys. J.* **C42**, 43 (2005), *Eur. Phys. J.* **C34**, 277 (2004).
14. B. A. Kniehl, G. Kramer and M. Maniatis, *Nucl. Phys.* **B711**, 345 (2005).
15. A. Daleo, D. de Florian and R. Sassot, *Phys. Rev.* **D71**, 034013 (2005).
16. B. A. Kniehl, these proceedings.
17. J. Kwiecinski, A. Martin and J. Outhwait, *Eur. Phys. J.* **C9**, 611 (1999).
18. S. Chekanov *et al.* [ZEUS Collaboration], *DESY*-05-17.
19. A. Aktas *et al.* [H1 Collaboration], *DESY*-05-135.
20. ZEUS Collaboration, XXII International Symposium on Lepton-Photon Interactions at High Energy , 2005, abstract 278.

22. L. Jönsson, Proceedings of New Trends in HERA Physics 2003.
23. J. Kwiecinski, S. C. Lang, A. Martin, *Phys. Rev.* **D54**, 1874 (1996).

LIGHT-HADRON ELECTROPRODUCTION AT NEXT-TO-LEADING ORDER AND IMPLICATIONS

BERND A. KNIEHL

II. Institut für Theoretische Physik, Universität Hamburg,
Luruper Chaussee 149, 22761 Hamburg, Germany
E-mail: kniehl@desy.de

We review recent results on the inclusive electroproduction of light hadrons at next-to-leading order in the parton model of quantum chromodynamics implemented with fragmentation functions and present updated predictions for HERA experiments based on the new AKK set. We also discuss phenomenological implications of these results.

1. Introduction

In the framework of the parton model of quantum chromodynamics (QCD), the inclusive production of single hadrons is described by means of fragmentation functions (FFs) $D_a^h(x,\mu)$. At lowest order (LO), the value of $D_a^h(x,\mu)$ corresponds to the probability for the parton a produced at short distance $1/\mu$ to form a jet that includes the hadron h carrying the fraction x of the longitudinal momentum of a. Analogously, incoming hadrons and resolved photons are represented by (non-perturbative) parton density functions (PDFs) $F_a^h(x,\mu)$. Unfortunately, it is not yet possible to calculate the FFs from first principles, in particular for hadrons with masses smaller than or comparable to the asymptotic scale parameter Λ. However, given their x dependence at some energy scale μ, the evolution with μ may be computed perturbatively in QCD using the timelike Dokshitzer-Gribov-Lipatov-Altarelli-Parisi (DGLAP) equations. Moreover, the factorization theorem guarantees that the $D_a^h(x,\mu)$ functions are independent of the process in which they have been determined and represent a universal property of h. This entitles us to transfer information on how a hadronizes to h in a well-defined quantitative way from e^+e^- annihilation, where the measurements are usually most precise, to other kinds of experiments, such as photo-, lepto-, and hadroproduction. Recently, FFs for light charged hadrons with complete quark flavour separation were determined[1] through

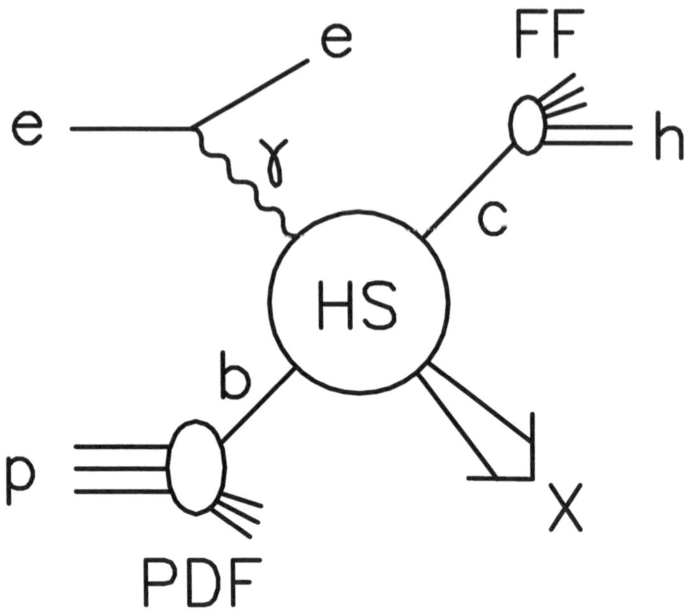

Figure 1. Parton-model representation of $ep \to eh + X$.

a global fit to e^+e^- data from LEP, PEP, and SLC including for the first time the light-quark tagging probabilities measured by the OPAL Collaboration at LEP,[2] thereby improving previous analyses.[3,4]

The QCD-improved parton model should be particularly well applicable to the inclusive production of light hadrons carrying large transverse momenta (p_T) in deep-inelastic lepton-hadron scattering (DIS) with large photon virtuality (Q^2) due to the presence of two hard mass scales, with $Q^2, p_T^2 \gg \Lambda^2$. In Fig. 1, this process is represented in the parton-model picture. The hard-scattering (HS) cross sections, which include colored quarks and/or gluons in the initial and final states, are computed in perturbative QCD. They were evaluated at LO more than 25 years ago.[5] Recently, the next-to-leading-order (NLO) analysis was performed independently by three groups.[6,7,8] A comparison between Refs. 7, 8 using identical input yielded agreement within the numerical accuracy.

The cross section of $e^+p \to e^+\pi^0 + X$ in DIS was measured in various distributions with high precision by the H1 Collaboration at HERA in the forward region, close to the proton remnant.[9,10] This measurement reaches down to rather low values of Bjorken's variable $x_B = Q^2/(2P\cdot q)$, where P and q are the proton and virtual-photon four-momenta, respectively, and $Q^2 = -q^2$, so that the validity of the DGLAP evolution might be challenged by Balitsky-Fadin-Kuraev-Lipatov (BFKL) dynamics.

In Ref. 7, the H1 data[9,10] were compared with NLO predictions evaluated with the KKP FFs.[3] In Sec. 2, we summarize the analytical calculation performed in Ref. 7. In Sec. 3, we present an update of this comparison based on the new AKK FFs.[1] Our conclusions are summarized in Sec. 4.

2. Analytical calculation

The partonic subprocesses contributing at LO are

$$\begin{aligned} &\gamma^* + q \to q + g, \\ &\gamma^* + q \to g + q, \\ &\gamma^* + g \to q + \overline{q}, \end{aligned} \tag{1}$$

where q represents any of the n_f active quarks or antiquarks and it is understood that the first of the final-state partons is the one that fragments into the hadron h.

At NLO, processes (1) receive virtual corrections, and real corrections arise through the partonic subprocesses

$$\begin{aligned} &\gamma^* + q \to q + g + g, \\ &\gamma^* + q \to g + q + g, \\ &\gamma^* + g \to q + \overline{q} + g, \\ &\gamma^* + g \to g + q + \overline{q}, \\ &\gamma^* + q \to q + q + \overline{q}, \\ &\gamma^* + q \to \overline{q} + q + q, \\ &\gamma^* + q \to q + q' + \overline{q}', \\ &\gamma^* + q \to q' + \overline{q}' + q, \end{aligned} \tag{2}$$

where $q' \neq q, \overline{q}$. The virtual corrections contain infrared (IR) singularities, both of the soft and/or collinear types, and ultraviolet (UV) ones, which are all regularized using dimensional regularization with $D = 4 - 2\epsilon$ space-time dimensions yielding poles in ϵ in the physical limit $D \to 4$. The latter arise from one-loop diagrams and are removed by renormalizing the strong-coupling constant and the wave functions of the external partons in the respective tree-level diagrams, while the former partly cancel in combination with the real corrections. The residual IR singularities are absorbed into redefinitions of the PDFs and FFs. We extract the IR singularities in the real corrections by performing the phase space integrations using the dipole subtraction formalism.[11]

3. Comparison with H1 data

We work in the modified minimal-subtraction ($\overline{\rm MS}$) renormalization and factorization scheme with $n_f = 5$ massless quark flavors and identify the renormalization and factorization scales by choosing $\mu^2 = \xi[Q^2 + (p_T^*)^2]/2$, where the asterisk labels quantities in the $\gamma^* p$ center-of-mass (c.m.) frame and ξ is varied between 1/2 and 2 about the default value 1 to estimate the theoretical uncertainty. At NLO (LO), we employ set CTEQ6M (CTEQ6L1) of proton PDFs,[12] the NLO (LO) set of AKK FFs,[1] and the two-loop (one-loop) formula for the strong-coupling constant $\alpha_s^{(n_f)}(\mu)$ with $\Lambda^{(5)} = 226$ MeV (165 MeV).[12]

The H1 data[9,10] were taken in DIS of positrons with energy $E_e = 27.6$ GeV on protons with energy $E_p = 820$ GeV in the laboratory frame, yielding a c.m. energy of $\sqrt{S} = 2\sqrt{E_e E_p} = 301$ GeV. The DIS phase space was restricted to $0.1 < y < 0.6$ and $2 < Q^2 < 70$ GeV^2, where $y = Q^2/(x_B S)$. The π^0 mesons were detected within the acceptance cuts $p_T^* > 2.5$ GeV (except where otherwise stated), $5° < \theta < 25°$, and $x_E > 0.01$, where θ is their angle with respect to the proton flight direction and $E = x_E E_p$ is their energy in the laboratory frame. The comparisons with our updated LO and NLO predictions are displayed in Figs. 2(a)–(d). The QCD correction (K) factors, i.e. the NLO to LO cross section ratios, are presented in the downmost frame of each figure.

Comparison of Figs. 2(a)–(d) with Figs. 3, 5(a), 6(c), and 7 in Ref. 7, where the KKP FFs[3] were used, reveals that the update of our FFs, from set KKP to set AKK,[1] has hardly any visible impact on the theoretical predictions considered here. This may be understood by observing that the OPAL light-quark tagging probabilities for charged pions,[2] included in the AKK analysis, agree well with the assumption made in the KKP one that $D_u^{\pi^\pm}(x, \mu_0) = D_d^{\pi^\pm}(x, \mu_0)$ at the starting scale μ_0 of the DGLAP evolution. In Figs. 3(a) and (b),[8] the H1 data[10] on $d\sigma/dp_T^*$ for $2 < Q^2 < 4.5$ GeV^2, $4.5 < Q^2 < 15$ GeV^2, or $15 < Q^2 < 70$ GeV^2 and on $d\sigma/dx_B$ for $p_T^* >$

3.5 GeV and $2 < Q^2 < 8$ GeV^2, $8 < Q^2 < 20$ GeV^2, or $20 < Q^2 < 70$ GeV^2, respectively, are compared with the LO and NLO predictions evaluated with the KKP FFs[3] or those by Kretzer (K).[4] While the LO predictions based on the KKP and K sets agree very well, the NLO predictions based on the K set appreciably undershoot those based on the KKP set. If it were not for the theoretical uncertainty, one might conclude that the H1 data prefer the KKP set at NLO.

From the downmost frames in Figs. 2(a)–(d), we observe that the K factors are rather sizeable, although the μ values are reasonably large. In Fig. 4,[8] the impact of the H1 forward-selection cuts on the K factor is studied for the case of $d\sigma/dx_B$ for $2 < Q^2 < 8$ GeV^2 and $p_T^* > 3.5$ GeV. Towards the lower end of the considered x_B range, the K factor reaches one order of magnitude if these cuts are imposed [see also Fig. 2(c)]. However, if the latter are removed, the K factor collapses to acceptable values of around 3. From this finding, we conclude that these cuts almost quench the LO cross section. In other words, in the extreme forward regime, the latter is effectively generated by the $2 \to 3$ partonic subprocesses of Eq. (2).

It is interesting to investigate the relative importance of the *tagged* partons, i.e. the one (a) that originates from the proton and the one (b) that fragments into the hadron. In Fig. 5,[8] the NLO contributions from the four most important ab channels to $d\sigma/dx_B$ for $2 < Q^2 < 8$ GeV^2 and $p_T^* > 3.5$ GeV with the H1 forward-selection cuts are shown together with the total LO contribution. We observe that the gg channel makes up approximately two thirds of the cross section in the low-x_B regime.

4. Conclusions

We calculated the cross section of $ep \to e\pi^0 + X$ in DIS for finite values of p_T^* at LO and NLO in the parton model of QCD[7] using the new AKK FFs[1] and compared it with a precise measurement by the H1 Collaboration at HERA.[9,10]

We found that our LO predictions always significantly fell short of the H1 data and often exhibited deviating shapes. However, the situation dramatically improved as we proceeded to NLO, where our default predictions, endowed with theoretical uncertainties estimated by moderate unphysical-scale variations, led to a satisfactory description of the H1 data in the preponderant part of the accessed phase space. In other words, we encountered K factors much in excess of unity, except towards the regime of asymptotic freedom characterized by large values of p_T^* and/or Q^2. This was unavoidably accompanied by considerable theoretical uncertainties. Both features suggest that a reliable interpretation of the H1 data within the QCD-improved parton model ultimately necessitates a full next-to-next-to-leading-order analysis, which is presently out of reach, however. For the time being, we conclude that the successful comparison of the H1 data with our NLO predictions provides a useful test of the universality and the scaling violations of the FFs, which are guaranteed by the factorization theorem and are ruled by the DGLAP evolution equations, respectively.

Significant deviations between the H1 data and our NLO predictions only occurred in certain corners of phase space, namely in the photoproduction limit $Q^2 \to 0$, where resolved virtual photons are expected to contribute, and in the limit $\eta \to \infty$ of the pseudorapidity $\eta = -\ln[\tan(\theta/2)]$, where fracture functions are supposed to enter the stage. Both refinements were not included in our analysis. Interestingly, distinctive deviations could not be observed towards the lowest x_B values probed, which indicates that the realm of BFKL dynamics has not actually been accessed yet.

Acknowledgments

The author thanks G. Kramer and M. Maniatis for their collaboration. This work was supported in part by BMBF Grant No. 05 HT1GUA/4.

References

1. S. Albino, B. A. Kniehl and G. Kramer, *Nucl. Phys.* **B725**, 181 (2005); *Nucl. Phys.* **B734**, 50 (2006).
2. OPAL Collaboration, G. Abbiendi et al., *Eur. Phys. J.* **C16**, 407 (2000).
3. B. A. Kniehl, G. Kramer and B. Pötter, *Nucl. Phys.* **B582**, 514 (2000); *Phys. Rev. Lett.* **85**, 5288 (2000); *Nucl. Phys.* **B597**, 337 (2001).
4. S. Kretzer, *Phys. Rev.* **D62**, 054001 (2000).
5. A. Mendez, *Nucl. Phys.* **B145**, 199 (1978).
6. P. Aurenche, R. Basu, M. Fontannaz and R. M. Godbole, *Eur. Phys. J.* **C34**, 277 (2004).
7. B. A. Kniehl, G. Kramer and M. Maniatis, *Nucl. Phys.* **B711**, 345 (2005); **B720**, 231(E) (2005).
8. A. Daleo, D. de Florian and R. Sassot, *Phys. Rev.* **D71**, 034013 (2005).
9. H1 Collaboration, C. Adloff et al., *Phys. Lett.* **B462**, 440 (1999).
10. H1 Collaboration, A. Aktas et al., *Eur. Phys. J.* **C36**, 441 (2004).
11. S. Catani and M. H. Seymour, *Nucl. Phys.* **B485**, 291 (1997); **B510**, 503(E) (1997).
12. J. Pumplin, D. R. Stump, J. Huston, H.-L. Lai, P. Nadolsky and W.-K. Tung, *JHEP* **0207**, 012 (2002).

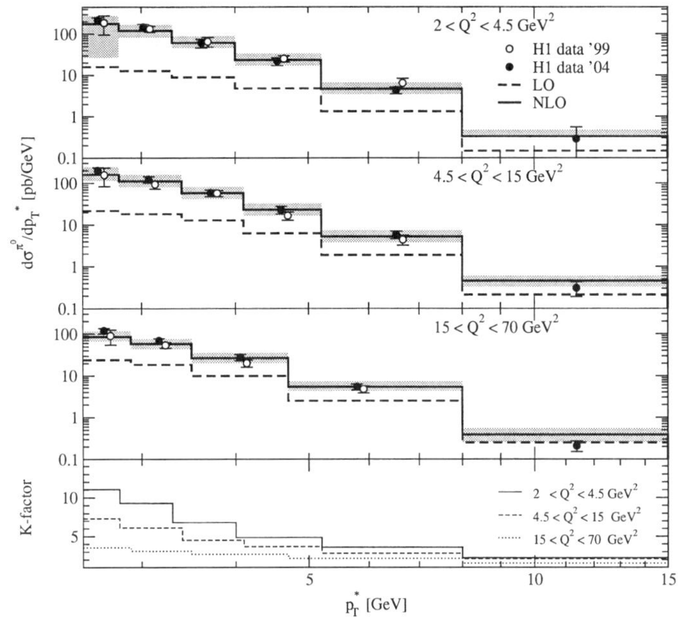

(a)

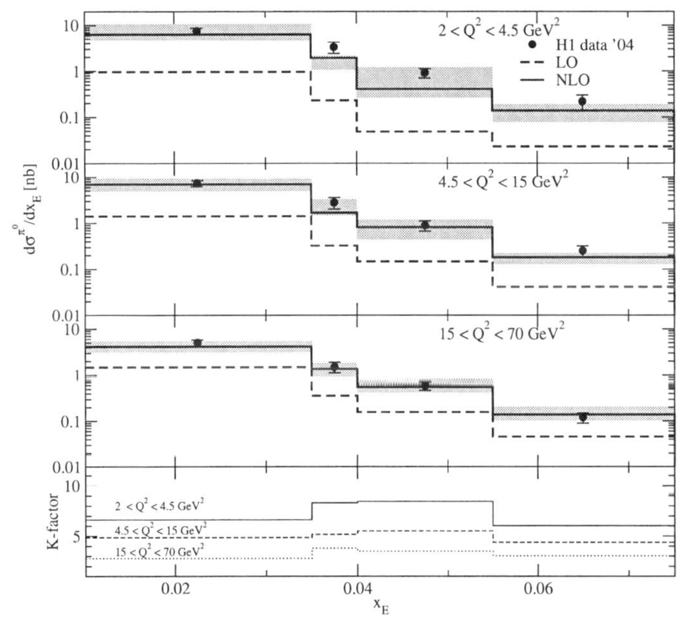

(b)

Figure 2. H1 data on (a) $d\sigma/dp_T^*$ and (b) $d\sigma/dx_E$ for $2 < Q^2 < 4.5\ \mathrm{GeV}^2$, $4.5 < Q^2 < 15\ \mathrm{GeV}^2$, or $15 < Q^2 < 70\ \mathrm{GeV}^2$, on (c) $d\sigma/dx_B$ for $p_T^* > 3.5\ \mathrm{GeV}$ and $2 < Q^2 < 8\ \mathrm{GeV}^2$, $8 < Q^2 < 20\ \mathrm{GeV}^2$, or $20 < Q^2 < 70\ \mathrm{GeV}^2$, and on (d) $d\sigma/dQ^2$ from Refs. 9 (open circles) and 10 (solid circles) are compared with our default LO (dashed histograms) and NLO (solid histograms) predictions including theoretical uncertainties (shaded bands). The K factors are also shown.

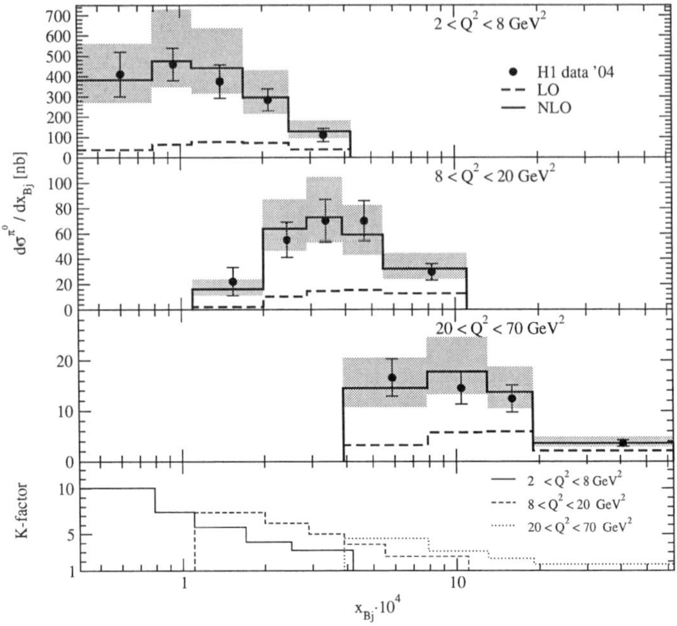

(c)

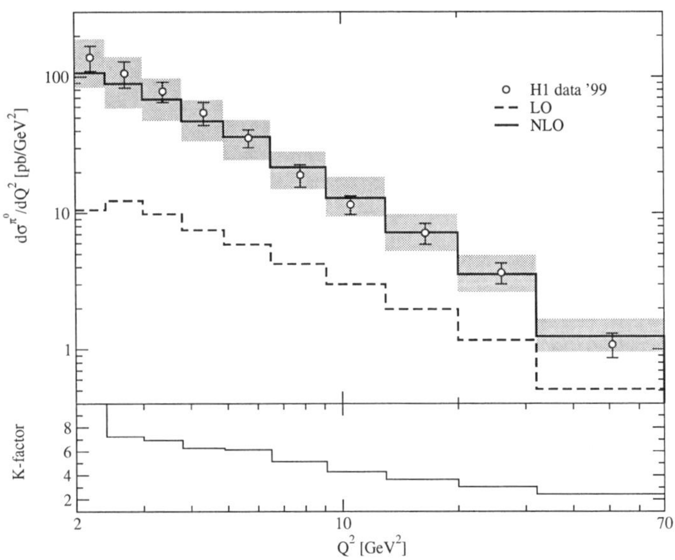

(d)

Figure 2. Continued.

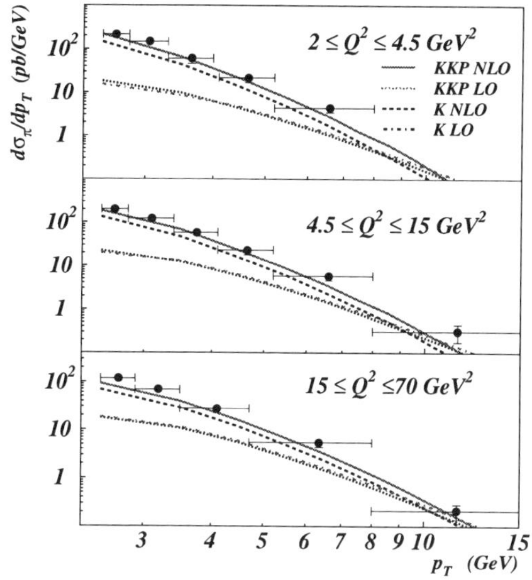

(a)

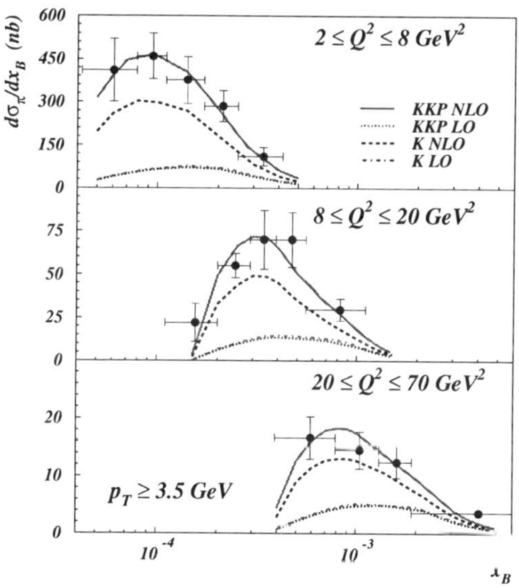

(b)

Figure 3. H1 data[10] on (a) $d\sigma/dp_T^*$ for $2 < Q^2 < 4.5$ GeV^2, $4.5 < Q^2 < 15$ GeV^2, or $15 < Q^2 < 70$ GeV^2 and on (b) $d\sigma/dx_B$ for $p_T^* > 3.5$ GeV and $2 < Q^2 < 8$ GeV^2, $8 < Q^2 < 20$ GeV^2, or $20 < Q^2 < 70$ GeV^2 are compared with the LO and NLO predictions evaluated with the KKP[3] or K[4] FFs (taken from Ref. 8).

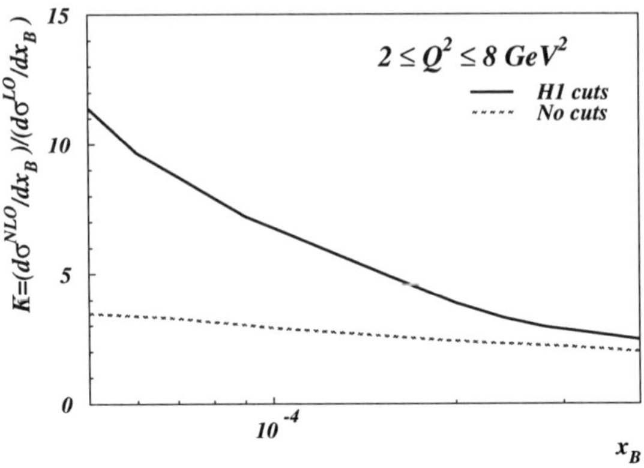

Figure 4. K factors of $d\sigma/dx_B$ for $2 < Q^2 < 8$ GeV^2 and $p_T^* > 3.5$ GeV with and without the H1 forward-selection cuts[10] (taken from Ref. 8).

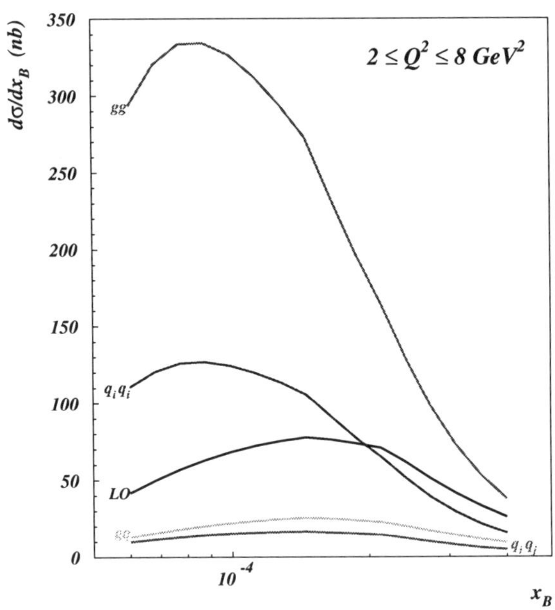

Figure 5. Total LO contribution and NLO contributions from the four most important ab channels, where a and b are the partons connected with the PDFs and FFs, respectively, to $d\sigma/dx_B$ for $2 < Q^2 < 8$ GeV^2 and $p_T^* > 3.5$ GeV with the H1 forward-selection cuts[10] (taken from Ref. 8).

PARTICLE PRODUCTION AND FRAGMENTATION

D. H. SAXON
Faculty of Physical Sciences,
University of Glasgow,
Glasgow, G12 8QQ, Scotland
E-mail: d.saxon@physics.gla.ac.uk

Recent results from HERA are presented on a range of topics: charged multiplicities, production of non-strange mesons and strange particles, charm fragmentation, baryons decaying to strange particles, antideuteron production, Bose-Einstein correlations, and new interpretations of results on prompt photon production in DIS.

1. Charged particle multiplicities in DIS and DDIS.

Historically, charged particle multiplicities have been measured in DIS in the current region of the Breit frame - defined as the frame in which there is no energy transfer in the collision. In zeroth order perturbative QCD this corresponds to a quark-parton jet with energy $Q/2$ and can be compared to a half-event in e^+e^- annihilation. Early ZEUS results covered the range $10 < Q^2 < 4000$ GeV^2 with rather large errors at the high end. The results lie mostly on top of e^+e^- annihilation but fall below those at $Q^2 < 50$ GeV^2. New analyses in the Breit frame using $2E_{current}$ as the scale instead of $Q/2$ fix this problem at low energy - see figure 1.

Using one hemisphere of the Breit frame offers a rapidity range of only $\ln(Q/m)$ where m is the pion mass. Instead we can work in the $\gamma^* p$ centre-of-mass frame using the hadronic energy W as the scale and access instead a larger rapidity range of $\ln(W/m)$ (recall $W^2 \simeq Q^2/x$.) These results are also shown in figure 1 and show excellent agreement between DIS and e^+e^-, and with calculations using LEPTO and ARIADNE.

ZEUS have also looked at an alternative approach, which is to use only the best part of the tracker ($20° < \theta < 160°$) where the tracking efficiencies are high and well-known and to plot the multiplicity (excluding the recoil electron) as a function of $M^2_{\text{eff}} = (\Sigma E)^2 - (\Sigma \mathbf{p})^2$. The results cover the range $5 < M_{\text{eff}} < 40$ GeV and agree well with ARIADNE. There is a weak Bjorken-x dependence across the range $0.006 < x < 0.1$.

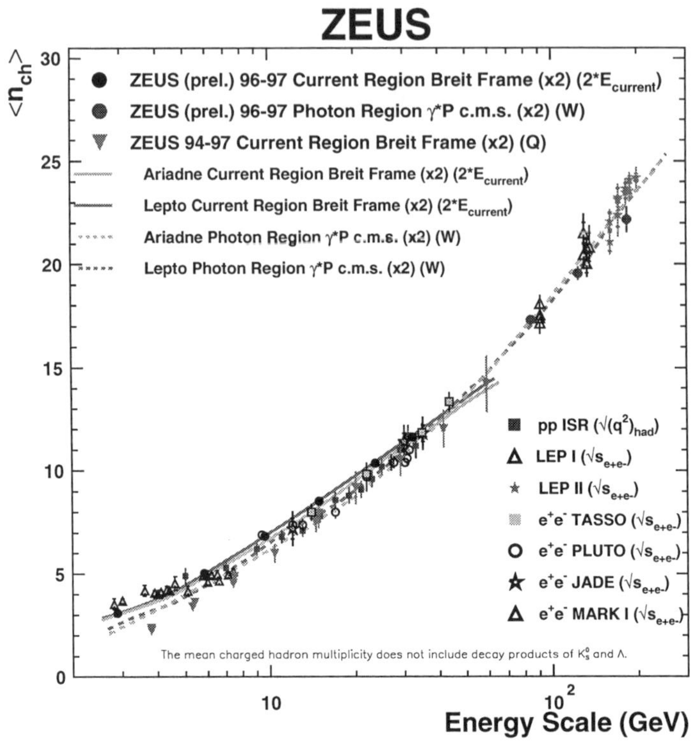

Figure 1. Charged particle multiplicities.

H1 have compared multiplicities in DIS and DDIS. The variable used to characterise the DDIS event is the mass of the diffractively produced hadronic system, M_x, giving a rapidity range of $\ln(M_x/m)$. Many comparison plots are made of mean multiplicity versus W or Q^2 for different M_x ranges. As M_x increases, the DDIS mean multiplicities, and their rapidity spectra, smoothly approach the DIS values from below. This is as one might expect naively in a colour-string approach where what matters is the number of units of rapidity of colour-string available to fragment into hadrons.

2. Non-strange inclusive meson photoproduction

H1 have analysed the $\pi^+\pi^-$ mass spectrum in photoproduction. Mass peaks for ρ^0, f_0 and f_2 are identified over a large and steeply falling background.

They plot (see figure 2) a universal curve of spin-weighted rate against $p_T + m$. Within the discrimination of a log-log plot, these mesons and η and π^+ lie on the same line and follow a power law over six orders of magnitude.

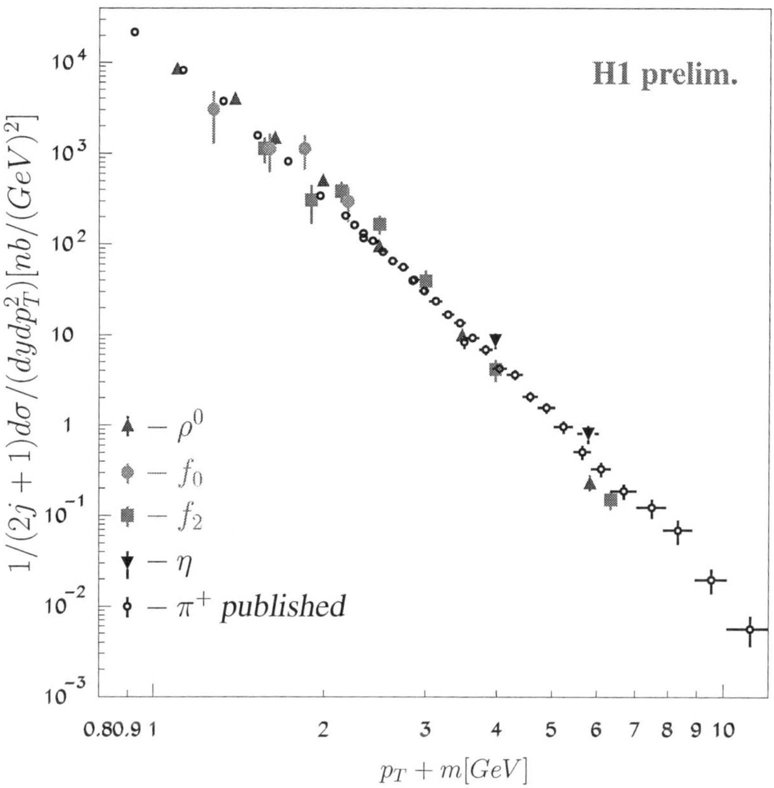

Figure 2. Meson production rates.

3. K^0, Λ^0 and $\overline{\Lambda}^0$ production in DIS.

ZEUS present clean mass peaks for these strange particles and plot the p_T, η, Q^2 and x distributions for K_s^0, $\Lambda^0 + \overline{\Lambda}^0$, and the ratios $(\Lambda^0 + \overline{\Lambda}^0)/K_s^0$, and $(\Lambda^0 - \overline{\Lambda}^0)/(\Lambda^0 + \overline{\Lambda}^0)$ for $Q^2 > 25$ GeV^2. The error bars (see figure 3) are small and there is mostly very pleasing agreement with the predictions of ARIADNE. ARIADNE tends to overestimate the K^0 production rate, suggesting that a lower value of $s/u \simeq 0.22$ may be appropriate. No significant difference is found between Λ^0 and $\overline{\Lambda}^0$. The Λ/K ratio is underestimated by ARIADNE for $x < 0.003$, as shown in figure 3.

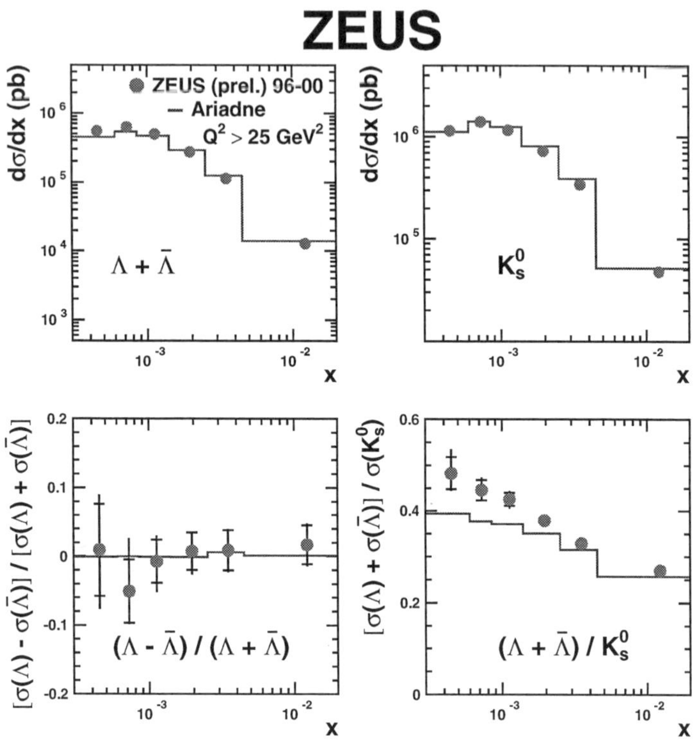

Figure 3. Strange particle production compared to predictions of ARIADNE.

Λ^0 and $\overline{\Lambda}^0$ polarisations are measured using the $\Lambda \rightarrow \pi^- p$ decay as analyser. In the Λ^0 rest frame the decay angular distribution with respect to a chosen axis is proportional to $(1 + \alpha P \cos\theta)$ with $\alpha = 0.642$. Using unpolarised $e^\pm$ beams no significant polarisations are observed either normal to the production plane (with 6% errors) or along the Λ flight path (with 12% errors). They have thus demonstrated the ability to measure these polarisations, with possible application to polarised $e^\pm$ running at HERA-II.

4. Charm fragmentation.

Results from H1 and ZEUS measure the production rates of D^{*+}, D^+, D^0, D_s^+ and Λ_c^+. D^{*+} are measured by the traditional decay chain $D^{*+} \to D^0\pi^+$, $D^0 \to K^-\pi^+$. H1 use vertex detector signatures to isolate the decays $D^+ \to K^-\pi^+\pi^+$ and $D^0 \to K^-\pi^+$ with relatively small backgrounds. They also use vertex tagging to measure $D_s^\pm$ production via the decay chain $D_s^\pm \to \phi\pi^\pm$, $\phi \to K^+K^-$. ZEUS observe the same channel and also find a signal of (1440 ± 220) Λ_c^+ events on a background of some 20000 events in the $pK^-\pi^+$ decay channel.

This allows them to compare fragmentation fractions such as $f(c \to D^0)$ in photoproduction, DIS (using both H1 and ZEUS data) and e^+e^- annihilation. There is good agreement on the fractions in all cases, though there is a hint that ZEUS find a somewhat lower value for $f(c \to D^{*+})$ than e^+e^- annihilation.

5. Baryons decaying to strange particles.

The search for pentaquarks has revived interest in baryons decaying to strange particles. We do not cover pentaquark issues here but look at other particles whose production has now been observed, often small but significant signals on large backgrounds. Both ZEUS and H1 identify protons and antiprotons by dE/dx information, restricted to momenta below 1.5 GeV/c. ZEUS demand that charged tracks emerge from the primary production vertex and identify K_s^0 via a $\pi^+\pi^-$ secondary vertex, selecting $p_T(K_s^0) > 0.3$ GeV/c and $-1.5 < \eta(K_s^0) < 1.5$ to use the best part of the tracker. Additional cuts exclude Dalitz pairs, γ-conversions and Λ^0 candidates. Their $(K_s^0 p)$ mass resolution is 2.4 MeV.

Three data samples are used: photoproduction, DIS for $Q^2 > 1$ GeV^2, and DIS for $Q^2 > 25$ GeV^2. The particle multiplicities, and hence the backgrounds, are highest in photoproduction and lowest in the high-Q^2 sample. In the $K_o^0 p(\overline{p})$ mass spectrum, as well as a pentaquark candidate, $\Theta(1530)$, in DIS a clear $\Lambda_c(2286)$ is seen in all three samples with a width of typically 5.3 ± 3.0 MeV (see figure 4.) The most prominent signal is (278 ± 67) events in low-Q^2 DIS. It is seen equally in $K_s^0 p$ and $K_s^0 \overline{p}$ $(162 \pm 36$ and 116 ± 38 events) and equally at forward and backward lab rapidity $(131 \pm 40$ and 145 ± 34 events.) This is consistent with expectations for $\gamma g \to c\bar{c}$ production.

A clear $\Lambda(1520)$ signal is seen equally in both K^-p and $K^+\overline{p}$ mass spectra in all three data samples: in photoproduction they observe $(13526 \pm$

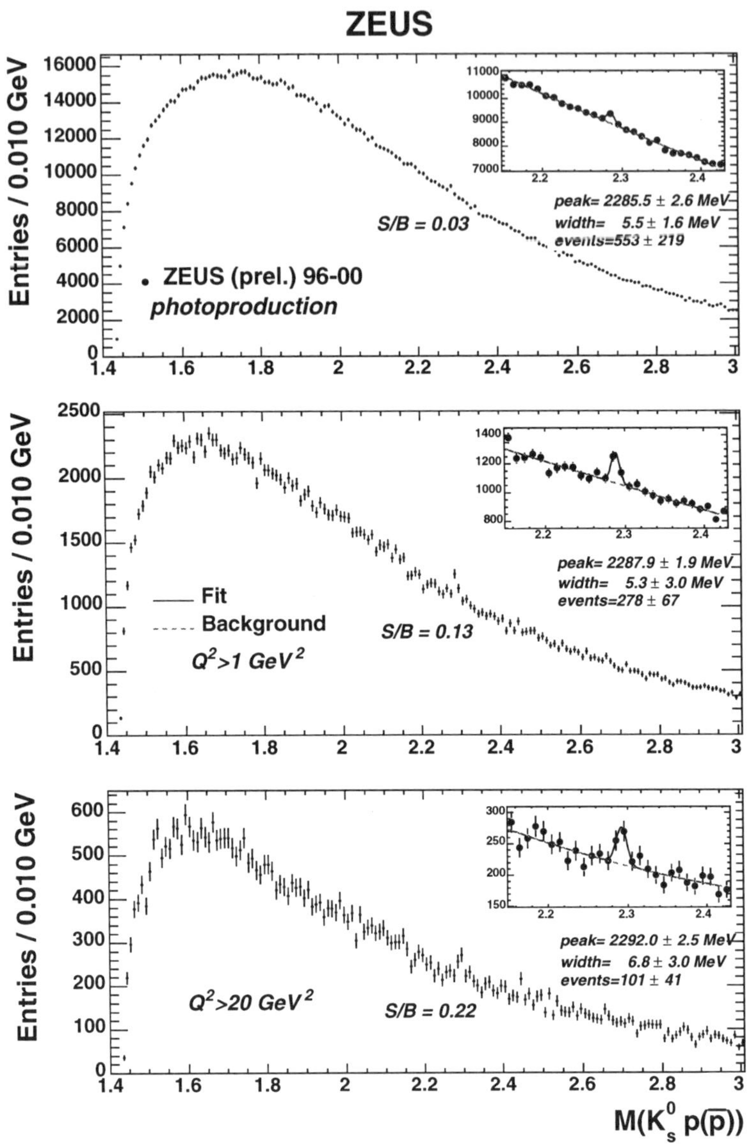

Figure 4. Observation of $\Lambda_c(2286)$ production.

561) events giving $S/B = 0.05$ with errors of 0.4 MeV on both mass and width. Again the production is the same at forward and backward rapidity again consistent with photon-gluon fusion.

In the $\Lambda\pi^\pm$ spectrum one searches for Ξ, Σ^* and pentaquark production. $\Xi(1320)$ and $\Sigma^*(1385)$ production are seen but there is no peak in the $\Theta^+(1530)$ region. There is a 4.6 σ peak near 1600 MeV in the $Q^2 > 1\text{GeV}^2$ sample, which might be $\Sigma(1580)$ or $\Sigma(1620)$ production.

Very clean peaks of (1561 ± 46) events in $\Xi^- \to \Lambda\pi^-$ and the same number in $\overline{\Xi}^+$ decay are seen after cuts are made to ensure a clean Λ decay vertex. These in turn lead to a clear narrow $\Xi^0(1530)$ signal in the $\Xi^-\pi^+$ (and antiparticle) mass spectrum.

6. Antideuteron production and heavy particle search.

H1 have used their excellent dE/dx resolution to measure the photoproduction of K, p, d, t and antiparticles for $\langle W(\gamma p)\rangle = 200$ GeV, $0.2 < p_T/M < 0.7$, $-0.4 < y_{lab} < 0.4$, using 5.5 pb^{-1} of data. Clear peaks are observed for K^+, p, d, t, K^- and $\overline{p}$ with 7% mass resolution, and a clean cluster of 45 antideuterons is observed, corresponding to a cross section of $(2.7 \pm 0.5 \pm 0.2$ nb). There are no candidates for heavier negative particles.

Antideuteron production is beyond standard fragmentation models. H1 compare their results to pp collisions at the ISR and to $AuAu$ collisions in STAR at RHIC as a function of p_T/M. Plotting the cross-section/total cross-section and the $\overline{d}/\overline{p}$ ratio the photoproduction and pp data are in good agreement but the $AuAu$ data are much higher. The data are also tested against a coalescence model, which favours d production if p and n are produced very close together. Heavy ion collisions have a much larger production volume and hence a much smaller chance of overlap. Again we find agreement between γp, pp and pA over a wide c.m. energy range, but very heavy ions $NeAu$, $AuPt$ and $PbPb$ give much lower coalescence probabilities, falling rapidly as the c.m. energy increases.

7. Bose-Einstein correlations in $K_s^0 K_s^0$ and $K^\pm K^\pm$.

ZEUS have performed a Bose-Einstein analysis for charged and neutral kaons. The correlation function used for two particles is $R(p_1, p_2) = \rho(p_1, p_2)/\rho(p_1)\rho(p_2)$. We set $Q_{12} = p_1 - p_2$ and fit the data to the form

$$R(Q_{12}) = \alpha(1 + \delta Q_{12})(1 + \lambda \exp[-r^2 Q_{12}^2])$$

where the physics interest lies in the source radius, r, and the incoherence parameter, $(0 < \lambda < 1)$. $R(Q_{12})$ is measured in the data by the double ratio

$$R = [P(\text{data})/P_{mix}(\text{data})]/[P(\text{MC})/P_{mix}(\text{MC})]$$

all evaluated at the same Q_{12}, where the suffix *mix* implies that the two particles are taken from different events. The Monte-Carlo has no Bose-Einstein correlations.

Looking at $K^{\pm}K^{\pm}$ in DIS events ($Q^2 > 2$ GeV^2) ZEUS find $r = (0.57 \pm 0.09 + 0.15 - 0.06)$ fm in good agreement with LEP and with charged pions. The λ value, $(0.31 \pm 0.06 + 0.09 - 0.06)$, is less than half that at LEP. The reason is unclear, though we note that the fragmentation may be different in the proton region of the c.m. frame and that $\phi(1020)$ decay may therefore play a different role.

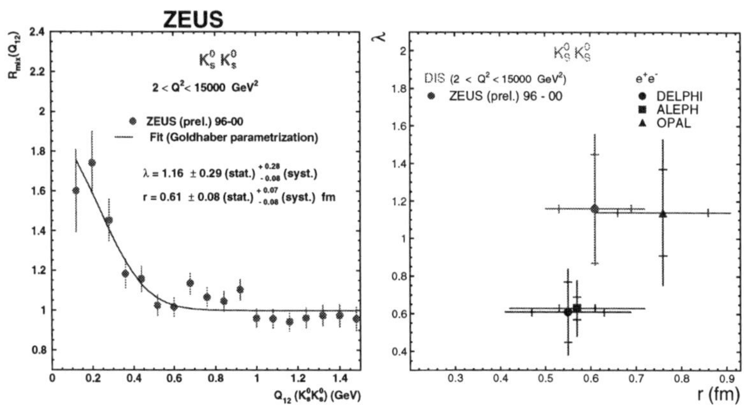

Figure 5. Bose-Einstein correlations.

In $K^0_sK^0_s$ ZEUS find a very similar value for r, in contrast with LEP results where there is a hierarchy, $r(\pi^{\pm}) > r(K^{\pm}) > r(\Lambda)$. (See figure 5.) ZEUS find $\lambda \simeq 1.2$, similar to OPAL but well above values from ALEPH and DELPHI who have removed $f_0(980) \to K^0_sK^0_s$ effects. Correction for this may bring the results into agreement.

8. Prompt photon emission.

We are looking here for hard photons emitted from the quark line in photoproduction or DIS. The experimental definition of photoproduction demands that the electron be lost down the beam pipe, so that a photon observed in the detector with $p_T > 5$ GeV must be far from it. We demand that the photon be isolated from all other outgoing particles by drawing a cone of radius 1 in $(\Delta\eta, \Delta\phi)$ around it and demanding that the electromagnetic cluster identified with the photon contain at least 90% of the energy found in the isolation cone. In DIS the photon is therefore far from both electron and quark.

π^0 and η^0 decay to photons provide major backgrounds to prompt photon signals. Both ZEUS and H1 have used transverse shower shape to discriminate these. In the ZEUS barrel calorimeter $(-0.7 < \eta < 0.9)$ the fine segmentation means that photons predominantly illuminate only 1 z-strip, π^0 decay mostly two strips amd η decay fills one to many strips. A photon signal is extracted statistically from fitting the shape distributions.

Photoproduction of prompt photons is rather well described by both PYTHIA (shape OK, normalisation a bit low) and by NLO QCD. H1 measurements of E_T^γ, η^γ and $\eta^{\rm jet}$ for accompanying jets are all described well. Corrections to NLO for multiple interactions (as estimated by PYTHIA) improve the fit, but there are no surprises.

Deep inelastic scattering turns out to be more of a challenge. ZEUS have published measurements of inclusive prompt photon emission and of the (photon+jet) final state, and they compared these to two Monte Carlo calculations (PYTHIA v6.206 and HERWIG v6.1) and to NLO, that is $O(\alpha^3\alpha_s)$, calculations by Kramer and Spiesberger. The data are shown in figure 6. PYTHIA and HERWIG get the normalisation wrong, by factors (in the inclusive case) of 2.4 and 8.3 respectively, and HERWIG gets too low a mean Q^2, while PYTHIA gets the slope of the rapidity spectrum wrong. A similar story holds in the case of prompt photon plus one jet.

Comparisons with NLO calculations, only available for the photon plus jet case, are more encouraging. The normalisation and mean Q^2 are good, (within 1.7 S.D. for cross-section) as are the photon rapidity and $E_T^{\rm jet}$ distribution. But the jet rapidity is predicted to be more forward peaked than the data and the photon E_T spectrum is more steeply falling than the data. More statistics will obviously help to clarify whether these issues are significant or not.

After these results were published, Martin, Roberts, Stirling and Thorne produced a different approach, in which the outgoing hard photon is emitted by the electron, which in turn scatters off a photon which is a constituent of the proton, produced by QED/QCD evolution of the proton structure function. The high Q^2 in the event (averaging 85 GeV^2) in the data is carried by the exchanged electron. MRST calculate the photon structure of the proton $x\gamma_p(x, Q^2)$ to be about 0.03 at $x \simeq 0.005$, (i.e. about 300 times

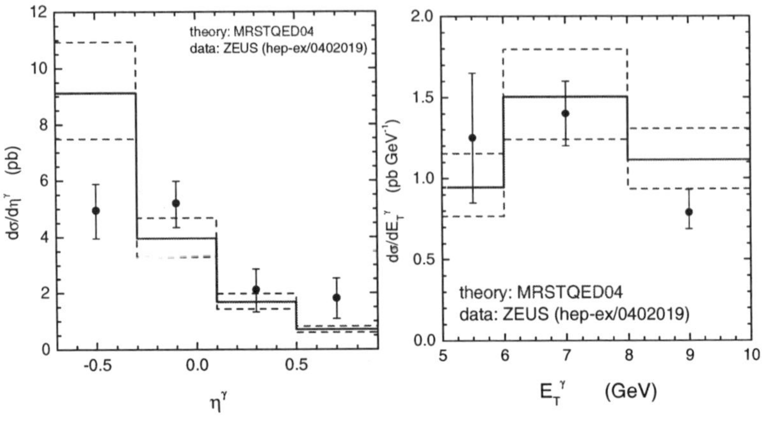

Figure 6. Inclusive prompt photon production in DIS compared to MRST predictions (absolute normalisation).

lower that the gluon content. This in turn gives a cross-section prediction similar to the data (well within 1 S.D.). The predictions for inclusive photon production (they do not calculate photon+jet) are overlaid on the data in figure 6.

The upshot is in some sense intriguing. Both the NLO and the structure approach are presumably valid and naively should be added together. Yet both explain to some extent the whole signal, in one case for inclusive photons and in the other in the sub-domain of photon plus jet. It is not clear how to reconcile these, and we await further developments.

References

Most of the work described here can be found at
http://www-zeus.desy.de/physics/phch/conf/lp05_eps05/ and
http://www-h1.desy.de/h1/www/general/home/intra_home.html.
The prompt photon results are found at
H1 Collab., Eur. Phys. J. **C38** (2005) 437-445,
ZEUS Collab., Phys. Lett. **B595** (2004) 86-100 and
A.D. Martin et al., Eur. Phys. J. **C39** (2005) 155-161.

SOFT GLUON LOGARITHMIC RESUMMATION AND HADRON MASS EFFECTS IN SINGLE HADRON INCLUSIVE PRODUCTION*

S. ALBINO

2nd Institute for Theoretical Physics, Hamburg University, Germany

We define a general scheme for the evolution of fragmentation functions which resums soft gluon logarithms in a manner consistent with fixed order evolution. We present an explicit example of our approach in which double logarithms are resummed using the Double Logarithmic Approximation. We show that this scheme reproduces the Modified Leading Logarithm Approximation in certain limits, and find that after using it to fit quark and gluon fragmentation functions to experimental data, a good description of the data from the largest x_p values to the peak region in $\xi = \ln(1/x_p)$ is obtained. In addition, we develop a treatment of hadron mass effects which gives additional improvements at large ξ.

1. Introduction

The extraction of fragmentation functions (FFs) at large and intermediate x from experimental data on single hadron inclusive production has been successfully performed [1] using the fixed order (FO) DGLAP [2] evolution to next-to-leading order (NLO) [3]. However, this formalism fails at small x due to unresummed soft gluon logarithms (SGLs), and a different analysis is required in which these SGLs are resummed. In this contribution we outline a general approach [4] which extends the FO evolution of FFs to smaller x values by resumming SGLs. We present explicit results for the case where the largest SGLs, being the double logarithms (DLs), are resummed using the Double Logarithmic Approximation (DLA) [5,6] in the LO DGLAP evolution, and use this scheme to describe, and to fit FFs to, experimental data. We also introduce a novel and simple approach to incorporate hadron mass effects.

*Work supported in part by DFG through Grant No. KN 365/3-1 and by BMBF through Grant No. 05 HT4GUA/4.

2. SGL Resummation in DGLAP Evolution

The DGLAP equation reads

$$\frac{d}{d\ln Q^2}D(x,Q^2)=\int_x^1\frac{dy}{y}P(z,a_s(Q^2))D\left(\frac{x}{y},Q^2\right), \tag{1}$$

where, for brevity, we omit hadron and parton labels. D is a vector containing the gluon FF D_g and the quark and antiquark FFs D_q and $D_{\overline{q}}$ respectively, in linear combinations according to the choice of basis, and P is the matrix of the splitting functions. We define $a_s=\alpha_s/(2\pi)$. We choose the simplest basis, consisting of valence quark FFs $D_q^-=D_q-D_{\bar{q}}$, non-singlet quark FFs D_{NS}, and $D=(D_\Sigma,D_g)$, where for n_f quark flavours the singlet quark FF is given by $D_\Sigma=\frac{1}{n_f}\sum_{q=1}^{n_f}D_q^+$.

Analytic operations are often simpler in Mellin space, where a function $f(x)$ becomes $f(\omega)=\int_0^1 dx x^\omega f(x)$, since the convolution in x space in Eq. (1) becomes the simple product

$$\frac{d}{d\ln Q^2}D(\omega,Q^2)=P(\omega,a_s(Q^2))D(\omega,Q^2). \tag{2}$$

Consider the formal expansion of $P(a_s)$ in a_s keeping x (or ω if we want to work in Mellin space) fixed, viz.

$$P(a_s)=\sum_{n=1}^{\infty}a_s^nP^{(n-1)} \tag{3}$$

where the $P^{(n-1)}$ are functions of x (or ω). Equation (3) truncated at some chosen (finite) n is known as the FO approach, and, in x space, is not valid at small x due to the presence of terms which in the limit $x\to 0$ behave like $(a_s^n/x)\ln^{2n-m-1}x$ for $m=1,...,2n-1$. Such logarithms are called SGLs, and m labels their class. As x decreases, these unresummed SGLs will spoil the convergence of the FO series for $P(x,a_s)$ once $\ln(1/x)=O(a_s^{-1/2})$. Consequently the evolution of $D(x,Q^2)$ will not be valid here, since the whole range $x\le y\le 1$ contributes in Eq. (1).

SGLs are defined to be all those terms of the form a_s^n/ω^{2n-m} only, where $m=1,...,2n$ and labels the class of the SGL, in the expansion about $\omega=0$ of Eq. (3) in Mellin space. For $m=1,...,2n-1$, this definition agrees with the form of the SGLs in x space given above, since $\omega^{-p}=-((-1)^p/p!)\int_0^1 dx x^\omega(1/x)\ln^{p-1}x$ for $\mathrm{Re}(\omega)>0$ and $p\ge 1$. Such terms spoil the convergence of the series in Eq. (3) as $\omega\to 0$. To construct a scheme valid for large *and* small ω (or x), we write P in the form $P=P^{\mathrm{FO}}+P^{\mathrm{SGL}}$, where P^{SGL} contains only and all the SGLs in P, so that P^{FO}

is completely free of SGLs. Since terms of the type $p = 0$ ($m = 2n$), which are included in our definition of SGLs, are non-singular, they may therefore be left unresummed. Thus we separate $P^{\rm SGL}$ into $P^{\rm SGL} = P^{\rm SGL}_{p\geq 1} + P^{\rm SGL}_{p=0}$. $P^{\rm SGL}_{p=0}$, which is independent of ω, is expanded as a series in a_s. On the other hand, by summing all SGLs in each class m, $P^{\rm SGL}_{p\geq 1}(\omega, a_s)$ is resummed in the form

$$P^{\rm SGL}_{p\geq 1}(\omega, a_s) = \sum_{m=1}^{\infty} \left(\frac{a_s}{\omega}\right)^m g_m\left(\frac{a_s}{\omega^2}\right), \tag{4}$$

and truncated for some finite m. The remaining FO contribution to P, $P^{\rm FO}(\omega, a_s)$, is expanded in a_s keeping ω fixed, $P^{\rm FO}(\omega, a_s) = \sum_{n=1}^{\infty} a_s^n P^{{\rm FO}(n-1)}(\omega)$, and truncated for some finite n. Finally, the result for $P(\omega, a_s)$ is inverse Mellin transformed to obtain $P(x, a_s)$, and then Eq. (1) is solved exactly. We shall call this the SGL+FO+FOδ scheme, where "+FOδ" means that the $p = 0$ terms, which are each proportional to $\delta(1-x)$ in x space, are left as a FO series in a_s. In summary, this scheme is the result of resumming all SGLs for which $m = 1, ..., 2n-1$ in the form of Eq. (4), and treating all remaining terms as in the FO approach.

In the region for which $a_s \ll 1$ and x is above values for which $\ln(1/x) = O(a_s^{-1/2})$, the SGL+FO(+FOδ) scheme should give a good description of the evolution for the following reasons. Firstly, Eq. (4) implies $P^{\rm SGL}_{p\geq 1}(x, a_s) = \frac{1}{x\ln x}\sum_{m=1}^{\infty}(a_s \ln x)^m f_m\left(a_s \ln^2 x\right)$. Since $a_s \ln x$ is always small, this series is a valid approximation when x is small. On the other hand, as $x \to 1$ the SGLs for the types $p \geq 1$ all vanish, and therefore so does each term in the series. The full contribution from the type $p = 0$ terms is just $P^{\rm SGL}_{p=0}(x, a_s) = \delta(1-x)\sum_{n=1}^{\infty} C_n a_s^n$, and the expansion of $P^{\rm FO}(x, a_s)$ in a_s, $P^{\rm FO}(x, a_s) = \sum_{n=1}^{\infty} a_s^n P^{{\rm FO}(n-1)}(x)$, is finite for all x.

3. DLA Improved DGLAP Evolution

From the DLA [6], if the evolution is rewritten in the form

$$\begin{aligned}\frac{d}{d\ln Q^2}D(x,Q^2) = &\int_x^1 \frac{dy}{y}\frac{2C_A}{y}Ay^{2\frac{d}{d\ln Q^2}}\left[a_s(Q^2)D\left(\frac{x}{y},Q^2\right)\right] \\ &+\int_x^1 \frac{dy}{y}\overline{P}(y,a_s(Q^2))D\left(\frac{x}{y},Q^2\right),\end{aligned} \tag{5}$$

then $\overline{P}(x, a_s)$ is free of DLs. Explicitly, $A = 0$ for the DL evolving parts of the components $D = D_q^-$ and $D = D_{NS}$, while

$$A = \begin{pmatrix} 0 & \frac{2C_F}{C_A} \\ 0 & 1 \end{pmatrix} \tag{6}$$

for the component $D = (D_\Sigma, D_g)$. Note that A is a projection operator, i.e. it obeys $A^2 = A$. $\overline{P}$ can be determined from P explicitly by expanding the operator $y^{2\frac{d}{d\ln Q^2}}$ in Eq. (5) in $\frac{d}{d\ln Q^2}$ and using Eq. (1).

We will examine what constraint Eq. (5) provides for P. In Mellin space, Eq. (5) becomes

$$\begin{aligned}\left(\omega + 2\frac{d}{d\ln Q^2}\right)\frac{d}{d\ln Q^2}D(\omega,Q^2) &= 2C_A a_s(Q^2) A D(\omega,Q^2)\\ &+\left(\omega + 2\frac{d}{d\ln Q^2}\right)\overline{P}(\omega,a_s(Q^2))D(\omega,Q^2).\end{aligned} \tag{7}$$

After substituting Eq. (2) into Eq. (7) and dividing out the overall factor of $D(\omega, Q^2)$, we obtain the following constraint on P:

$$\left(\omega + 2\frac{d}{d\ln Q^2}\right)(P-\overline{P}) + 2\left(P-\overline{P}\right)P - 2C_A a_s A = 0. \tag{8}$$

We now use the fact that $\overline{P}$ is free of DLs to obtain an explicit constraint for $P^{\rm DL}$. We first make the replacement $P = \widetilde{P} + P^{\rm DL}$ in Eq. (8). Expanding Eq. (8) as a series in a_s/ω keeping a_s/ω^2 fixed and extracting the first, $O((a_s/\omega)^2)$, term gives

$$2(P^{\rm DL})^2 + \omega P^{\rm DL} - 2C_A a_s A = 0. \tag{9}$$

Equation (9) gives two solutions for each component of P. Since P is never larger than a 2×2 matrix in the basis consisting of singlet, gluon, non-singlet and valence quark FFs, there are four solutions. We choose the solution

$$P^{\rm DL}(\omega,a_s) = \frac{A}{4}\left(-\omega + \sqrt{\omega^2 + 16C_A a_s}\right), \tag{10}$$

since, in the component $D = (D_\Sigma, D_g)$, the expansion of the result in Eq. (10) in a_s to $O(a_s^2)$ keeping ω fixed,

$$P^{\rm DL}(\omega,a_s) = \begin{pmatrix} 0 & a_s\frac{4C_F}{\omega} - a_s^2\frac{16C_F C_A}{\omega^3} \\ 0 & a_s\frac{2C_A}{\omega} - a_s^2\frac{8C_A^2}{\omega^3}\end{pmatrix} + O(a_s^3), \tag{11}$$

agrees with the DLs in the literature, while in the components $D = D_q^-$ and $D = D_{NS}$, $P^{\rm DL} = 0$. The other possibilities do not give these results and/or cannot be expanded in a_s, i.e. they are non-perturbative. Equation (10) agrees with previous results [6,7].

At small ω, Eq. (10) implies that

$$D_{q,\overline{q}} = \frac{C_F}{C_A}D_g, \tag{12}$$

obtained after integrating Eq. (2) and neglecting the constant, which is valid at large Q. We will use Eq. (12) to partially constrain our choice of parameterization at low x in the next section.

The $O(a_s)$ single logarithm (SL) contribution to P is a type $p = 0$ term (see Sec. 2), and is given by

$$P^{\mathrm{SL}(0)}(\omega) = \begin{pmatrix} 0 & -3C_F \\ \frac{2}{3}T_R n_f & -\frac{11}{6}C_A - \frac{2}{3}T_R n_f \end{pmatrix}, \tag{13}$$

Approximating $\overline{P}$ by $a_s P^{\mathrm{SL}(0)}$ Eq. (7) can be regarded as a generalized version of the Modified Leading Logarithm Approximation (MLLA) [6,7,8] equation to include quarks, in the sense that the g component of this latter equation for $D = (D_\Sigma, D_g)$ when Eq. (12) is invoked is precisely the MLLA equation. We therefore conclude Eq. (7) is more complete than the MLLA equation.

$P^{\mathrm{DL}}(x, a_s)$ is obtained by inverse Mellin transform of Eq. (10), which yields $P^{\mathrm{DL}}(x, a_s) = \frac{A\sqrt{C_A a_s}}{x \ln \frac{1}{x}} J_1\left(4\sqrt{C_A a_s} \ln \frac{1}{x}\right)$, where $J_1(y)$ is the Bessel function of the first kind, given by $J_1(y) = \frac{1}{\pi}\int_0^\pi d\theta \cos(y \sin\theta - \theta)$.

In the next section we shall consider the DL+LO+LOδ scheme, in which we take $P^{\mathrm{SGL}} \approx P^{\mathrm{DL}}$, and take P^{FO} (and $P^{\mathrm{SGL}}_{p=0}$) to $O(a_s)$ only.

In Fig. 1, we see that $P_{gg}(x, a_s)$ in the DL+LO scheme, which is equal to the DL+LO+LOδ scheme when $x \neq 1$, interpolates well between its $O(a_s)$ approximation in the FO approach at large x and $P^{\mathrm{DL}}_{gg}(x, a_s)$ at small x (the small difference here comes from $P^{\mathrm{FO}(0)}(x)$ at small x). DL resummation clearly makes a large difference to P at small x.

4. Comparisons with Data

In this section, we study the outcome of fitting FFs evolved in the LO DGLAP approach and in our DL+LO+LOδ scheme to experimental data on the normalized differential cross section for light charged hadron production in the process $e^+e^- \to (\gamma, Z) \to h + X$, where h is the observed hadron and X is anything else. These data exist for different center-of-mass (COM) energies $\sqrt{s}$ and values of $x_p = 2p/\sqrt{s}$, where p is the momentum of the observed hadron, which constrain the FFs in the region of x for which $x_p \leq x \leq 1$. We fit to data for which $\xi < \ln(\sqrt{s}/1\mathrm{GeV})$, where $\xi = \ln(1/x_p)$. At LO in the coefficient functions, these data are described in terms of the evolved FFs by

$$\frac{1}{\sigma(s)}\frac{d\sigma}{dx_p}(x_p, s) = \frac{1}{n_f \langle Q(s) \rangle} \sum_q Q_q(s) D_q^+(x_p, Q^2), \tag{14}$$

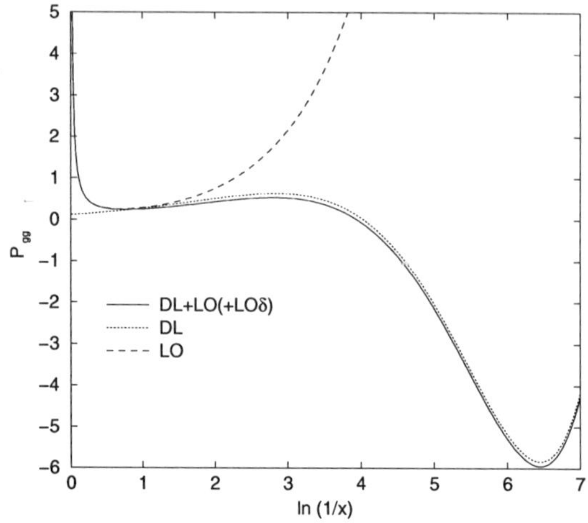

Figure 1. (i) $P_{gg}(x, a_s)$ calculated in the DL+LO(+LOδ) scheme, (ii) $P_{gg}(x, a_s)$ calculated to $O(a_s)$ in the FO approach (labelled "LO"), and (iii) $P_{gg}^{\rm DL}(x, a_s)$ (labelled "DL"). $a_s = 0.118/(2\pi)$.

where Q_q is the electroweak charge of a quark with flavour q and $\langle Q \rangle$ is the average charge over all flavours. We will take $n_f = 5$ in all our calculations. Since we sum over hadron charges, we set $D_{\overline{q}} = D_q$. These data depend on the FFs in the combinations $f_{uc}(x, Q_0^2) = \frac{1}{2}\left(u(x, Q_0^2) + c(x, Q_0^2)\right)$, $f_{dsb}(x, Q_0^2) = \frac{1}{3}\left(d(x, Q_0^2) + s(x, Q_0^2) + b(x, Q_0^2)\right)$ and the gluon $g(x, Q_0^2)$. For each of these three FFs, we choose the parameterization

$$f(x, Q_0^2) = N \exp[-c \ln^2 x] x^\alpha (1-x)^\beta, \tag{15}$$

since at intermediate and large x the FF is constrained to behave like $f(x, Q_0^2) \approx N x^\alpha (1-x)^\beta$, which is the standard parameterization used in global fits at large x, while at small x (where $(1-x)^\beta \approx 1$) the FF is constrained to behave like $\lim_{x \to 0} f(x, Q_0^2) = N \exp\left[-c \ln^2 \frac{1}{x} - \alpha \ln \frac{1}{x}\right]$, which for $c > 0$ is a Gaussian in $\ln(1/x)$, as predicted by the DLA for sufficiently large Q_0. We use Eq. (12) to remove four free parameters by imposing the constraints $c_{uc} = c_{dsb} = c_g$ and $\alpha_{uc} = \alpha_{dsb} = \alpha_g$. The relation

$$N_{uc} \approx N_{dsb} \approx \frac{C_F}{C_A} N_g \tag{16}$$

implied by Eq. (12) is not imposed, since we want to describe large x data as well. We also fit $\Lambda_{\rm QCD}$. We choose $Q^2 = s$, although it is only important that the latter two quantities are kept proportional, since the constant of proportionality has no effect on the final FF parameters and the description of the data (and therefore the quality of the fit). This implies that there will be an overall theoretical error on our fitted values for $\Lambda_{\rm QCD}$ of a factor of $O(1)$. Since all data will be at $\sqrt{s} \geq 14$ GeV, we choose $Q_0 = 14$ GeV. The evolution is performed by numerically integrating Eq. (1).

4.1. *Fixed Order Evolution*

We first perform a fit using standard LO DGLAP evolution. We obtain $\chi^2_{\rm DF} = 3.0$, and the results are shown in Fig. 2 and Table 1. The result for $\Lambda_{\rm QCD}$ is quite consistent with that of other analyses, at least within the theoretical error. It is clear that FO DGLAP evolution fails in the description of the peak region. The unphysical negative value of β for the gluon is because the gluon FF is weakly constrained, since it couples to the data only through the evolution (see Eq. (14)).

Table 1. Parameter values for the FFs at $Q_0 = 14$ GeV parameterized as in Eq. (15) from a fit to all data listed in the text using DGLAP evolution in the FO approach to LO. $\Lambda_{\rm QCD} = 388$ MeV.

FF \ Parameter	N	β	α	c
g	0.22	−0.43	−2.38	0.25
$u+c$	0.49	2.30	[−2.38]	[0.25]
$d+s+b$	0.37	1.49	[−2.38]	[0.25]

4.2. *Incorporation of Soft Gluon Resummation*

We now redo the previous fit, but now evolving in the DL+LO+LOδ scheme. The results are shown in Table 2 and Fig. 3. We obtain $\chi^2_{\rm DF} = 2.1$,

Table 2. Parameter values for the FFs at $Q_0 = 14$ GeV parameterized as in Eq. (15) from a fit to all data listed in the text using DGLAP evolution in the DL+LO+LOδ scheme. $\Lambda_{\rm QCD} = 801$ MeV.

FF \ Parameter	N	β	α	c
g	1.60	5.01	−2.63	0.35
$u+c$	0.39	1.46	[−2.63]	[0.35]
$d+s+b$	0.34	1.49	[−2.63]	[0.35]

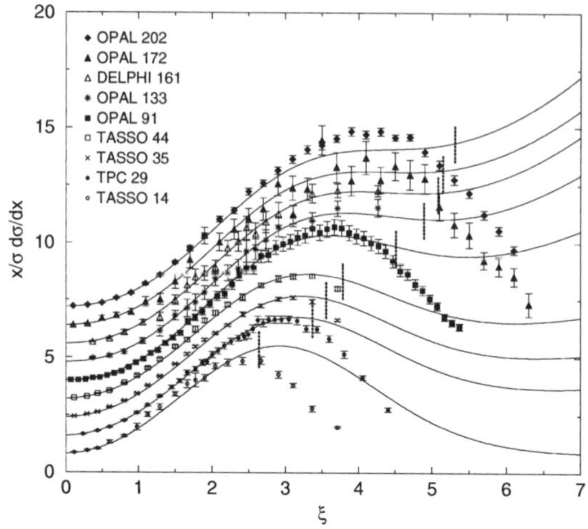

Figure 2. Fit to data as described in Table 1. Some of the data sets used for the fit are shown, together with their theoretical predictions from the results of the fit. Data to the right of the vertical dotted lines have not been used in the fit. Each curve is shifted up by 0.8 for clarity.

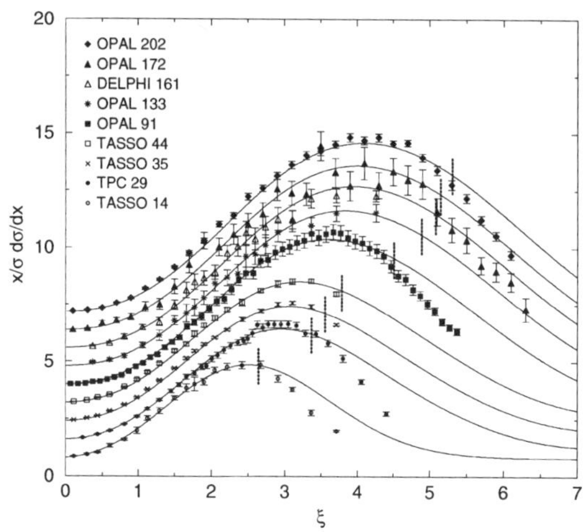

Figure 3. Fit to data as described in Table 2.

a significant improvement to the fit above with FO DGLAP evolution. In particular, the data around the peak are now much better described.

We note that had we made the usual DLA (MLLA) choice $Q = \sqrt{s}/2$

instead of our choice $Q = \sqrt{s}$ which is usually employed in analyses using the DGLAP equation, we would have obtained half the result $\Lambda_{\rm QCD} \approx 800$ MeV. N_g is too large by a factor of about 2 relative to its prediction in Eq. (16). However, as noted before, the initial gluon FF is weakly constrained in our fits.

4.3. *Incorporation of Hadron Mass Effects*

To attempt to improve the description beyond the peak, we now study hadron mass effects, which are important at small x_p. It is helpful to work with light cone coordinates, in which any 4-vector V is written in the form $V = (V^+, V^-, \mathbf{V}_T)$ with $V^\pm = \frac{1}{\sqrt{2}}(V^0 \pm V^3)$ and $\mathbf{V}_T = (V^1, V^2)$. In the COM frame, the momentum of the electroweak boson takes the form $q = \left(\frac{\sqrt{s}}{\sqrt{2}}, \frac{\sqrt{s}}{\sqrt{2}}, \mathbf{0}\right)$. The momentum of the hadron in the COM frame is chosen as $p_h = \left(\frac{\eta\sqrt{s}}{\sqrt{2}}, \frac{m_h^2}{\sqrt{2}\eta\sqrt{s}}, \mathbf{0}\right)$, where $\eta = p_h^+/q^+$ is the light cone scaling variable. Therefore the relation between the two scaling variables in the presence of hadron mass is $x_p = \eta\left(1 - \frac{m_h^2}{s\eta^2}\right)$. As a generalization of the massless case, we assume the cross section we have been calculating is $(d\sigma/d\eta)(\eta, s)$, i.e. $\frac{d\sigma}{d\eta}(\eta, s) = \int_\eta^1 \frac{dy}{y}\frac{d\sigma}{dy}(y, s, Q^2) D\left(\frac{\eta}{y}, Q^2\right)$, which is related to the measured observable $(d\sigma/dx_p)(x_p, s)$ via

$$\frac{d\sigma}{dx_p}(x_p, s) = \frac{1}{1 + \frac{m_h^2}{s\eta^2(x_p)}}\frac{d\sigma}{d\eta}(\eta(x_p), s). \tag{17}$$

Although the data are for light charged hadrons, the vast majority of particles are pions, so we will assume all particle masses are equal. We now perform the DL+LO+LOδ fit again but with m_h included in the list of free parameters. We obtain the results in Table 3. The parameters are not substantially different to those in Table 2. The result for m_h is reasonable for light charged hadrons. We find $\chi^2_{\rm DF} = 2.03$, i.e. no significant improvement to the quality of the fit, and the comparison with data is similar to that in Fig.3. However, treatment of mass effects renders the value of $\Lambda_{\rm QCD}$ more reasonable.

5. Conclusions

Using the approach in this contribution gives a much better fit to all the data than the FO approach and the MLLA [9] do, even if the fit is still not in the acceptable range. Further improvement in the large ξ region can

Table 3. As in Table 2, but incorporating mass effects in the fit. $\Lambda_{\text{QCD}} = 399$ MeV and $m_h = 252$ MeV.

FF \ Parameter	N	β	α	c
g	1.59	7.80	−2.65	0.33
$u + c$	0.62	1.43	[−2.65]	[0.33]
$d + s + b$	0.74	1.60	[−2.65]	[0.33]

be expected from the inclusion of higher order SGLs. Our scheme allows a determination of quark and gluon FFs over a wider range of data than previously achieved, and should be used to extend the NLO global fits of FFs to lower x_p values.

Acknowledgments

This work was supported in part by the Deutsche Forschungsgemeinschaft through Grant No. KN 365/5-1 and by the Bundesministerium für Bildung und Forschung through Grant No. 05 HT4GUA/4.

References

1. B. A. Kniehl, G. Kramer and B. Pötter, Nucl. Phys. **B582** (2000) 514; S. Kretzer, Phys. Rev. **D62** (2000) 054001; L. Bourhis, M. Fontannaz, J. P. Guillet and M. Werlen, Eur. Phys. J. **C19** (2001) 89; S. Albino, B. A. Kniehl and G. Kramer, Nucl. Phys. **B725** (2005) 181.
2. L. N. Lipatov, Yad. Fiz. **20** (1974) 181 [Sov. J. Nucl. Phys. **20** (1975) 94]; V. N. Gribov and L. N. Lipatov, Yad. Fiz. **15** (1972) 781 [Sov. J. Nucl. Phys. **15** (1972) 438]; G. Altarelli and G. Parisi, Nucl. Phys. **B126** (1977) 298; Yu. L. Dokshitzer, Zh. Eksp. Teor. Fiz. **73** (1977) 1216 [Sov. Phys. JETP **46** (1977) 641].
3. G. Curci, W. Furmanski and R. Petronzio, Nucl. Phys. **B175** (1980) 27; W. Furmanski and R. Petronzio, Phys. Lett. **B97** (1980) 437.
4. S. Albino, B. A. Kniehl, G. Kramer and W. Ochs, Phys. Rev. Lett. (*in press*), arXiv:hep-ph/0503170; arXiv:hep-ph/0510319.
5. A. Bassetto, M. Ciafaloni, G. Marchesini and A. H. Mueller, Nucl. Phys. **B207** (1982) 189; V. S. Fadin, Yad. Fiz. **37** (1983) 408 [Sov. J. Nucl. Phys. **37** (1983) 245].
6. Y. L. Dokshitzer, V. A. Khoze, A. H. Mueller and S. I. Troian, Basics of Perturbative QCD (Editions Frontières, Gif-sur-Yvette, 1991).
7. A. H. Mueller, Nucl. Phys. **B213** (1983) 85.
8. Y. L. Dokshitzer and S. I. Troian, Proc. 19th Winter School of the LNPI, Vol. 1, p. 144 (Leningrad, 1984); Y. L. Dokshitzer and S. I. Troian, LNPI-922 preprint (1984).
9. S. Albino, B. A. Kniehl, G. Kramer and W. Ochs, Eur. Phys. J. **C36** (2004) 49.

4 Heavy-Flavour Production

HEAVY-FLAVOR PHOTO- AND ELECTROPRODUCTION AT NLO

I. SCHIENBEIN
Southern Methodist University
Department of Physics
104 Fondren Science Building
3215 Daniel Avenue
Dallas, Texas 75275-0175, USA
E-mail: schien@physics.smu.edu

We review one-particle inclusive production of heavy-flavored hadrons in a framework which resums the large collinear logarithms through the evolution of the FFs and PDFs and, at the same time, retains the full dependence on the heavy-quark mass without additional theoretical assumptions. The main focus is on the production of $D^\star$ mesons in deep inelastic electron–proton scattering at HERA. We show results, neglecting for the time being the heavy-quark mass terms, for deep inelastic $D^\star$ meson production at finite transverse momenta. Work to implement this process in the above mentioned massive QCD framework is in progress.

1. Introduction

One-particle inclusive production processes provide extensive tests of perturbative quantum chromodynamics (pQCD). In contrast to fully inclusive processes, it is possible to study distributions in the momentum of the final state particle and to apply kinematical cuts which come close to the experimental situation. On the other hand, contrary to even more exclusive cases, QCD factorization theorems [1,2] still hold stating that this class of observables can be computed as convolutions of *universal* parton distribution functions (PDFs) and fragmentation functions (FFs) with perturbatively calculable hard scattering cross sections. As is well-known, it is due to the factorization property that the parton model of QCD has predictive power. Hence, tests of the universality of the PDFs and FFs are of crucial importance for validating this QCD framework. At the same time, lowest order expressions for the hard scattering cross sections are often not sufficient for meaningful tests and the use of higher order computations is mandatory.

The perturbative analysis is becoming more involved and interesting if the observed final state hadron contains a heavy (charm or bottom) quark. In this case, the heavy-quark mass m_h enters as an additional scale. Clearly, the conventional massless formalism, also known as zero-mass variable-flavor-number scheme (ZM-VFNS), can also be applied to this case, provided the hard scale Q of the process is much bigger than the heavy-quark mass such that terms m_h/Q are negligible. However, at present collider energies, most of the experimental data lie in the kinematic region $Q \gtrsim m_h$ and it is necessary to take the power-like mass terms into account in a consistent framework.

The subject of this article is the theoretical description of one-particle inclusive production of heavy-flavored hadrons $X_h = D, B, \Lambda_c, \ldots$ in a massive variable-flavor-number scheme (GM-VFNS). In such a scheme the large collinear logarithms of the heavy-quark mass $\ln \mu/m_h$ are subtracted from the hard scattering cross sections and resummed through the evolution of the fragmentation functions (FFs) and parton distribution functions (PDFs). At the same time, finite non-logarithmic mass terms m_h/Q are kept in the hard part and fully taken into account.

In order to test the pQCD formalism, in particular the universality of the FFs, it is important to provide a description of all relevant processes in a coherent framework. Therefore, it is important to work out the GM-VFNS at next-to-leading order (NLO) of QCD for all the relevant processes. Previously, the GM-VFNS has been applied to the following processes: $\gamma + \gamma \to D^{\star+} + X$ (direct part) [3], $\gamma + \gamma \to D^{\star+} + X$ (single resolved part) [4], $\gamma + p \to D^{\star+} + X$ (direct part) [5], $p + \bar{p} \to (D^0, D^{\star+}, D^+, D_s^+) + X$ [6,7,8], where the latter results for hadron–hadron collisions also constitute the resolved contribution to the photoproduction process $\gamma + p \to X_h + X$.

In this contribution, we will review the production of heavy-flavored hadrons X_h in hadron–hadron, photon–proton and deep inelastic electron–proton collisions where the main focus will be on the electroproduction case. We will show results, neglecting for the time being the m_h/Q mass terms, for deep inelastic $D^\star$ meson production at finite transverse momenta.

2. Theoretical Framework

2.1. *GM-VFNS*

The differential cross sections for inclusive heavy-flavored hadron production can be computed in the GM-VFNS according to the familiar factorization formulae, however, with heavy-quark mass terms included in the hard scattering cross sections [9]. Generically, the physical cross sections are expressed as convolutions of PDFs for the incoming hadron(s), hard scattering cross sections, and FFs for the fragmentation of the outgoing parton into the observed hadron. All possible partonic subprocesses are taken into account. The massive hard scattering cross sections are constructed in a way that in the limit $m_h \to 0$ the conventional ZM-VFNS is recovered. A more detailed discussion of the GM-VFNS and the construction of the massive hard scattering cross sections can be found in Refs. 6, 7 and the conference proceedings [10,11].

2.2. *Fragmentation Functions*

A crucial ingredient entering these calculation are the non-perturbative FFs for the transition of the final state parton into the observed hadron X_h. For charm-flavored mesons, X_c, three sets of FFs have been employed in our analyses: (i) BKK98-D [12]: For $X_c = D^{*+}$, such FFs were extracted at LO and NLO in the $\overline{\text{MS}}$ factorization scheme with $n_f = 5$ massless quark flavors several years ago [12] from the scaled-energy (x) distribution $d\sigma/dx$ of the cross section of $e^+e^- \to D^{*+} + X$ measured by the ALEPH [13] and OPAL [14] Collaborations at CERN LEP1. (ii) KK05-D [15]: Recently, the BKK98-D analysis was extended [15] to include $X_c = D^0, D^+, D_s^+, \Lambda_c^+$ by exploiting appropriate OPAL data [16]. (iii) KK05-D2: In Refs. [12,15], the starting scales μ_0 for the DGLAP evolution of the $a \to X_c$ FFs in the factorization scale μ'_F have been taken to be $\mu_0 = 2m_c$ for $a = g, u, \overline{u}, d, \overline{d}, s, \overline{s}, c, \overline{c}$ and $\mu_0 = 2m_b$, with $m_b = 5$ GeV, for $a = b, \overline{b}$. The FFs for $a = g, u, \overline{u}, d, \overline{d}, s, \overline{s}$ were assumed to be zero at $\mu'_F = \mu_0$ and were generated through the DGLAP evolution to larger values of μ'_F. For consistency with the $\overline{\text{MS}}$ prescription for PDFs, we repeated the fits of the X_c FFs for the choice $\mu_0 = m_c, m_b$. This changes the c-quark FFs only marginally, but has an appreciable effect on the gluon FF, which is important at Tevatron energies, as was found for D^{*+} production in Ref. [6]. These new FFs will be presented elsewhere.

2.3. *Input Parameters*

If not stated otherwise, the following parameters have been chosen for the numerical results presented below. For the proton PDFs we have employed the CTEQ6M/CTEQ6.1M PDFs from the CTEQ Collaboration [17] and the KK05-D2 FFs. We have set $m_c = 1.5$ GeV, $m_b = 5$ GeV and have used the two-loop formula for $\alpha_s^{(n_f)}(\mu_R)$ in the $\overline{\text{MS}}$ scheme with $\alpha_s^{(5)}(m_Z) = 0.118$. Our theoretical predictions depend on three scales, the renormalization scale μ_R, and the initial- and final-state factorization scales μ_F and μ'_F,

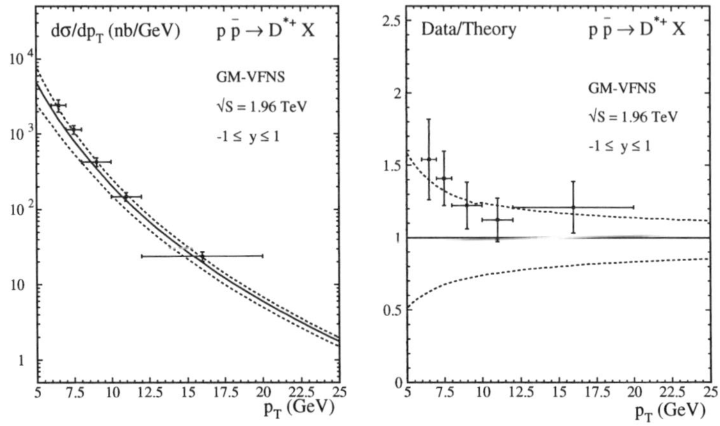

Figure 1. Comparison of the CDF data [18] with our NLO predictions for D^{*+}. The solid line represents our default prediction obtained with $\mu_R = \mu_F = \mu'_F = m_T$, while the dashed lines indicate the scale uncertainty estimated by varying μ_R, μ_F, and μ'_F independently within a factor of 2 up and down relative to the central values. The right figure shows the data-over-theory representation with respect to our default prediction.

respectively. Our default choice for hadro- and photoproduction has been $\mu_R = \mu_F = \mu'_F = m_T$, where $m_T = \sqrt{p_T^2 + m_h^2}$ is the transverse mass. The scale choice in the electroproduction case will be specified below.

3. Hadroproduction

Recently, the CDF collaboration has published first cross section data for inclusive production of D^0, D^+, D^{*+}, and D_s^+ mesons in $p\bar{p}$ collisions [18] obtained in Run II at the Tevatron at center-of-mass energies of $\sqrt{S} = 1.96$ TeV. The data come as distributions $d\sigma/dp_T$ with y integrated over the range $|y| \leq 1$ and the particle and antiparticle contributions are averaged.

Our theoretical predictions in the GM-VFNS are compared with the CDF data for $D^\star$ mesons on an absolute scale in Fig. 1 (left) and in the data-over-theory representation with respect to our default results in Fig. 1 (right). We find good agreement in the sense that the theoretical and experimental errors overlap where the experimental results are gathered on the upper side of the theoretical error band, corresponding to a small value of μ_R and large values of μ_F and μ'_F, the μ_R dependence being dominant in the upper p_T range. As is evident from Fig. 1 (right), the central data points tend to overshoot the central QCD prediction by a factor of about 1.5

at the lower end of the considered p_T range, where the errors are largest, however. This factor is rapidly approaching unity as the value of p_T is increased. The tendency of measurements of inclusive hadroproduction in Tevatron run II to prefer smaller renormalization scales is familiar from single jets, which actually favor $\mu_R = p_T/2$ [19].

For more details and a comparison with the data for the D^0, D^+, and D_s^+ mesons we refer to Ref. 8. A comparison of NLO predictions for $p+\bar{p} \to B^+ + X$ in the GM-VFNS with recent CDF data [20] is in preparation [21].

4. Photoproduction

Inclusive photoproduction of $D^\star$ mesons, $\gamma + p \to D^\star + X$, has been studied in Ref. 5 where the direct part has been computed in the GM-VFNS whereas the resolved part has been included in the ZM-VFNS. In this analysis the BKK98-D FFs have been utilized and, for the resolved contribution, the GRV92 photon PDFs [22]. The other parameters have been chosen as specified in Sec. 2.3. In Fig. 6 of Ref. 5, the central numerical predictions for the transverse momentum (p_T) distributions of the $D^\star$ meson have been compared with preliminary ZEUS data [23]. There exist similar data by the H1 collaboration [24] which have not been used in this analysis. As can be seen in this figure, the agreement of the p_T-distributions with the data is quite good down to $p_T \simeq 2m_c$ and the mass effects turn out to be small. In order to extend the range of applicability of the GM-VFNS into the region $p_T < 3$ GeV more work on the matching to the 3-fixed flavor theory would be needed. The Figs. 7 – 9 of Ref. 5, showing results for the rapidity (y), invariant mass (W) and inelasticity ($z(D^\star)$) distributions, have to be taken with a grain of salt since they receive large contributions from the transverse momentum region $1.9 < p_T < 3$ GeV which is outside the range of validity of the present theory.

With the work in Ref. 6, it is now possible to include also the resolved part in the GM-VFNS. It will be interesting to compare the complete GM-VFNS framework at NLO of QCD, combined with updated FFs, in more detail with ZEUS and H1 photoproduction data once they are finalized.

5. Electroproduction

Recently, the single inclusive electroproduction of light hadrons at finite transverse momenta has attracted quite a lot of interest, where the outgoing hadron is required to carry a non-vanishing transverse momentum (p_T^*) in the center-of-mass system (CMS) of the virtual photon and the incoming proton. The following partonic subprocesses contribute at leading order (LO) and are of the order $O(\alpha_s)$: $\gamma^\star + q \to q + g$ and $\gamma^\star + g \to q + \bar{q}$. Very recently, the NLO ($O(\alpha_s^2)$) corrections to this process have been accomplished by three independent groups [25,26,27]. Suffice to say here that the NLO corrections increase the LO results in certain kinematical regions by large factors and are essential for bringing theory in agreement with the experimental results. For more details see Ref. 28.

Endowed with appropriate fragmentation functions, the computation in Ref. 27 has been employed to obtain predictions in the ZM-VFNS for the production of heavy-flavored hadrons in electron–proton collisions at HERA, $e+p \to e+X_h+X$. The electron and proton energies have been set to $E_e = 27.5$ GeV and $E_p = 820$ GeV (Fig. 2) resp. 920 GeV (Figs. 3, 4) in the laboratory frame. Furthermore, the renormalization and factorization scales have been chosen [27] as $\mu_R^2 = \mu_F^2 = {\mu'_F}^2 = \xi \frac{Q^2+(p_T^*)^2}{2}$. The dimensionless parameter ξ has been varied between 1/2 and 2 about the default value 1 in order to estimate the scale uncertainties of the theoretical predictions. The other input parameters have been chosen as described in Sec. 2.3. The following cuts have been imposed on the numerical results in Figs. 2–4: $2 < Q^2 < 100$ GeV^2, $0.05 < y < 0.7$, $1.5 \leq p_{T,\mathrm{Lab}}(D^\star) \leq 15$ GeV, and $|\eta_{\mathrm{Lab}}(D^\star)| < 1.5$ where the momentum transfer Q^2, the inelasticity y, and the pseudorapidity η are defined as usual. Furthermore, in Figs. 3–4 we have asked for the additional constraint $p_T^*(D^\star) > 2$ GeV where $p_T^*(D^\star)$ is the transverse momentum of the $D^\star$ meson in the γ^*p-CMS. This cut is essential to avoid the collinear singularities as $p_T^* \to 0$. The results have been calculated with $n_f = 5$ flavors and include the contribution where a bottom quark fragments into the $D^\star$ meson via $b \to B \to D^\star$. No distinction is made between $D^{\star+}$ and $D^{\star-}$. Figure 2 shows the p_T^* spectrum in comparison with H1 data [29], collected in the years 1994 to 1996 with 27.5 GeV positrons colliding with 820 GeV protons at CMS energies of $\sqrt{S} = 300$ GeV. The data points are reasonably well described if one keeps in mind that the theory is expected to work for larger $p_T^* \gtrsim 2$ GeV. Furthermore, in the ZM-VFNS the heavy-quark mass terms are missing which are potentially important in the region of small p_T^*. These terms will be included as soon as ongoing work to implement this process in the GM-VFNS has been completed.

Finally, results for other distributions ($p_{T,\mathrm{Lab}}$, η_{Lab}, W, z, Q^2, and x_{Bj}) are presented in Figs. 3–4.

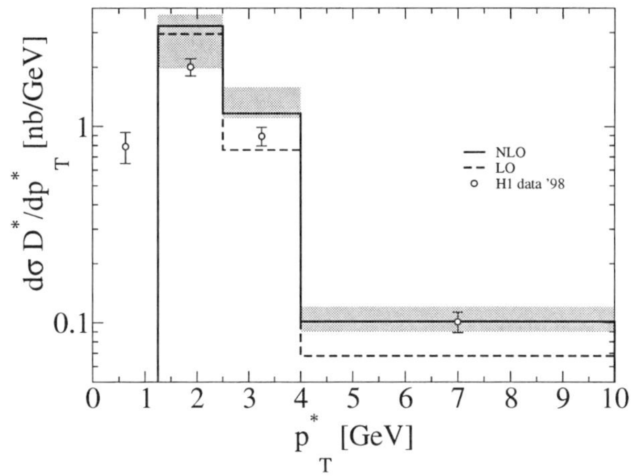

Figure 2. NLO QCD predictions in the ZM-VFNS for electroproduction of $D^\star$ mesons in dependence of the transverse momentum of the $D^\star$ meson in the $\gamma^\star p$ CMS. H1 data [29] are shown for comparison. See the text for further details.

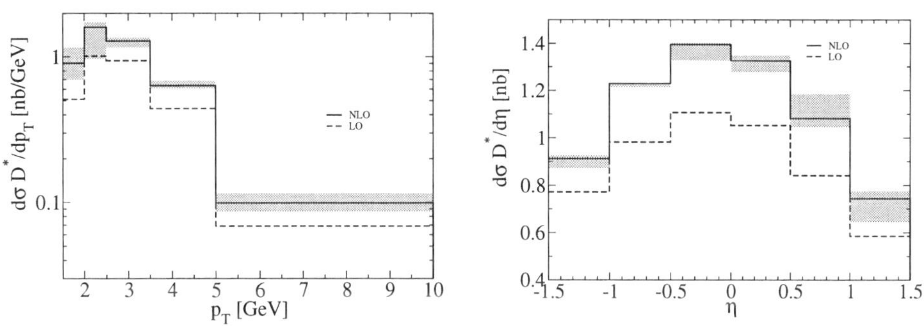

Figure 3. NLO QCD predictions in the ZM-VFNS for electroproduction of $D^\star$ mesons in dependence of the transverse momentum (left) and the pseudo rapidity (right) of the $D^\star$ meson in the laboratory frame. For further details on the kinematical cuts and the uncertainty band see the text.

6. Summary

We have discussed one-particle inclusive production of heavy-flavored hadrons in hadron–hadron, photon–proton, and electron–proton collisions in a massive variable-flavor-number scheme (GM-VFNS). The importance of a unified treatment of all these processes, based on QCD factorization theorems, has been emphasized, in order to provide meaningful tests of the

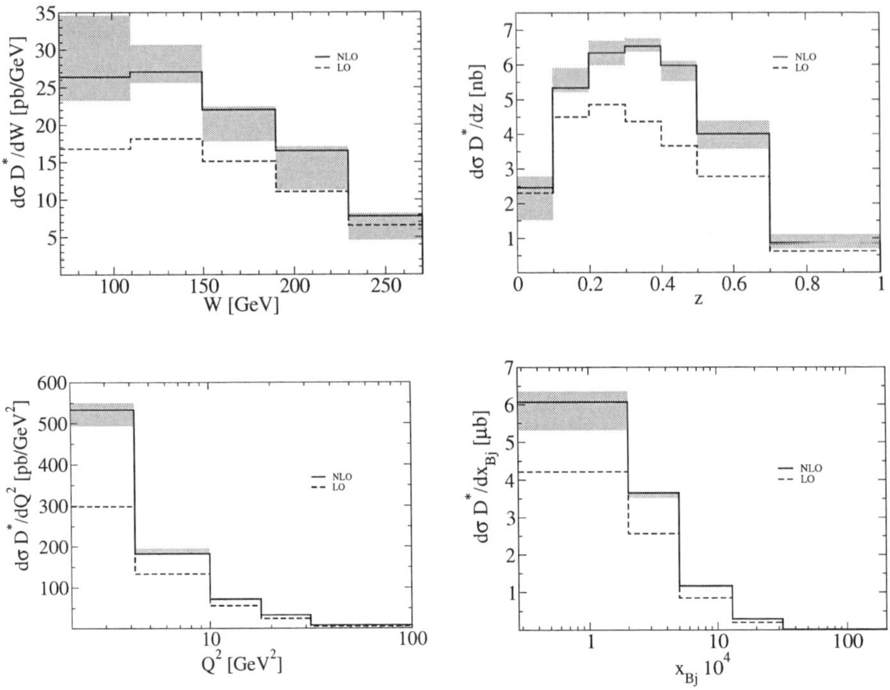

Figure 4. As in Fig. 3 for the $\gamma^\star p$ invariant mass W, the inelasticity z, the momentum transfer Q^2 and Bjorken-x.

universality of the FFs and hence of QCD. At the same time, it is necessary to incorporate heavy-quark mass effects in the formalism since many of the present experimental data points lie in a kinematical region where the hard scale of the process is not much larger than the heavy-quark mass. This is achieved in the GM-VFNS, which includes heavy-quark mass effects in a rigorous way and still relies on QCD factorization. We have discussed numerical results for the three reactions. In general, the description of the transverse momentum spectra is quite good down to transverse momenta $p_T \simeq 2m_h$. Extending the range of applicability of our scheme to smaller p_T would require more work on the matching to the corresponding theories in the fixed flavor number schemes. Our ZM-VFNS results for the electroproduction of $D^\star$ mesons indicate that future experimental results by the H1 collaboration can be nicely described. All this leads us to the expectation that a good overall description of the data can be reached in the future.

Acknowledgments

The author would like to thank the organizers of the Ringberg workshop on *New Trends in HERA Physics 2005* for the kind invitation, B. A. Kniehl, G. Kramer and H. Spiesberger for their collaboration, and M. Maniatis for providing the figures 2 – 4 and for useful discussions.

References

1. J. C. Collins, D. E. Soper, and G. Sterman, in *Perturbative Quantum Chromodynamics*, edited by A. H. Mueller (World Scientific, 1989).
2. J. C. Collins and D. E. Soper, Ann. Rev. Nucl. Part. Sci. **37**, 383 (1987).
3. G. Kramer and H. Spiesberger, Eur. Phys. J. **C22**, 289 (2001).
4. G. Kramer and H. Spiesberger, Eur. Phys. J. **C28**, 495 (2003).
5. G. Kramer and H. Spiesberger, Eur. Phys. J. **C38**, 309 (2004).
6. B. A. Kniehl, G. Kramer, I. Schienbein, and H. Spiesberger, Phys. Rev. **D71**, 014018 (2005).
7. B. A. Kniehl, G. Kramer, I. Schienbein, and H. Spiesberger, Eur. Phys. J. **C41**, 199 (2005).
8. B. A. Kniehl, G. Kramer, I. Schienbein, and H. Spiesberger, Phys. Rev. Lett. **96**, 012001 (2006).
9. J. C. Collins, Phys. Rev. **D58**, 094002 (1998).
10. I. Schienbein, *Open heavy-flavour photoproduction at NLO*, Proceedings of the Ringberg Workshop, *New Trends in HERA Physics 2003*, edited by G. Grindhammer, B. A. Kniehl, G. Kramer and W. Ochs, World Scientific, 2004, p. 197; I. Schienbein, Proceedings of the 12th International Workshop on Deep Inelastic Scattering (DIS 2004), Strbske Pleso, Slovakia, 14-18 Apr 2004, hep-ph/0408036; B. A. Kniehl, G. Kramer, I. Schienbein, and H. Spiesberger, Proceedings of the 13th International Workshop on Deep Inelastic Scattering (DIS 2005), Madison, Wisconsin, USA, April 27 - May 1, 2005, hep-ph/0507068.
11. J. Baines et al, Summary report of the heavy-flavour working group of the HERA-LHC Workshop, hep-ph/0601164.
12. J. Binnewies, B. A. Kniehl, and G. Kramer, Phys. Rev. **D58**, 014014 (1998).
13. R. Barate *et al.*, ALEPH Collaboration, Eur. Phys. J. **C16**, 597 (2000).
14. K. Ackerstaff *et al.*, OPAL Collaboration, Eur. Phys. J. **C1**, 439 (1998).
15. B. A. Kniehl and G. Kramer, Phys. Rev. **D71**, 094013 (2005).
16. G. Alexander *et al.*, OPAL Collaboration, Z. Phys. **C72**, 1 (1996).
17. J. Pumplin *et al.*, JHEP **07**, 012 (2002); D. Stump *et al.*, JHEP **10**, 046 (2003).
18. D. Acosta *et al.*, CDF Collaboration, Phys. Rev. Lett. **91**, 241804 (2003).
19. R. Field, for the CDF Collaboration, in Proceedings of the XIIIth International Workshop on Deep Inelastic Scattering (DIS05), Madison, Wisconsin, 2005 (American Institute of Physics, Melville, to be published); B. Davies, for the D0 Collaboration, *ibid.*
20. D. Acosta *et al.*, CDF Collaboration, Phys. Rev. **D71**, 032001 (2005).

21. B. A. Kniehl, G. Kramer, I. Schienbein, and H. Spiesberger, (work in progress).
22. M. Glück, E. Reya, and A. Vogt, Phys. Rev. **D46**, 1973 (1992).
23. S. Chekanov *et al.*, ZEUS Collaboration, 31st International Conference on High Energy Physics, ICHEP02, July 24–31, 2002, Amsterdam, Abstract 786; see also: ZEUS Collaboration, `http://www-zeus.de/public_plots`, Measurement of $D^\star$ photoproduction at HERA, 2003.
24. H1 Collaboration, International Europhysics Conference on High Energy Physics, EPS03, July 17–23, 2003, Aachen, Abstract 097, DESY-H1prelim-03-071 and earlier H1 paper given in this reference.
25. P. Aurenche, R. Basu, M. Fontannaz, and R. M. Godbole, Eur. Phys. J. **C34**, 277 (2004).
26. A. Daleo, D. de Florian, and R. Sassot, Phys. Rev. **D71**, 034013 (2005).
27. B. A. Kniehl, G. Kramer, and M. Maniatis, Nucl. Phys. **B711**, 345 (2005); Erratum-ibid. **B720**, 231 (2005).
28. B. A. Kniehl, these proceedings.
29. C. Adloff *et al.*, H1 Collaboration, Nucl. Phys. **B545**, 21 (1999).

PHYSICS WITH CHARM QUARKS AT HERA

J. H.LOIZIDES
University College London,
Gower street,
London WC1E 6BT England
E-mail: jhl@hep.ucl.ac.uk

Measurements of charm production in deep inelastic scattering (DIS) and photoproduction have been carried out by the ZEUS and H1 collaborations at HERA. Protons at 920 GeV are collided with positrons of 27.5 GeV. Results using integrated luminosities up to 120 pb^{-1} for HERA I and 35 pb^{-1} for HERA II are presented. Single and double differential cross sections are compared to perturbative QCD (pQCD) predictions. Charm DIS and photoproduction cross sections are in reasonable agreement with pQCD calculations. Charm and jet angular correlations show regions of phase space where NLO is not sufficient to describe these data sets, indicating the need for improved calculations such as MC@NLO or NNLO predictions. Charm fragmentation ratios and fractions generally support the hypothesis that fragmentation proceeds independently from the hard sub-process in e^+e^- and ep collisions. This test of charm universality is confirmed by the measured D meson ground states at HERA and LEP. Stringent tests of QCD have been performed; theoretical errors dominate or are equivalent to experimental errors in the comparisons.

1. Introduction

The electron/positron-proton collider HERA at the DESY laboratory is a unique facility to test Quantum Chromo Dynamics (QCD) in a variety of measurements involving charm. The transition region between soft and hard interactions allows the possibility to better understand the present limits of applicability of perturbative QCD (pQCD).

When Q^2, the negative squared four momentum exchange at the electron/positron vertex, is small the reaction can be considered as a photoproduction process. Charm quarks in photoproduction and in deep inelastic scattering (DIS) have been extensively studied at HERA [1,2,3,4,5,6]. These measurements are consistent with pQCD calculations indicating boson-gluon fusion (BGF) as the dominant mechanism of charm production. Charm is mainly tagged in the 'golden' decay channel of the D^* meson

$D^{*+} \to K^- \pi^+ \pi^+ (+\text{c.c.})$. More advanced instrumentation using secondary vertex tagging has been used to measure other charm cross sections at H1 [7]. ZEUS has first results in tagging of D mesons using the newly installed microvertex chamber from the recent running phase of HERA II. For more detailed information on heavy quark production such as BGF, resolved, 'massive' & 'massless' schemes see [8,9,10,12,13].

2. Inclusive charm production

The production cross sections of the charmed mesons $D^\pm$, D^0, $D_s^\pm$, $D^{*\pm}$ have been measured by H1 [7]. The differential production cross sections for all four D mesons measured in the same kinematic region show similar dependence on transverse momentum $P_T(D)$ and the pseudo-rapidity $\eta(D)$ of the charmed hadron and Q^2. This similarity implies that the fragmentation fractions are independent of kinematics and can be measured from the integrated D meson cross sections. ZEUS has measured differential $D^{*\pm}$ meson cross sections as a function of Q^2 (fig. 1) across the transition region ($Q^2 \approx 1$ GeV^2) from DIS to photoproduction and compared to NLO predictions; there is good agreement throughout the entire range.

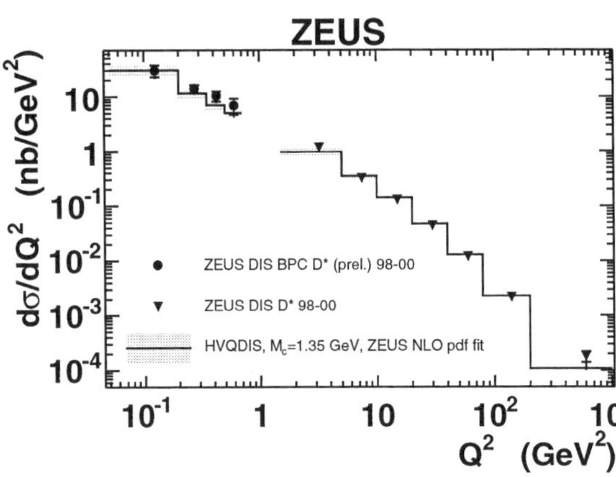

Figure 1. Differential D^* meson cross section as a function of Q^2 for low Q^2 and from previous results on D^* production at high Q^2 compared to the NLO predictions from the HVQDIS program.

Correspondingly ratios of the total production rates are used to test the isospin invariance of the fragmentation process and to extract the strangeness suppression factor γ_s (fig. 2), the fraction P_V of D-mesons produced in a vector state (fig. 3), and the ratio of neutral and charged D-meson production rates $R_{u/d}$ (fig. 4). The results are compared with values measured at e^+e^- colliders and allow tests of the assumed universality of the charm fragmentation process [7,14]. Table 1 shows the results of the branching fractions $f(c \to D^+)$, $f(c \to D^0)$, $f(c \to D_s^+)$, $f(c \to D^{*+})$ and $f(c \to \Lambda_c^+)$ (see fig. 5 for the invariant mass distribution of Λ_c candidates). The values at the different colliders are in good agreement, with competitive errors for γp; sometimes better than e^+e^- collision measurements. So the assumption that charm fragmentation fractions are universal is confirmed.

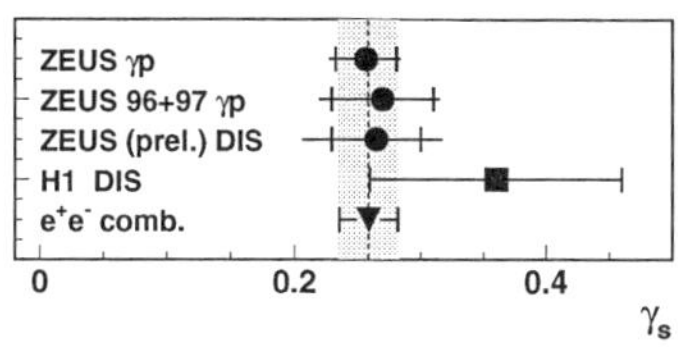

Figure 2. The strangeness-suppression factor in charm fragmentation, γ_s.

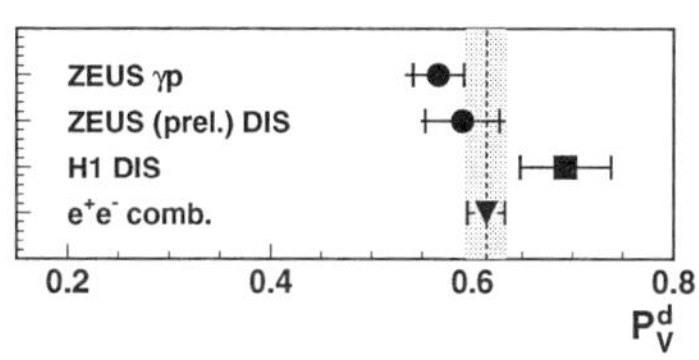

Figure 3. The fraction of charged D mesons produced in a vector state, P_V^d.

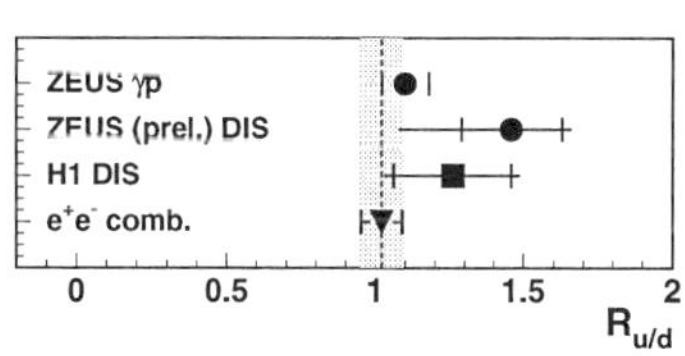

Figure 4. The ratio of neutral to charged D-meson production rates, $R_{u/d}$.

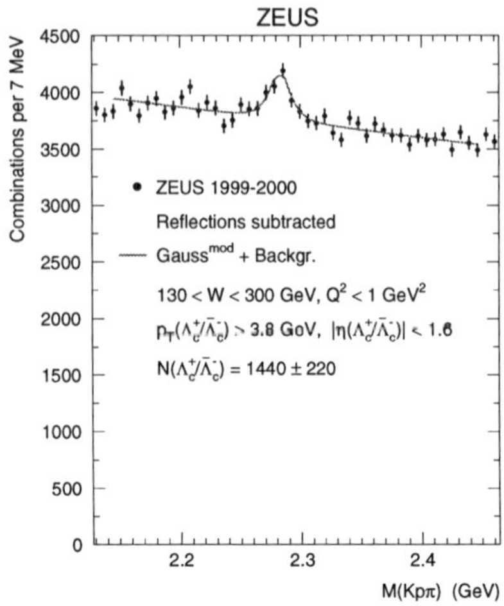

Figure 5. The $M(Kp\pi)$ distribution for the $\Lambda_c^+/\bar{\Lambda}_c^-$ candidates (dots). The solid curve represents a fit to the sum of a modified Gaussian function and a linear background function.

Table 1. The fractions of c quarks hadronising as a particular charm hadron, $f(c \to D, \Lambda_c)$. The fractions are shown for the D^+, D^0, D_s^+ and Λ_c^+ charm ground states and for the D^{*+} state.

	ZEUS (γp) $p_T(D,\Lambda_c) > 3.8\,\text{GeV}$ $\lvert\eta(D,\Lambda_c)\rvert < 1.6$		Combined e^+e^- data [21]		H1 (DIS) [22]
	stat.	syst. br.	stat.⊕ syst.	br.	total
$f(c \to D^+)$	0.217 ± 0.014	${}^{+0.013}_{-0.005}\,{}^{+0.014}_{-0.016}$	0.226 ± 0.010	${}^{+0.016}_{-0.014}$	0.203 ± 0.026
$f(c \to D^0)$	0.523 ± 0.021	${}^{+0.018}_{-0.017}\,{}^{+0.022}_{-0.032}$	0.557 ± 0.023	${}^{+0.014}_{-0.013}$	0.560 ± 0.046
$f(c \to D_s^+)$	0.095 ± 0.008	${}^{+0.005}_{-0.005}\,{}^{+0.026}_{-0.017}$	0.101 ± 0.009	${}^{+0.034}_{-0.020}$	0.151 ± 0.055
$f(c \to \Lambda_c^+)$	0.144 ± 0.022	${}^{+0.013}_{-0.022}\,{}^{+0.037}_{-0.025}$	0.076 ± 0.007	${}^{+0.027}_{-0.016}$	
$f(c \to D^{*+})$	0.200 ± 0.009	${}^{+0.008}_{-0.006}\,{}^{+0.008}_{-0.012}$	0.238 ± 0.007	${}^{+0.003}_{-0.003}$	0.263 ± 0.032

3. Charm jet cross sections in photoproduction

Differential inclusive D^* meson cross sections are not well described by pQCD predictions [18,19]. Therefore it is interesting to compare pQCD predictions for charm jets.

The predictions available are the 'massive' [9] scheme, where charm is produced only dynamically and there is no a priori charm in the proton or photon, and the 'massless' scheme [10], where charm is assumed to be massless and treated in the same manner as a light quark. Cross sections were measured double differentially as a function of E_T^{jet} and η^{jet} [20,11]. Both predictions describe the data well, with a slight excess seen at high values of E_T^{jet} compared to the 'massive' scheme. As the charm mass becomes small compared to the scale E_T^{jet}, this excess is not visible for the 'massless' calculation. Dijet correlation cross-sections $d\sigma/d\Delta\phi^{\rm jj}$ (fig. 6) show a large deviation from the massive NLO QCD prediction at low $\Delta\phi^{\rm jj}$ for the resolved-enriched ($x_\gamma^{\rm obs} < 0.75$) sample. These regions are expected to be particularly sensitive to higher-order effects. The HERWIG MC model which incorporates leading-order matrix elements followed by parton showers and hadronisation describes the shape of the measurements well. This indicates that for the precise description of charm dijet photoproduction higher-order calculations or the implementation of additional parton showers in current NLO calculations are needed.

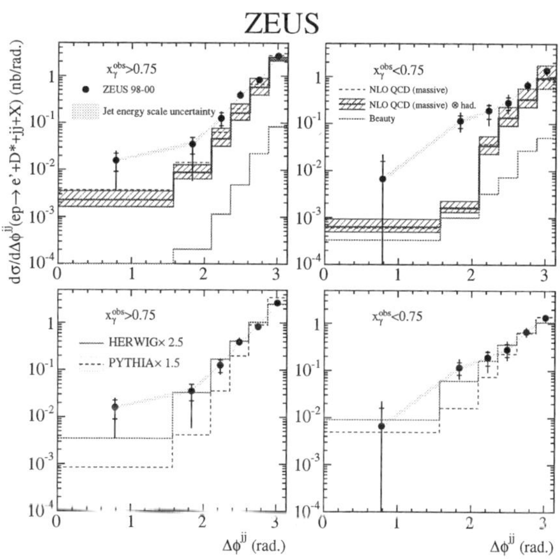

Figure 6. Cross-section $d\sigma/d\Delta\phi^{\rm jj}$ for $(ep \to e' + D^* + {\rm jets} + X)$, separated into direct enriched ($x_\gamma^{\rm obs} > 0.75$) and resolved enriched ($x_\gamma^{\rm obs} < 0.75$). The data (solid dots) are compared (top plots) to the massive QCD prediction with (solid line) and without (dotted line) hadronisation corrections applied. The theoretical uncertainties (hatched band) come from the change in scales simultaneously with the change in charm mass. The beauty component is also shown (upper histogram). The data are also compared (bottom plots) with HERWIG (solid line) and PYTHIA (dashed line) MC predictions are multiplied by the indicated factors.

4. Charm fragmentation functions

The hadronisation of a charm quark into a jet of particles consisting of a charmed meson and other mesons is not calculable in pQCD. The transition of a charm quark into a charmed meson is usually described by a phenomenological and non-perturbative fragmentation function, which is expected to be universal. One considerable source of uncertainty in comparing pQCD calculations in next-to-leading order (NLO) with data is due to the choice of fragmentation function and their parameters. Two widely used functions are the Peterson et al. (eq. 1) and the Kartvelishvili et al. function (eq. 2).

$$D_Q^H = \frac{N}{z[1 - (1/z) - \epsilon(1-z)]^2} \,, \tag{1}$$

$$D_Q^H = N z^\alpha (1-z) \,. \tag{2}$$

The free parameters ϵ and α determine the 'hardness' of the respective fragmentation function. In these parameterisations, z describes the fraction of the charm quark's energy carried by the heavy meson.

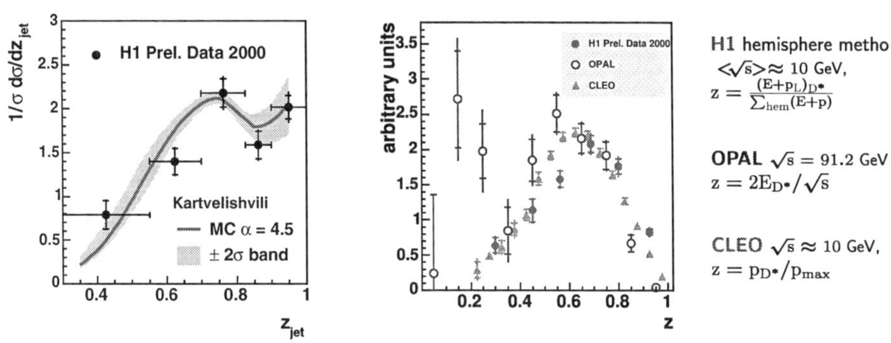

Figure 7. Measured D^* fragmentation function for jet method compared with RAPGAP model prediction using the Kartvelishvili parametrisation for non-perturbative fragmentation function (left plot). Comparison of the z-distributions from CLEO, OPAL and ALEPH with the one from the hemisphere method from H1. All distributions are normalised to unit area from $z = 0.4$ to $z = 1$ (right plot).

There are many observables sensitive to the non-perturbative fragmentation function, both in e^+e^- and ep collisions. In the case of e^+e^-, the natural choice is to measure the dependence of the D^* meson production cross section as a function of the variable z, defined as $z = E(D^*)/(\sqrt{s}/2)$.

In ep collisions this is not so obvious; H1 have considered two methods. Figure 7 shows the jet method where the energy and direction of the charm quark is approximated by reconstructing the jet which contains the D^* meson, where $z_{jet} = \frac{(E+p_L)_{D^*}}{(E+p)_{jet}}$. Another method is called the hemisphere method which allows the event to be split into two hemispheres. One hemisphere will contain the charm quark and the other the anti-charm quark, here z is defined as $z_{hem} = \frac{(E+p_L)_{D^*}}{\sum_{hem}(E+p)}$, where the denominator is the sum of all particles with momentum projections in the D^* hemisphere. More detailed information can be found elsewhere [23].

Table 2. The H1 extracted fragmentation parameters for Peterson and Kartvelishvili parametrisations, for RAPGAP/PYTHIA MC.

Parametrisation	Hemisphere Method	Jet Method	Range
Peterson	$0.018^{+0.004}_{-0.004}$	$0.030^{+0.006}_{-0.005}$	$0.014 < \epsilon < 0.036$
Kartvelishvili	$5.9^{+0.9}_{-0.6}$	$4.5^{+0.5}_{-0.5}$	$4 < \alpha < 6.8$

These data are compared to recent measurements in e^+e^- annihilations and are in rough agreement with the recent data from CLEO at a similar centre-of-mass energy for $c\bar{c}$ pair production. The difference at low z values in e^+e^- annihilations from OPAL is due to gluon splitting into $c\bar{c}$ pairs. The extracted values of ϵ and α span a large range (table 2); more data and a more rigorous approach is needed using NLO programs as well as theoretical developments such as MC@NLO in order to be able to attain a better understanding of fragmentation functions.

5. Charm Contribution to proton structure function F_2

The charm structure function F_2^{cc} [1,2] was extracted from:

$$F^{c\bar{c}}_{2,\mathrm{meas}}(x_i, Q^2) = \frac{\sigma_{i,\mathrm{meas}}(ep \to D^* X)}{\sigma_{i,\mathrm{theo}}(ep \to D^* X)} F^{c\bar{c}}_{2,\mathrm{theo}}(x_i, Q^2) \quad (3)$$

Latest results [17] are based on the impact parameter in the transverse plane, of tracks to the primary vertex, as measured by the H1 vertex detector. The ratio $F_2^{c\bar{c}}/F_2$ rises from 10% to 30% as Q^2 increases and x decreases. Figure 8 shows the $F_2^{c\bar{c}}$ results from ZEUS and H1 compared to the ZEUS NLO QCD fit.

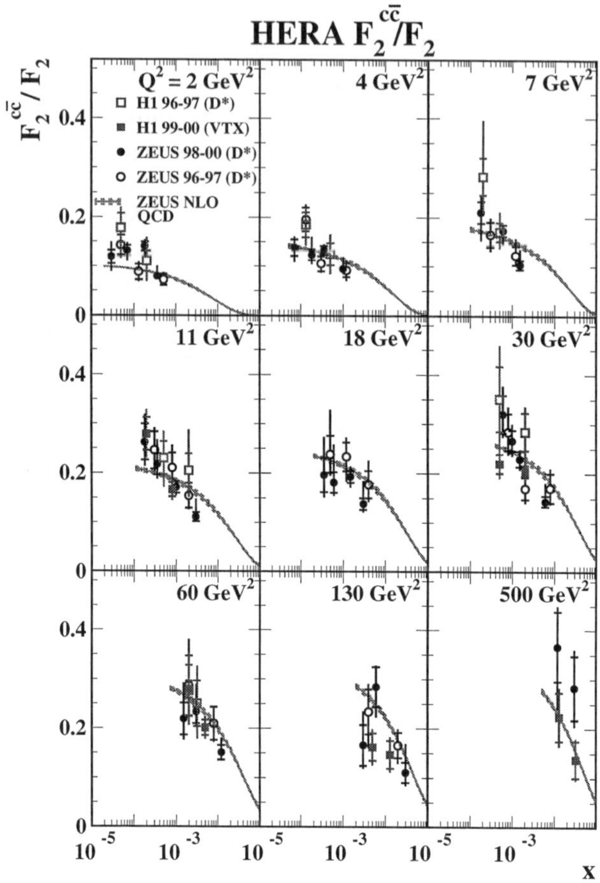

Figure 8. The $F_2^{c\bar{c}}$ results from H1, ZEUS are shown as a function of x for different bins of Q^2. The solid line represents the ZEUS NLO QCD fit with the uncertainties being the dashed lines.

Reduced experimental errors will allow these measurements to be used together with the inclusive F_2 measurements to additionally constrain the gluon density.

6. First look at HERA II data

In the most recently published ZEUS measurement [1], made using 65 pb^{-1} e^+p data and 17 pb^{-1} e^-p data, it was observed that the D^* cross section was systematically higher in e^-p than in the e^+p data set, and rising with Q^2. Such a difference was not expected from any physics process, and so the phenomenon was treated as a statistical fluctuation. Figure 9 shows these results using the 1998-2000 data set and the newly collected 2003-2005 HERA II data where the e^-p luminosity is doubled with respect to the 1998-2000 data set. There is no evidence of any production rate anomaly in the HERA II data set.

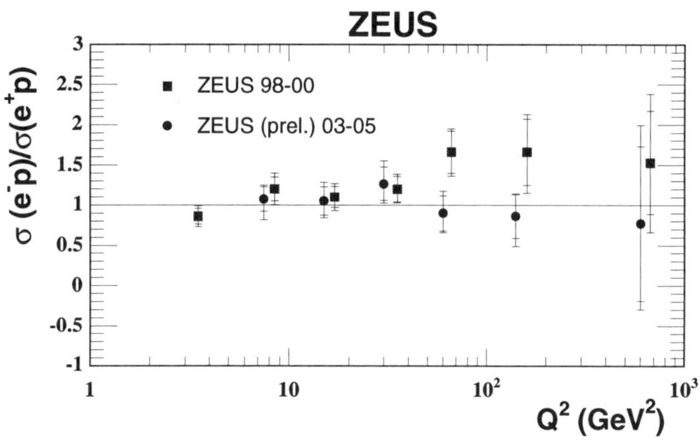

Figure 9. The ratio $\sigma(e^-p)/\sigma(e^+p)$ for the process $ep \to e' + D^* + X$, where $D^{*\pm} \to D^0, \pi_s^\pm$, and $D^0 \to K^\mp, \pi^\pm$ in the 2004-2005 running period (solid dots), compared to the ZEUS measurement for the 1998-2000 running period (solid squares).

7. Conclusions

There has been a wealth of charm physics produced at HERA from H1 and ZEUS providing stringent tests of pQCD. Charm fragmentation ratios and fractions generally support the hypothesis that fragmentation proceeds independently of the hard sub-process in ep and e^+e^- collisions. Fragmentation functions for D^* mesons have been measured and various models have been fitted to these distributions. Unfortunately, it is still unclear how one derives meaningful and applicable fragmentation functions to apply to theoretical predictions with the variation of these parameterisations being one of the largest theoretical uncertainties. Charm jet correlations show there are regions of phase space which cannot be described by NLO predictions, showing the need for improved theoretical predictions such as NNLO or MC@NLO. For these charm measurements, the theoretical errors dominate or are at best equivalent to experimental errors. The new additional instrumentation for ZEUS and higher luminosity being delivered by HERA will allow for more precise and new measurements to be made, with the aim of gaining further understanding of the production dynamics of heavy quarks in electron/positron-proton interactions.

References

1. [ZEUS Collaboration], S. Chekanov *et al.*, Phys. Rev. **D 69**, 012004 (2004).
2. [H1 Collaboration], C. Adloff et al., Phys. Lett. **B 528**, 199 (2002).
3. [ZEUS Collaboration], J. Breitweg et al., Phys. Lett. **B 481**, 213 (2000).
4. [ZEUS Collaboration], J. Breitweg et al., Eur. Phys. J. **C 12**, 35 (2000).
5. [H1 Collaboration], C. Adloff et al., Nucl. Phys. **B 545**, 21 (1999).
6. [ZEUS Collaboration], J. Breitweg et al., Eur. Phys. J. **C 6**, 67 (1999).
7. [H1 Collaboration], A. Aktas *et al.*, Eur. Phys. J. C **38**, 447 (2005), hep-ex/0408149.
8. B. W. Harris and J. Smith, Phys. Rev. **D 57**, 2806 (1998).
9. S. Frixione et al., Adv. Ser. Direct. High Energy Phys. **15**, 609 (1998).
10. G. Heinrich and B. Kniehl, Phys. Rev. **D 70**, 094035 (2004), hep-ph/0409303.
11. [H1 Collaboration],"Photoproduction of $D^{*\pm}$ Mesons Associated with a Jet at HERA," presented at DIS2005, Spring 2005.
12. M. Cacciari, S. Frixione and P. Nason, JHEP **0103**, 006 (2001), hep-ph/0102134.
13. S. Frixione and P. Nason, JHEP **0203**, 053 (2002), hep-ph/0201281.
14. [ZEUS Collaboration],"Measurement of charm fragmentation ratios and fractions in γp collisions at HERA", Lepton Photon 2005, PAPER-263.
15. [H1 Collaboration], "D^* Mesons and Associated Jet Production in Deep Inelastic Scattering," *Prepared for 32nd International Conference on High-Energy Physics (ICHEP 04), Beijing, China, 16-22 Aug 2004.*
16. [ZEUS Collaboration], S. Chekanov *et al.*, Eur. Phys. J. C **44**, 351 (2005), hep-ex/0508019.
17. [H1 Collaboration], A. Aktas *et al.*, Eur. Phys. J. C **40**, 349 (2005), hep-ex/0411046.
18. [ZEUS Collaboration], "Measurement of D^* photoproduction at HERA", submitted to ICHEP2002.
19. [H1 Collaboration], "Photoproduction of D^* Mesons at HERA", submitted to ICHEP2004.
20. [ZEUS Collaboration], S. Chekanov *et al.*, Nucl. Phys. B **729**, 492 (2005), hep-ex/0507089.
21. L. Gladilin, Preprint hep-ex/9912064 (1999).
22. [H1 Collaboration], A. Aktas et al., Eur. Phys. J. **C 38**, 447 (2005).
23. [H1 Collaboration],"The Charm Fragmentation Function in DIS," presented at DIS2005, Spring 2005.

BEAUTY PRODUCTION AT HERA

O. BEHNKE

Physik. Institut Uni Heidelberg
Philosophenweg 12, 69120 Heidelberg
E-mail: obehnke@mail.desy.de

A review is given on beauty production at HERA. Many new results have recently become available based on various tagging techniques and covering altogether a wide kinematic phase space in photon virtualities and b quark transverse momenta. Differential beauty production cross sections are presented and compared to perturbative QCD predictions. The first measurements of the structure function $F_2^{b\bar{b}}$ are shown.

1. Introduction

The dominant beauty production mechanism at HERA is photon gluon fusion (PGF), which is shown in the left diagram in figure 1. The total

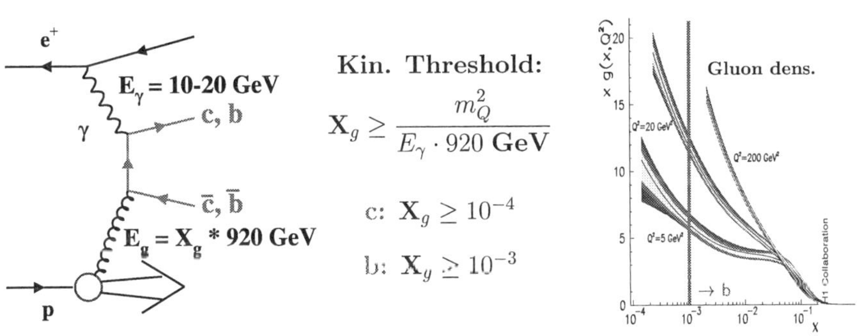

Figure 1. Comparison of kinematical thresholds for charm and beauty production at HERA in terms of the minimum proton momentum fraction x_g of the gluon that enters the hard interaction. The left plot shows the PGF process where the energies are indicated that enter the heavy quark pair production process. The threshold formula for x_g in the middle of the figure is calculated from $(x_g p + q)^2 \approx 2x_g pq = 4x_g E_\gamma E_p \geq (2m_Q)^2$, where p and q are the four vectors of the proton and photon, respectively. The right plot shows the gluon density in the proton as determined from scaling violations of F_2. The vertical line indicates the minimum fraction $X_g = 10^{-3}$ for beauty production at HERA.

production rates of light, charm and beauty quarks at HERA scale roughly like

$$\sigma_{uds} : \sigma_c : \sigma_b \sim 2000 : 200 : 1.$$

The strong suppression of beauty events is mostly due to the limited kinematic phase space, as illustrated in figure 1. To produce two beauty quarks in the PGF process a minimal proton momentum fraction of the gluon of 10^{-3} is needed, while for charm the threshold is one order of magnitude lower. A further suppression factor of *four* for beauty production relative to charm production is due to the smaller electric charge of down type quarks compared to up type quarks.

Since the last Ringberg workshop in 2003 a wealth of new beauty production measurements have become available at HERA, which will be summarised here. The results are based on a variety of tagging methods and cover in total a large kinematic phase space of photon virtualities $0 < Q^2 < 1000$ GeV^2 and b quark transverse momenta $0 < p_T < 30$ GeV.

2. Theory

Since the large b mass provides a hard scale, rendering a small α_s, it would be expected that beauty production can be accurately calculated using perturbative QCD (pQCD). Calculations are available in Next to Leading Order (NLO) in the massive scheme, where the b quark mass is fully taken into account. For photoproduction the program FMNR [1] is used and for deep inelastic scattering (DIS) HVQDIS [2]. In the massive scheme, u, d and s are the only active flavours in the proton and the photon, and charm and beauty are produced dynamically in the hard scattering as shown in figure 1. These predictions are expected to be reliable for the largest part of the kinematic phase space at HERA. However, at very large transverse momenta $p_{t,b} \gg m_b$ or photon virtualities $Q^2 \gg m_b^2$, the predictions of the massive scheme are expected to become unreliable due to neglected higher order terms in the perturbation series of the form $[\alpha_s \ln(p_{t,b}^2/m_b^2)]^n$ or $[\alpha_s \ln(Q^2/m_b^2)]^n$, which appear to any order n and represent collinear gluon radiation from the heavy quark lines. In this kinematic range, the massless scheme can be used, in which charm and beauty are treated as active flavours in both the proton and also in the hadronic structure of the photon, in addition to u, d and s. The kinematics of the heavy quarks is treated massless in this scheme, mass effects are taken into account as effective cutoffs for the dynamic evolution of the heavy quarks in the proton and photon parton density functions. In this scheme the above higher order terms are resummed to all orders. Unfortunately there are at the present time no results from massless scheme calculations available which could be compared to the HERA beauty measurements. Results are available from mixed scheme calculations, which apply the massive (massless) scheme for small (large) high transverse momenta and photon virtualities with suitable interpolations in intermediate regions. The measurements of the proton structure function $F_2^{b\bar{b}}$, presented below, are compared to the mixed scheme predictions from MRST [3] and CTEQ [4] in NLO and to the first Next to Next to Leading Order (NNLO) calculation, provided by MRST [5].

3. Overview of measurements and the tagging techniques used

The table below gives an overview of the beauty measurements presented in this review. The analyses are ordered with respect to the minimum b quark transverse momentum p_T which is probed and the covered range of the photon virtuality Q^2. Note, that the numbers in the table are only approximative values.

Q^2 →

p_T ↓	Photoprod. $Q^2 \approx 0$	DIS $Q^2 \geq 1$ GeV^2	DIS $Q^2 \geq 150$ GeV^2
> 0 GeV	$\mu\mu$ $D^*\mu$	Incl. Lifet.	Incl. Lifet.
> 6 GeV	μ+Jets	μ+Jet	μ+Jet
> 11 GeV	Incl. Lifet.		

The following tagging techniques are applied:

- (μ+Jets): The measurements [6,7,8] use events with a muon from semileptonic b decay associated to a jet. The separation of beauty events from charm and light quark background is based on the large b quark mass, leading to large transverse momenta p_t^{rel} of the muon with respect to the axis of the associated jet. In [8] the long lifetime of the b quark is additionally exploited, leading to relatively large displacements of the muon track from the primary vertex.
- (Incl. Lifet.) The inclusive lifetime measurements [9,10,11] are based on the displacements of charged tracks from beauty decays from the primary vertex. Events are selected with at least one charged track measured with high quality in the vertex detector.

- ($\mu\mu$, $D^*\mu$) The double tag measurements use events with two muons [12] or with one muon and a fully reconstructed D^{*+} [13,14]. The separation of the beauty events from the background exploits the charge correlations ($\mu\mu$) or charge and angular correlations ($D^*\mu$) between the two particles.

All presented measurements are based on HERA I data. The achieved experimental accuracies are $\geq 20\%$ for inclusive measurements and for differential bins sometimes only $\sim 50\%$.

4. Results

The results section is ordered as follows: In 4.1 a selection of differential results of the μ+Jets analyses is presented. In 4.2 summary plots of all recent HERA measurements are shown, as a function of photon virtuality Q^2 and separately for photoproduction as a function of the b quark transverse momentum p_T. In 4.3 the results are concluded with the first measurements of the structure function $F_2^{b\bar{b}}$.

4.1. *μ+Jets analyses:*

Photoproduction: For the H1 and ZEUS photoproduction analyses [8,6] ($Q^2 < 1$ GeV^2) at least two jets are required in the final state with transverse momenta $p_t^{jet_{1(2)}} > 7(6)$ GeV. Figure 2 shows the differential cross sections as a function of the muon pseudorapidity (left) and transverse momentum (right). The H1 and ZEUS measurements agree well in the overlapping region. The data are also compared to a massive scheme NLO calculation, based on the program [1]. The estimated errors of the theory prediction are dominated by the uncertainties of the renormalisation and factorisation scales and of the b quark mass. The data tend to lie slightly above this calculation, however, within the errors the calculation describes all data points. The measured cross sections as a function of the muon transverse momentum are compared in figure 2 (right) to the NLO predictions in the respective kinematic ranges of the H1 and ZEUS measurements. In the lowest bin from 2.5 to 3.3 GeV the H1 measurement exceeds the prediction by a factor of ~ 2.5, while at higher transverse momenta a better agreement is observed. Such an excess is not seen in the ZEUS data. This discrepancy needs to be clarified in the future.

Deep inelastic scattering: For the H1 and ZEUS DIS analyses [8,7] ($Q^2 >$ 2 GeV^2) the jet algorithm is applied in the Breit frame and at least one jet

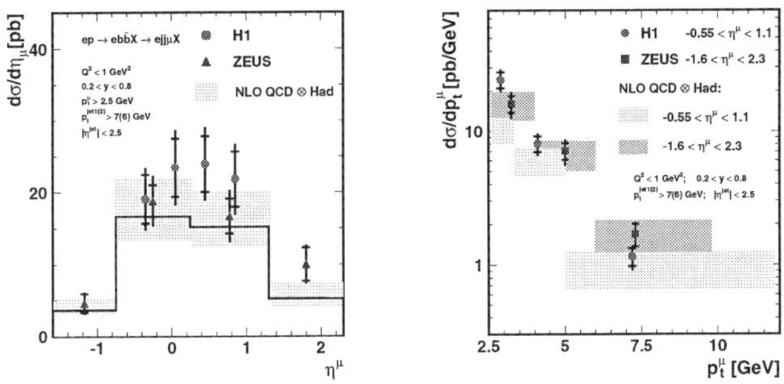

Figure 2. Differential beauty cross sections in photoproduction, for dijet events with a muon associated to one of the jets, as a function of (left) muon pseudorapidity and (right) muon transverse momentum. The H1 and ZEUS data are compared to predictions from a massive scheme NLO calculation.

with transverse momentum $p_{t,jet}^{Breit} > 6$ GeV is required. Figure 3 shows the differential cross sections of the H1 (top) and ZEUS (bottom) measurements as a function of (left) jet transverse momentum in the Breit frame, (middle) muon transverse momentum and (right) muon pseudorapidity. The data are compared to a massive scheme NLO calculation using the program [2]. The H1 and ZEUS measurements are performed in similar kinematic regions and also most of the observations are similar:

(1) An excess of data over NLO prediction by a factor ~ 2 is observed towards smaller muon transverse momenta below 4 GeV.
(2) A rise of the differential cross sections is observed towards more positive muon pseudorapidities, (i.e. more close to the proton direction) which is not reproduced by the NLO calculation.

The excess seen in the H1 data for lower muon transverse momenta is accompanied by an excess for lower jet transverse momenta while for ZEUS an excess is observed for higher jet momenta. More precise measurements are needed to clarify these different findings.

4.2. *Summary of results as a function of Q^2 and p_T*

Figure 4 shows a summary of the data/theory comparison for HERA beauty results as a function of Q^2. For the measurements sensitive to b quarks with $p_T^b \sim m_b$ or lower (black points) there is a trend that the massive NLO QCD predictions [1,2] tend to underestimate the b production rate at very

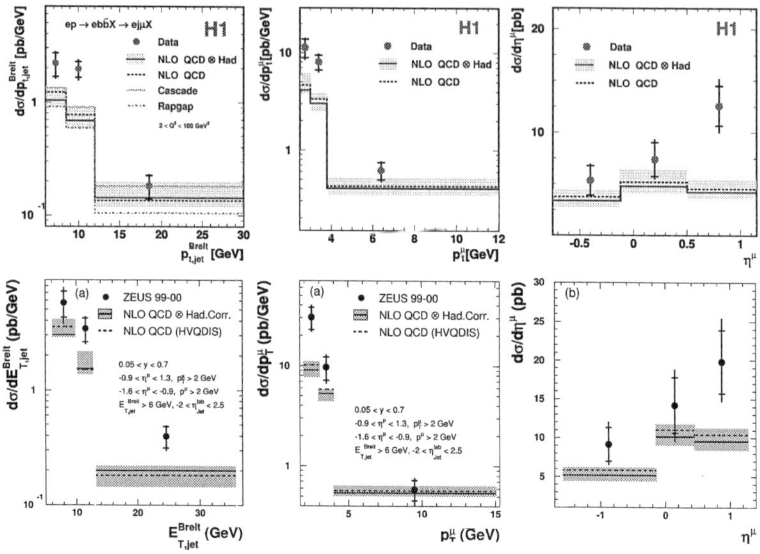

Figure 3. H1 (top) and ZEUS (bottom) measured differential beauty cross sections in DIS, for events with a muon and an associated jet, as a function of (left) jet transverse momentum in the Breit frame, (middle) muon transverse momentum and (right) muon pseudorapidity. The data are compared to predictions from a massive scheme NLO calculation.

low Q^2. For the higher p_T measurements (red/grey points), no clear trend is observed. Note that the estimated theoretical errors, which are typically of order 30%, are not shown. For the DIS data ($Q^2 > 2$ GeV^2) the curves from two mixed scheme NLO calculations are also shown in figure 4. These predictions from CTEQ [4] and MRST [3] are specifically provided for the $F_2^{b\bar{b}}$ inclusive structure function measurements which are shown here as the triangle points and which are discussed in 4.3. While the CTEQ curve is close to the the massive scheme calculation the MRST prediction is up to a factor of two higher. The data are not yet precise enough to separate between the three different predictions.

Figure 5 shows a similar compilation for all HERA measurements in photoproduction ($Q^2 < 1$ GeV^2), now as a function of the b quark p_T. Some measurements agree well with the massive scheme predictions but in general the data tend to be higher than the calculations. There are indications that the data exceed the predictions more significantly towards low p_T. Note, that several measurements appear in both summary figures.

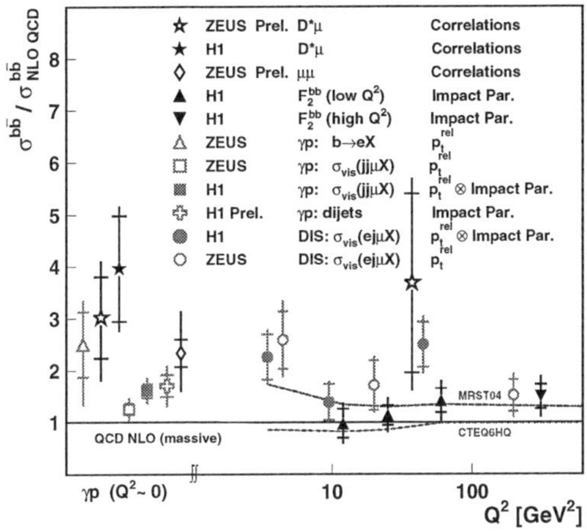

Key	Ref.	Signature	p_T cuts
Photoproduction			
◊	12	$\mu\mu$	low
☆	14	$D^*\mu$	low
★	13	$D^*\mu$	low
△	15	2 jets+e	medium
□	6	2 jets+μ	medium
■	8	2 jets+μ	medium
✙	11	tracks	high
DIS			
☆	14	$D^*\mu$	low
▲	10	tracks	low
▼	9	tracks	low
●	8	1 jet+μ	medium
○	7	1 jet+μ	medium

Figure 4. Ratio of beauty production cross section measurements at HERA to NLO QCD predictions in the massive scheme as a function of the photon virtuality Q^2. The predictions from the mixed scheme NLO calculations by MRST and CTEQ for the DIS kinematic regime $Q^2 > 2$ GeV^2 are also presented (valid for comparison with the measurements shown as triangles). Since theoretical errors are different for each point, they are not included in this plot.

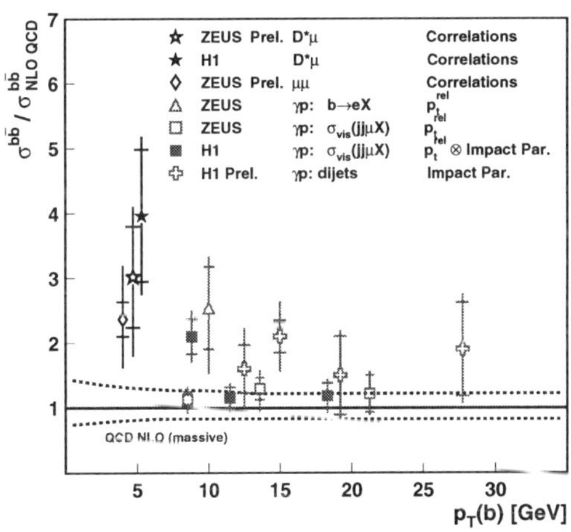

Figure 5. Ratio of beauty production cross section measurements in photoproduction at HERA to NLO QCD predictions in the massive scheme as a function of the transverse momentum of the b quark p_T. The dashed line gives an indication of the size of the estimated theoretical uncertainties.

4.3. *First measurement of structure function $F_2^{b\bar{b}}$*

Recently the first measurements [9,10] of the beauty contributions to the inclusive deep inelastic *ep* scattering have been made. These analyses are based on inclusive lifetime tagging. Figure 6 shows in the left plot the results obtained for the structure function $F_2^{b\bar{b}}$ as a function of Q^2 for various values of *Bjorken* x. The data exhibit positive scaling violations, i.e. rises of $F_2^{b\bar{b}}$ with Q^2 for fixed x. The data are compared to mixed scheme NLO calculations from CTEQ [4] and MRST [3]. The difference between the two calculations, which reaches a factor two at the lowest Q^2 and x, arises mainly from the different treatments of threshold effects by MRST and CTEQ. However, within the current experimental errors, these differences cannot yet be resolved and both calculations describe the data well. The data are also compared to the mixed scheme NNLO predictions from MRST [5], which is somewhat lower than the NLO prediction from the same group, but also agrees with the data. In the right plot of figure 6 the fractional contribution of beauty events to deep inelastic *ep* scattering is shown as expressed by the ratio $f^{b\bar{b}} = F_2^{b\bar{b}} / F_2$. The beauty contributions rises strongly from a few permille at small $Q^2 = 12$ GeV^2 $< m_b^2$ to about 3% at the largest $Q^2 = 500$ GeV$^2 \gg m_b^2$. This reflects the kinematic threshold behaviour, i.e. at small Q^2 and x the invariant mass of the gluon-photon system barely exceeds the minimal required mass of $2m_b$. For comparison the corresponding fractional charm contribution is also shown, which is in the covered phase space rather flat with values of about 20-30%.

5. Outlook

The ongoing data taking at HERA II will collect until summer 2007 at least five times more statistics than collected at HERA I. This together with the upgraded H1 and ZEUS detectors will improve beauty measurements, reaching a precision of about 10% for the total and 20% for the differential cross sections. This will allow to clarify whether the data indeed exceed the perturbative QCD calculations where currently indicated. Improved measurements of the structure function $F_2^{b\bar{b}}$ will allow to distinguish between the available calculations in massive and mixed pQCD schemes whose predictions differ up to a factor two.

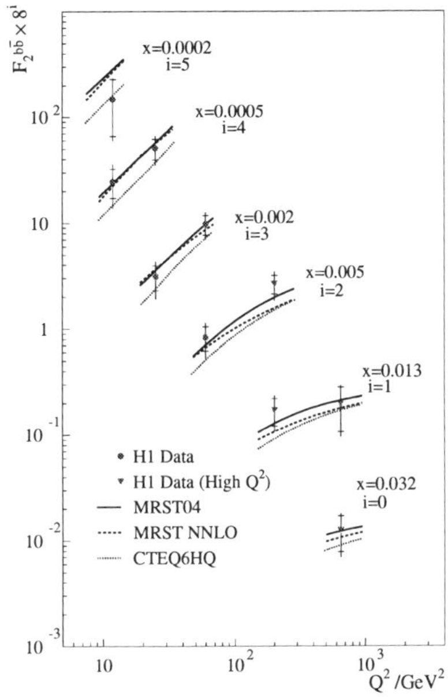

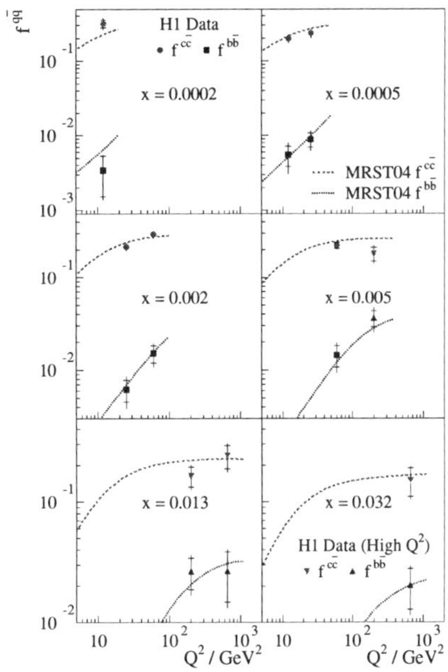

Figure 6. The first results for the beauty contribution $F_2^{b\bar{b}}$ to the inclusive structure function F_2 . The left plot shows $F_2^{b\bar{b}}$ as a function of Q^2 for various x. The right plot shows the observed relative beauty contributions to the total cross section $f^{b\bar{b}} = F_2^{b\bar{b}} / F_2$. The results for the relative charm contribution are also shown. The data are compared to different perturbative QCD calculations.

References

1. S. Frixione, P. Nason and G. Ridolfi, Nucl. Phys. B **454** (1995) 3 [hep-ph/9506226].
2. B. W. Harris and J. Smith, Nucl. Phys. B **452** (1995) 109 [hep-ph/9503484].
3. A. D. Martin, R. G. Roberts, W. J. Stirling and R. S. Thorne, Eur. Phys. J. C **39** (2005) 155 [hep-ph/0411040].
4. S. Kretzer, H. L. Lai, F. I. Olness and W. K. Tung, Phys. Rev. D **69** (2004) 114005 [hep-ph/0307022].
5. R. S. Thorne, AIP Conf. Proc. **792** (2005) 847 [hep-ph/0506251].
6. S. Chekanov *et al.* [ZEUS Collaboration], Phys. Rev. D **70** (2004) 012008 [hep-ex/0312057].
7. S. Chekanov *et al.* [ZEUS Collaboration], Phys. Lett. B **599** (2004) 173 [hep-ex/0405069].
8. A. Aktas *et al.* [H1 Collaboration], Eur. Phys. J. C **41** (2005) 453 [hep-ex/0502010].
9. A. Aktas *et al.* [H1 Collaboration], Eur. Phys. J. C **40** (2005) 349 [hep-

ex/0411046].
10. A. Aktas *et al.* [H1 Collaboration], Eur. Phys. J. C **45** (2006) 23 [hep-ex/0507081].
11. H1 Collaboration, contributed paper 405, XXII International Symposium on Lepton-Photon Interactions at High Energy, Uppsala, Sweden, 2005.
12. ZEUS Collaboration, contributed paper 269, XXII International Symposium on Lepton-Photon Interactions at High Energy, Uppsala, Sweden, 2005.
13. A. Aktas *et al.* [H1 Collaboration], Phys. Lett. B **621** (2005) 56 [hep-ex/0503038].
14. S. Chekanov *et al.* [ZEUS Collaboration], contributed paper 575, International Europhysics Conference on High Energy Physics (EPS 2003), Aachen, Germany, 2003.
15. J. Breitweg *et al.* [ZEUS Collaboration], Eur. Phys. J. C **18** (2001) 625 [hep-ex/0011081].

J/Ψ PHOTOPRODUCTION AT NEXT-TO-LEADING ORDER

L. N. MIHAILA
Institut für Theoretische Teilchenphysik, Universität Karlsruhe, 76131 Karlsruhe, Germany

In this paper, we report on the calculation of the cross section of J/Ψ inclusive production in direct $\gamma\gamma$ collisions at next-to-leading order within the factorization formalism of nonrelativistic quantum chromodynamics (NRQCD). Theoretical predictions for the future e^+e^- international linear collider ILC are also presented.

1. Introduction

Heavy quarkonium played an important role in establishing the asymptotic freedom of QCD, as its mass is much larger than the low-energy scale of the strong interactions Λ_{QCD}. On the other hand, the compensation of colour in heavy quarkonium gives new insights into the confinement mechanism, a nonperturbative process which is still not fully understood.
The factorization approach based on the nonrelativistic QCD (NRQCD) [1,2] provides a rigorous and systematic framework for calculating perturbative and relativistic corrections to heavy quarkonium production and annihilation rates. This formalism implies a separation of short-distance coefficients, which can be expressed as perturbative expansions in the strong coupling α_s, from the long-distance matrix elements (MEs), which must be extracted from the experiment. The MEs are predicted to scale with a definite power of the heavy quark relative velocity v. In this way, the theoretical calculations are organized as double expansions in α_s and v.
Within the NRQCD formalism, the colour-octet (CO) processes contribute to the production and decay rates at some level. Their quantitative significance was assessed for inclusive charmonium hadroproduction observed at the Fermilab Tevatron [3]. However, it is necessary to test the universality of the MEs in other kinds of production processes, such as $\gamma\gamma$ collisions. This process was studied at LEP2, where the photons originated from hard initial-state bremsstrahlung. Hopefully, the future e^+e^- linear collider will

be built, for which an additional source of hard photons would be provided by beamstrahlung.
However, the leading-order theoretical predictions to be compared with the experimental data suffer from considerable uncertainties, mostly from the dependences on the renormalization and factorization scales and from the lack of information on the nonperturbative MEs. In this paper, we report on the calculation of the next-to-leading order (NLO) corrections to the inclusive production of J/Ψ mesons in $\gamma\gamma$ collisions [4]. This is a first step in a comprehensive programme concerning the study of charmonium production at NLO within the NRQCD framework.

2. NLO corrections to J/Ψ production in direct photon-photon collisions

In the following, we take into consideration the process $\gamma\gamma \to J/\Psi + j + X$, where X denotes the hadronic remnant possibly including a second jet. We restrict the analysis to j and X free of charm quarks and to finite values for the J/Ψ transverse momentum p_T. Two-photon processes can be modelled by assuming that each photon either interacts directly (direct photoproduction) or fluctuates into hadronic components (resolved photoproduction). Thus, the above mentioned process receives contributions from the direct, single-resolved, and double-resolved channels. Furthermore, we concentrate on the NLO calculation of the inclusive cross section for direct photoproduction.
At LO in α_s, there is only one partonic subprocess that scales like v^2, namely

$$\gamma\gamma \to c\bar{c}[^3S_1^{(8)}] + g. \tag{1}$$

The representative Feynman diagrams together with the analytic expression of the LO cross section for the process (1) can be found in Ref. 5.
In the calculation of the production rates beyond LO in α_s, ultraviolet (UV), infrared (IR), and Coulomb divergences arise and need to be regularized. One of the most convenient methods to handle the UV and IR singularities which are present in the short-distance coefficients as well as in the MEs is dimensional regularization. We introduce a 't Hooft mass μ and a factorization mass M as unphysical scales, and formally distinguish between UV and IR poles. The Coulomb singularities are regularized by assigning a small relative velocity between the c and $\bar{c}$ quarks.
For the extraction of the short-distance coefficients we apply the projector formalism of Refs. 6, 7.

In the following, we study virtual and real radiative corrections separately, as their calculation requires different approaches.

2.1. *Virtual Corrections*

Feynman diagrams that generate the virtual corrections to the process (1) can be collected into two classes. The first one is obtained by attaching one virtual gluon in all possible ways to the tree-level diagrams. They include self-energy, triangle, box, and pentagon diagrams. Loop insertions in external gluon or c-quark lines are taken into consideration in the respective wave-function renormalization constants. The self-energy and triangle diagrams are in general UV divergent; the triangle, box and pentagon diagrams are in general IR divergent. The pentagon diagrams comprising only abelian gluon vertices also contain Coulomb singularities which cancel out similar poles in the radiative corrections to the operator $\left\langle O^H\left[{}^3S_1^{(8)}\right]\right\rangle$. For the analytical treatment of the abelian five-point functions, we refer to Ref. 8.

The diagrams of the second class comprise light-quark loops. The individual contributions arising from the triangle diagrams are equal to zero according to Furry's theorem [9]. The box diagrams contain UV and IR singularities, but their sum is finite.

The UV divergences comprised in self-energy and triangle diagrams are cancelled upon renormalization of the QCD gauge coupling $g_s = \sqrt{4\pi\alpha_s}$, the charm-quark mass m and field Ψ, and the gluon field A. We adopt the on-mass-shell (OS) scheme to renormalize m, Ψ, and A, while for g_s we employ the modified minimal-subtraction ($\overline{\text{MS}}$) scheme.

2.2. *Operator Renormalization*

For a consistent NLO analysis, one should also consider higher-order corrections in α_s to the four-quark operator $\left\langle O^H\left[{}^3S_1^{(8)}\right]\right\rangle$ within NRQCD. This bare d-dimensional operator has mass dimension $d-1$ while its renormalized version, which is extracted from the experimental data, has mass dimension 3. We thus introduce the 't Hooft mass scale of NRQCD, λ, to compensate for the difference between bare and renormalized operators. The corresponding tree-level and one-loop diagrams are depicted in Fig. 5 of Ref. 7. Using dimensional regularization and the NRQCD Feynman rules in the quarkonium rest frame and performing the integrations over the loop momenta after a Taylor expansion of the integrands in $1/m$, we obtain the unrenormalized one-loop result

$$\left\langle O^H\left[{}^3S_1^{(8)}\right]\right\rangle_1 = \left\langle O^H\left[{}^3S_1^{(8)}\right]\right\rangle_0 \times \left[1+\left(C_F-\frac{C_A}{2}\right)\frac{\pi\alpha_s}{2v}\right] + \frac{4\alpha_s}{3\pi m^2}\left(\frac{4\pi\mu^2}{\lambda^2}\right)^\epsilon \exp(-\epsilon\gamma_E)\left(\frac{1}{\epsilon_{\rm UV}}-\frac{1}{\epsilon_{\rm IR}}\right) \times \left(\sum_{J=0}^{2} C_F\left\langle O^H\left[{}^3P_J^{(1)}\right]\right\rangle + \sum_{J=0}^{2} B_F\left\langle O^H\left[{}^3P_J^{(8)}\right]\right\rangle\right), \quad (2)$$

where the subscript 0 labels the tree-level quantity and μ is the 't Hooft mass scale of QCD.

The presence of UV divergences indicates that the $\left\langle O^H\left[{}^3S_1^{(8)}\right]\right\rangle$ operator needs renormalization. For this, we choose the $\overline{\rm MS}$ scheme so that the $1/\epsilon_{UV}$ pole in Eq. (2) is cancelled out by the $1/\epsilon_{UV} + ln(4\pi) - \gamma_E$ pole in the coefficient of the counterterm ME. The term proportional with $1/v$ represents the Coulomb singularity, which compensates a similar term in the virtual corrections. After the renormalization of the NRQCD ME and cancellation of the Coulomb singularity, an IR counterterm at $O(\alpha_s)$ is generated from Eq. (2), which is indispensable to render the overall NLO result finite. This feature will be discussed in some detail in the next section.

2.3. *Real Corrections*

The real corrections to the process (1) arise from the partonic subprocesses

$$\gamma(k_1) + \gamma(k_2) \to c\bar{c}[n](p) + g(k_3) + g(k_4), \quad (3)$$

where $n = {}^3P_J^{(1)}, {}^1S_0^{(1)}, {}^1S_0^{(8)}, {}^3S_1^{(8)}, {}^3P_J^{(8)}$, and

$$\gamma(k_1) + \gamma(k_2) \to c\bar{c}[n](p) + q(k_3) + \bar{q}(k_4), \quad (4)$$

where $n = {}^1S_0^{(8)}, {}^3S_1^{(8)}, {}^3P_J^{(8)}$. For the respective Feynman diagrams we refer to Fig. 2 and Fig. 3 of Ref. 5. Note that the colour-singlet states $n = {}^3S_1^{(1)}$ in process (3) and $n = {}^3P_J^{(1)}, {}^1S_0^{(1)}$ in process (4) are forbidden by the Furry's theorem [9] and colour conservation, respectively.

Integrating the squared MEs of the processes (3) and (4) over the three-particle phase space while keeping the value of p_T finite, we encounter IR singularities, which can be of the soft and/or collinear type. In order to systematically extract these singularities, it is useful to slice the phase space by introducing infinitesimal dimensionless cut-off parameters δ_i and δ_f, which are connected with the initial and final states, respectively [10].

In the case of process (3) we distinguish *soft*, *final-state collinear* and *hard* regions of the phase space. In case of process (3) we differentiate *initial-state collinear* and *hard* regions of the phase space. While the individual contributions depend on the cut-off parameters δ_i or δ_f, their sum must be independent of them. We used the numerical verification of the cut-off independence as a check for our calculation.

The contributions originating from process (3) with $n = {}^3S_1^{(8)}$ contain IR (soft and final-state collinear) singularities which cancel out singularities of the same type comprised in the virtual corrections, as required by the Kinoshita-Lee-Nauenberg theorem [11]. The subprocesses with $n = {}^3P_J^{(1)}, {}^3P_J^{(8)}$ generate contributions containing only soft divergences. These are cancelled out by the IR poles left over in the renormalized $\left\langle O^H \left[{}^3S_1^{(8)}\right]\right\rangle$ operator, which organise themselves as coefficients of the $\left\langle O^H \left[{}^3P_J^{(1,8)}\right]\right\rangle$ operators.

According to the mass factorization theorem [12], the form of the collinear singularities present in the squared amplitudes of processes (4), which are related with an incoming-photon leg, is universal and the pole can be absorbed into the bare PDF of the antiquark $\bar{q}$ inside the resolved photon. As a consequence, the real MEs acquire an explicit dependence on the mass factorization scale M. In turn, the resolved-photon contribution is evaluated with the renormalized photon PDF, which is also M dependent. In the sum of these two contributions, the M dependence cancels up to terms beyond NLO.

2.4. *Academic Study*

For the purpose of the following technical study, it is sufficient to consider a typical kinematic situation. We thus choose as our reference quantity the differential cross section $d^2\sigma/dp_T\,dy$ at $p_T = 5$ GeV and $y = 0$.

In Fig. 1(a), the NLO (solid line) and LO (dashed line) contributions due to the $c\bar{c}$ Fock state $n = {}^3S_1^{(8)}$ are shown as functions of the renormalization scale μ, while the NRQCD renormalization scale λ and the factorization scale M are kept fixed at their reference values $\lambda = m$ and $M = m_T$. We focus on this particular channel because it is the only one already open at LO, so that a compensation of the μ dependence can occur. Notice that μ is varied by more than one order of magnitude, from $m_T/4$ to $4m_T$. Passing from LO to NLO, the μ dependence is appreciably reduced, reflecting the partial compensation of the μ dependence. In fact, the related theoretical uncertainty amounts to $^{+70}_{-30}\%$ at LO and to $\pm18\%$ at NLO.

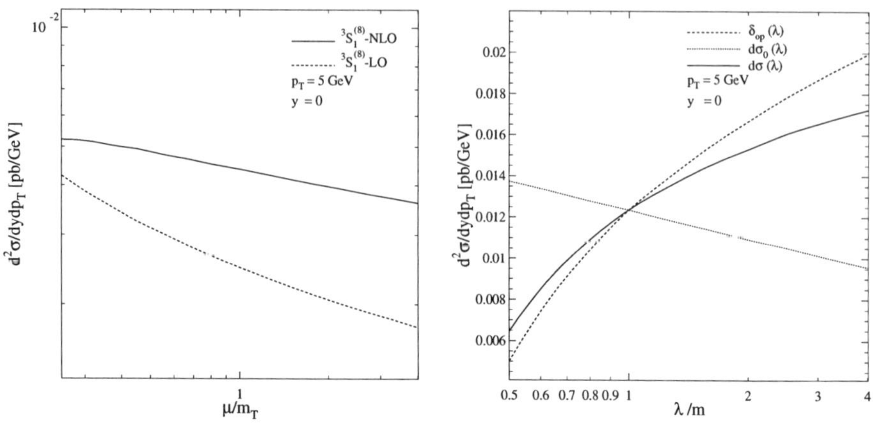

(a) (b)

Figure 1. (a) Differential cross section $d^2\sigma/dp_T\,dy$ in pb/GeV of $e^+e^- \to e^+e^- J/\psi + X$ in direct photoproduction at ILC with $\sqrt{s} = 500$ GeV for prompt J/ψ mesons with $p_T = 5$ GeV and $y = 0$. The NLO (solid line) and LO (dashed line) contributions due to the $c\bar{c}$ Fock state $n = {}^3S_1^{(8)}$ are shown as functions of μ. (b) The NLO result is shown as a function of λ, (i) keeping λ in the partonic cross sections fixed (dashed line), (ii) keeping λ in $\left\langle O^H\left[{}^3S_1^{(8)}\right]\right\rangle_r(\lambda)$ fixed (dotted line), and (iii) varying all occurrences of λ simultaneously (solid line).

In Fig. 1(b), the NLO result is shown as a function of λ (solid line), while μ and M are kept fixed at their reference values. Notice that λ is varied by almost one order of magnitude, from $m/2$ to $4m$. The related theoretical uncertainty amounts to $\pm 50\%$. The λ-dependent terms are $d\sigma_0$, the NLO cross section, and $\delta_{\rm op}$, the counterterm for the NRQCD operator$\langle O^3S_1^{(8)}\rangle$ renormalization. In order to exhibit the partial compensation in λ dependence between these two terms, we also include in Fig. 1(b) the results that are obtained by only varying λ in $d\sigma_0$ (dashed line) or $\delta_{\rm op}$ (dotted line) at a time. The dotted line is straight, reflecting the fact that $\delta_{\rm op}$ depends on λ through a single logarithm.

3. Phenomenological Study

In the following, we study the phenomenological significance of the process $\gamma\gamma \to J/\Psi + j + X$ at ILC in its e^+e^- mode, with $\sqrt{s} = 500$ GeV and the photon sources bremsstrahlung and beamstrahlung coherently superposed. For the energy spectrum of the photons we refer to Eq. (27) of Ref. 13

and Eq. (2.14) of Ref. 14, and for the designed experimental parameters to Ref. 15 and Ref. 16, respectively.

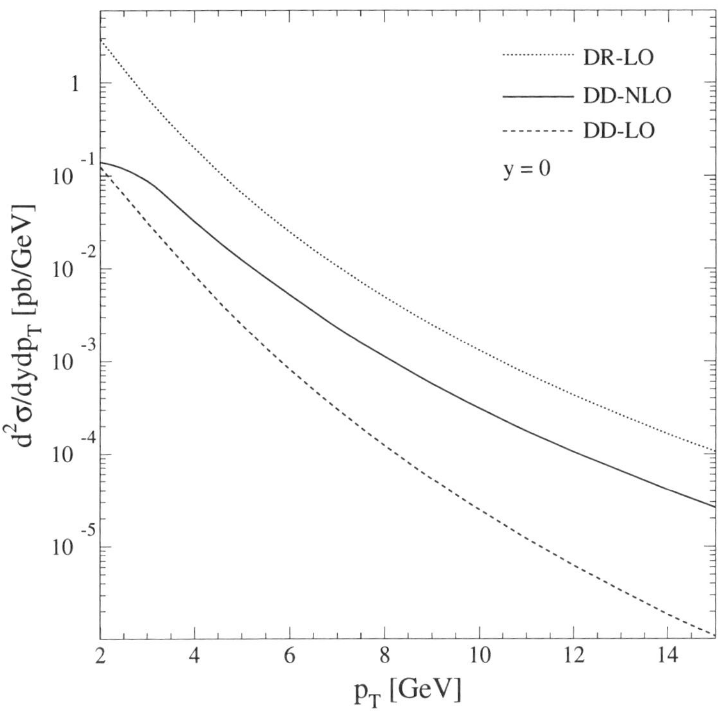

Figure 2. Differential cross section $d^2\sigma/dp_T\,dy$ of $e^+e^- \to e^+e^- J/\psi + X$ at ILC with $\sqrt{s} = 500$ GeV for $y = 0$ as a function of p_T.

We use $m = 1.5$ GeV, $\alpha = 1/137.036$, and the two-loop formula for $\alpha_s^{(n_f)}(\mu)$ [17] with $n_f = 3$ active quark flavours and $\Lambda_{\mathrm{QCD}}^{(3)} = 299$ MeV [18]. As for photon PDFs, we employ the NLO set from Glück, Reya, and Schienbein (GRS) [18], which are implemented in the fixed-flavour-number scheme, with $n_f = 3$. Our default choice of renormalization and factorization scales is $\mu = M = m_T$ and $\lambda = m$. As for J/ψ, χ_{cJ}, and ψ' MEs, we adopt the LO sets determined in Ref. 19 using the LO set of proton PDFs from Martin, Roberts, Stirling, and Thorne (MRST98LO) [20].

In Fig. 2 and Fig. 3 we study $d^2\sigma/dp_T\,dy$ for $y = 0$ as a function of p_T

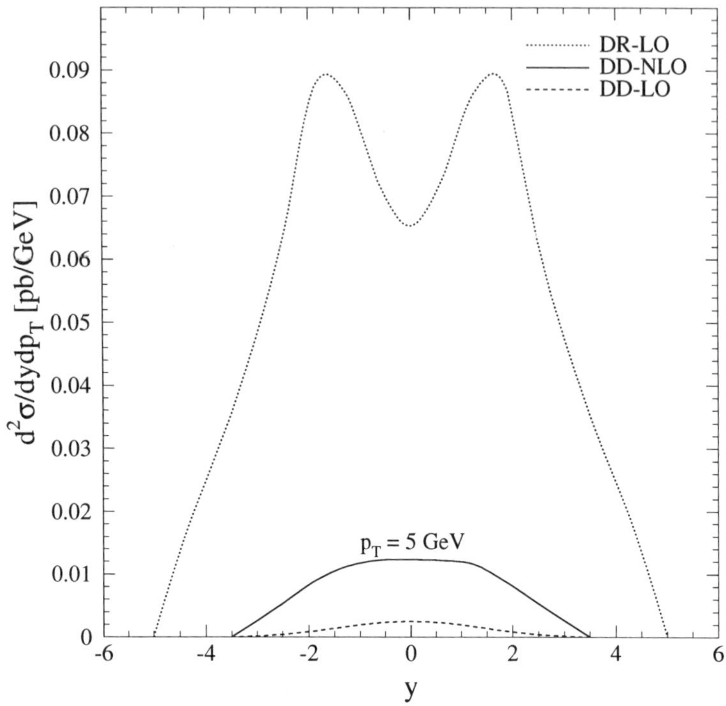

Figure 3. Differential cross section $d^2\sigma/dp_T\,dy$ of $e^+e^- \to e^+e^- J/\psi + X$ at ILC with $\sqrt{s} = 500$ GeV for $p_T = 5$ GeV as a function of y.

and for $p_T = 5$ GeV as a function of y, respectively. The solid and dashed lines represent NLO and LO results, respectively, for the direct photoproduction. The dotted lines correspond to the LO result of the single-resolved photoproduction evaluated with the LO versions of the $\alpha_s^{(n_f)}(\mu)$ and the photon PDFs. Notice that we do not consider p_T values smaller than 2 GeV, where additional IR singularities occur.

From Fig. 2, we observe that the NLO result of direct photoproduction falls off with increasing p_T more slowly than the LO one. This feature may be accounted for by the *fragmentation-prone* partonic subprocesses that contribute at NLO, while they are absent at LO. Such subprocesses contain a gluon with a small virtuality that splits into a $c\bar{c}$ pair in the Fock state $n = {}^3S_1^{(8)}$. They generally provide dominant contributions at $p_T \gg 2m$ due

to the presence of a large gluon propagator. In single-resolved photoproduction, a fragmentation-prone partonic subprocess already contributes at LO. This explains why the solid and dotted curves in Fig. 2 run parallel in the upper p_T range. In the lower p_T range, the fragmentation-prone partonic subprocesses do not matter, as the gluon propagator becomes finite and the Fock state $n = {}^3S_1^{(8)}$ is already present at LO.
As one can notice from Fig. 3, at $p_T = 5$ GeV, single-resolved photoproduction is still overwhelming. The two pronounced maxima of the LO single-resolved result may be explained by the occurrence of a virtual gluon in a t channel that can be almost collinear with the incoming quark q or by the one of a virtual quark in the u channel that can become almost collinear with the incoming photon. The NLO result increases towards forward and backward directions due to the finite remainders of the initial-state collinear singularities that were absorbed into the photon PDFs.

4. Conclusions

The experimental verification of the NRQCD factorization hypothesis is of great importance especially for charmonium, because the charm quark mass might not be large enough to justify the nonrelativistic approximation.
In this paper, we studied at NLO the inclusive production of prompt J/Ψ mesons with finite values of p_T. This is the first time that an inclusive $2 \to 2$ production process was treated at NLO within the NRQCD factorization formalism. A complete NLO calculation of prompt J/Ψ production in $\gamma\gamma$ collisions allows a NLO treatment of photo- and hadroproduction as well. This will provide a solid basis for the attempt to accommodate the NRQCD theoretical predictions with the experimental data to be taken at high-energy colliders, such as the Tevatron (RunII) and HERA (HERA-II).

Acknowledgments

The author thanks M. Klasen, B.A. Kniehl, and M. Steinhauser for a fruitful collaboration and B. Kniehl for the kind invitation to this workshop.

References

1. W.E. Caswell, G.P. Lepage, *Phys. Lett.* **B167** 437 (1986).
2. G.T. Bodwin, E. Braaten, G.P. Lepage, *Phys. Rev.* **D51** 1125 (1995); *ibid.* **55** 5853 (1997), Erratum.
3. CDF Collaboration, F. Abe et al., *Phys. Rev. Lett.* **69** 3704 (1992); *ibid.* **71** 2537 (1993); *ibid.* **79** 572 (1997); *ibid.* **79** 578 (1997) ;

D0 Collaboration, S. Abachi et al., *Phys. Lett.* **B370** 239 (1996);
D0 Collaboration, B. Abbott et al., *Phys. Rev. Lett.* **82** 35 (1999).

4. M. Klasen, B.A. Kniehl, L.N. Mihaila, M. Steinhauser, *Nucl. Phys.* **B713** 487 (2005).
5. M. Klasen, B.A. Kniehl, L.N. Mihaila, M. Steinhauser, *Nucl. Phys.* **B609** 518 (2001) 518.
6. J. Kaplan, J.H. Kühn, *Phys. Lett.* **B78** 252 (1978).
7. A. Petrelli, M. Cacciari, M. Greco, F. Maltoni, M.L. Mangano, *Nucl. Phys.* **B514** 245 (1998).
8. W. Beenakker, S. Dittmaier, M. Krämer, B. Plümper, M. Spira, P.M. Zerwas, *Nucl. Phys.* **B653** 151 (2003).
9. W.H. Furry, *Phys. Rev.* **51** 125 (1937).
10. K. Fabricius, G. Kramer, G. Schierholz, I. Schmitt, *Z. Phys.* **C11** 315 (1981); B.F. Harris, J.F. Owens, *Phys. Rev.* **D65** 094032 (2002).
11. T. Kinoshita, *J. Math. Phys.* **3** 650 (1962); T.D. Lee, M. Nauenberg, *Phys. Rev.* **B133** 1549 (1964).
12. R.J. DeWitt, L.M. Jones, J.D. Sullivan, D.E. Willen, H.W. Wyld, Jr., *Phys. Rev.* **D19** 2046 (1979); *ibid.* **20** 1751 (1979), Erratum.
13. S. Frixione, M.L. Mangano, P. Nason, G. Ridolfi, *Phys. Lett.* **B319** 339 (1993).
14. P. Chen, T.L. Barklow, M.E. Peskin, *Phys. Rev.* **D49** 3209 (1994).
15. J. A. Aguilar-Saavedra et al., in: R.-D. Heuer, D. Miller, F. Richard, P. Zerwas (Eds.), TESLA Technical Design Report, Part III, p. III-166.
16. J. Andruszkow et al., in: R. Brinkmann, K. Flöttmann, J. Rossbach, P. Schmüser, N. Walker, H. Weise (Eds.), TESLA Technical Design Report, Part II, p. II-1.
17. Particle Data Group, K. Hagiwara et al., *Phys. Rev.* **D66** 010001 (2002).
18. M. Glück, E. Reya, I. Schienbein, *Phys. Rev.* **D60** 054019 (1999); *ibid.* **62** 019902 (1999), Erratum.
19. E. Braaten, B.A. Kniehl, J. Lee, *Phys. Rev.* **D62** 094005 (2000).
20. A.D. Martin, R.G. Roberts, W.J. Stirling, R.S. Thorne, *Eur. Phys. J.* **C4** 463 (1998).

J/ψ PHOTO-PRODUCTION AT LARGE Z IN SOFT COLLINEAR EFFECTIVE THEORY

SEAN FLEMING

Physics Department
University of Arizona,
Tucson, AZ 85721, USA
E-mail: fleming@physics.arizona.edu

ADAM K. LEIBOVICH

Department of Physics and Astronomy
University of Pittsburgh
Pittsburgh, PA 15260, USA
E-mail: akl2@pitt.edu

THOMAS MEHEN

Department of Physics
Duke University
Durham, NC 27708, USA
and
Jefferson Laboratory
12000 Jefferson Ave.
Newport News, VA 23606. USA
E-mail: mehen@phy.duke.edu

One of the outstanding problems in J/ψ physics is a systematic understanding of the differential photo-production cross section $d\sigma/dz(\gamma + p \to J/\psi + X)$, where $z = E_\psi/E_\gamma$ in the proton rest frame. The theoretical prediction based on the non relativistic QCD (NRQCD) factorization formalism has a color-octet contribution which grows rapidly in the endpoint region, $z \to 1$, spoiling perturbation theory. In addition there are subleading operators which are enhanced by powers of $1/(1-z)$ and they must be resummed to all orders. Here an update of a systematic analysis is presented. The approach used to organize the endpoint behavior of the photo-production cross section is based on a combination of NRQCD and soft collinear effective theory. While a final result is not yet available, an intermediate result indicates that better agreement between theory and data will be achieved in this framework.

1. Introduction

The production of J/ψ in high energy electron-proton interactions has proven to be a rich subject with great potential for furthering our understanding of the perturbative and non-perturbative regimes of QCD. In particular a number of different kinematic regions are accessible in *ep* collisions so that a range of different J/ψ production phenomena can be explored with a single experiment. The H1 and ZEUS collaborations have gathered a large amount of data on J/ψ production and study a diverse set of topics including processes where the scattered electron is in the forward region so that the photon exchanged with the proton is real.

The theoretical framework within which the production of non-relativistic bound states can be systematically treated is the NRQCD factorization formalism [1]. This approach relies on a non-relativistic effective theory of QCD [1,2,3], which is a systematic expansion of QCD about the limit of vanishing quark velocity: $v \to 0$. The resulting effective field theory consists of an infinite number of operators each scaling as a definite power of v, however, at a given order in v there are only a finite number of terms. As a consequence working to a specified numerical accuracy only requires including those contributions that affect the calculation at the level of significance desired.

Ref. 1 postulates that the J/ψ photo-production cross section in *ep* collisions is of the form

$$d\sigma(\gamma+p \to J/\psi+X) = f_{i/p} \otimes d\hat{\sigma}(i+\gamma \to c\bar{c}[n]+\hat{X}) \otimes F(c\bar{c}[n] \to J/\psi)\,, \quad (1)$$

where $d\hat{\sigma}(i + \gamma \to c\bar{c}[n] + \hat{X})$ is the short-distance scattering cross section for producing a $c\bar{c}$ pair with non-relativistic relative momentum in a color and angular momentum state indexed by n. The parton distribution function $f_{i/p}$ gives the probability for finding parton i in the proton, and $F(c\bar{c}[n] \to J/\psi)$ gives the probability for the $c\bar{c}$ pair with quantum numbers n to hadronize into a J/ψ. There is an implied sum over quantum numbers n, with each $F(c\bar{c}[n] \to J/\psi)$ scaling as $v^{\gamma(n)}$, with $\gamma(n)$ the scaling dimension of the operator. The short-distance coefficient $d\hat{\sigma}(i + \gamma \to c\bar{c}[n] + \hat{X})$ scales with powers of the strong coupling constant α_s, and can be enhanced by kinematic factors as well. As a result there is a competition between suppression by $v^{\gamma(n)}$ and enhancements of $d\hat{\sigma}$, and extra care must be taken when deciding which terms to keep in Eq. (1). In particular there are two effects that can greatly enhance the short-distance cross section: fragmentation contributions [4,5] and color-octet contributions [6].

In Refs. 7, 8, 9 the total J/ψ photo-production cross section $\sigma(\gamma +$

$p \to J/\psi + X)$ is calculated and compared to HERA data as well as other experimental data. The conclusion is that theory and data agree as long as the non-perturbative NRQCD production matrix elements are allowed to be negative. However, given the interpretation of the NRQCD matrix elements as the probability for a $c\bar{c}$ pair to fragment into a J/ψ and any number of soft particles there is a prejudice against negative values of the matrix elements. Furthermore there is uncertainty whether the NRQCD factorization formalism is valid for the total cross section since the J/ψ is not necessarily produced with large transverse momentum. Refs. 7, 9 also considered the differential cross section $d\sigma/dz(\gamma + p \to J/\psi(z) + X)$ where $z = p_p \cdot p_\psi / p_p \cdot p_\gamma$ for J/ψ transverse momentum greater than 1 GeV. The results of Ref. 7 compared to H1 data [10] are shown in Fig. 1. The theoretical

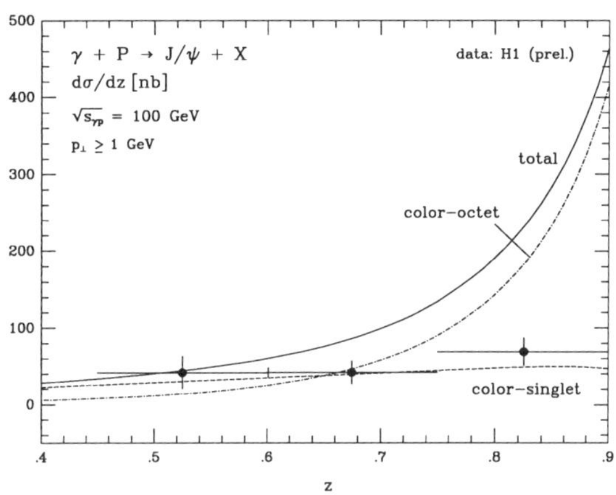

Figure 1. The J/ψ energy distribution $d\sigma/dz$ at the photon-proton center-of-mass energy $\sqrt{s} = 100$ GeV integrated in the range $p_\perp \geq 1$ GeV from Ref. 7.

curve is an increasing function of z while the experimental points indicate a flat distribution, and it is clear that the color-octet contribution dominates the differential cross section at larger values of z. The rise in the cross section is due to terms of the form

$$\left(\frac{\ln(1-z)}{1-z}\right)_+, \qquad \left(\frac{1}{1-z}\right)_+, \tag{2}$$

which arise at next-to-leading order in the color-octet differential cross section. These plus-distributions become arbitrarily large as $z \to 1$, and cause a breakdown of perturbation theory. As a consequence such terms must be resummed to make a sensible comparison of theory to data in the large z region.

In addition to the breakdown of perturbation theory, the non-perturbative NRQCD expansion breaks down when $z \to 1$, because of a set of subleading operators which scale as $v^{2n}/(1-z)^n$. This was first explored in Refs. 11, 12, where the infinite tower of enhanced operators was resummed into a shape function. In Ref. 13 an analysis of the differential cross section $d\sigma/dz$ including the shape function was carried out. The spectrum compared to H1 and ZEUS data [14] is shown in Fig. 2. The introduction of

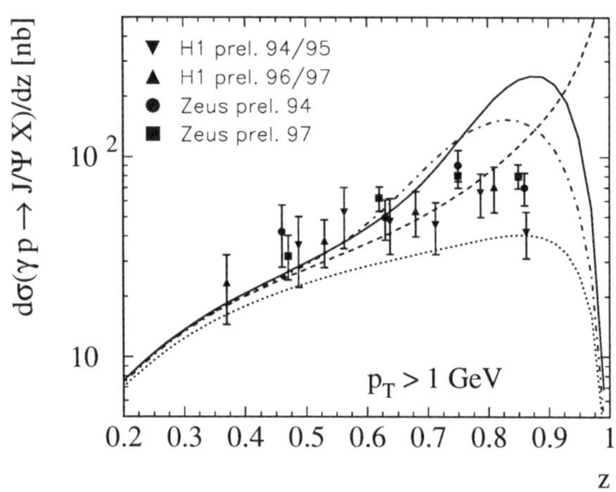

Figure 2. The J/ψ energy spectrum at $\sqrt{s} = 100$ GeV integrated in the range $p_\perp \geq 1$ GeV compared to HERA data. Solid, dashed-dotted, and dashed lines are the color-singlet contribution plus the color-octet shape function contribution for three different values of a shape function parameter. The dotted line is the color-singlet contribution alone. Plot from Ref. 13.

a shape function tames the endpoint divergences of the color-octet contribution at NLO, however, agreement with data is not great. In particular the shoulder in the theoretical prediction is too large and peaked too far to the right. This is due to the inconsistent treatment of the plus-distribution terms which also need to be resummed before a comparison of theory and data can be made. In Ref. 15 it was shown that in $e^+e^- \to J/\psi + X$ at the endpoint perturbative resummation in addition to the shape function significantly broadens the differential distribution relative to either including only the shape function or only carrying out a perturbative resummation without shape function.

2. The breakdown of the NRQCD expansion

It is helpful to study the kinematics of $\gamma + p \to J/\psi + X$ as $z \to 1$ to gain a better understanding of the physics that leads to the breakdown of NRQCD factorization in this kinematic regime. In the proton-photon center-of-mass frame

$$p_\gamma^\mu = \frac{1}{2}\sqrt{s}\bar{n}^\mu, \qquad p_p^\mu = \frac{1}{2}\sqrt{s}n^\mu, \qquad p_{c\bar{c}}^\mu = Mv^\mu + k^\mu\,, \tag{3}$$

where $n^\mu = (1,0,0,-1)$, $\bar{n}^\mu = (1,0,0,1)$, $M = 2m_c$, and v^μ and k^μ are the 4-velocity of the J/ψ and the residual momentum of the $c\bar{c}$ pair in the J/ψ respectively. In terms of the scaling variable z the J/ψ velocity is

$$p_\psi^\mu = M_\psi v^\mu = \frac{z\sqrt{s}}{2}\bar{n}^\mu + p_\perp^\mu + \frac{M_\perp^2}{2z\sqrt{s}}n^\mu \tag{4}$$

where $M_\perp^2 = M_\psi^2 - p_\perp^2$, with $p_\perp^2 = -\mathbf{p}_\perp^2$. By momentum conservation

$$\begin{aligned} p_X^\mu &= p_\gamma^\mu + p_p^\mu - p_{c\bar{c}}^\mu \\ &= \frac{1}{2}\frac{\sqrt{s}M}{M_\psi}\left(\frac{M_\psi}{M} - z\right)\bar{n}^\mu + \frac{1}{2}\sqrt{s}\left(1 - \frac{MM_\perp^2}{szM_\psi}\right)n^\mu - \frac{M}{M_\psi}p_\perp^\mu - k^\mu\,. \end{aligned} \tag{5}$$

Setting $k = 0$ the invariant mass squared of the final state is

$$p_X^2 = \frac{sM}{M_\psi}\left(\frac{M_\psi}{M} - z\right)\left(1 - \frac{MM_\perp^2}{szM_\psi}\right) + \frac{M^2}{M_\psi^2}p_\perp^2\,. \tag{6}$$

This is much larger than $\Lambda \sim 1\,,\mathrm{GeV}$ as long as $M_\psi/M - z \sim 1$. Note the requirement for $p_X^2 \geq 0$ translates into $\mathbf{p}_\perp^2 \leq (sz - MM_\psi)(M_\psi/M - z)$ so $p_\perp$ is forced to be small at the endpoint. Thus away from the endpoint the decay products can be integrated out and NRQCD factorization holds. However if $z \to M_\psi/M$ the invariant mass of the decay products becomes much smaller than the $\bar{n}\cdot p_X$ momentum component and the final state consists of a J/ψ and a jet in the direction of the incoming hadron. In this kinematic regime NRQCD factorization no longer holds, and an effective field theory which describes the light-like final state must be used. The appropriate theoretical framework is a combination of NRQCD to describe the J/ψ and soft collinear effective theory (SCET) [16,17,18,19] to describe the light-like decay products. A combination of SCET and NRQCD has already been used to study a variety of Quarkonium decay and production processes in kinematically restricted regions [20,21,22,15,23,24,25,26].

Before diving into a detailed description of SCET it is useful to determine the region in z where SCET is applicable. To do this return to Eq. (6) and include the residual momentum of the $c\bar{c}$ pair introduced in Eq. (3). The scaling of the residual momentum k^μ can be found by boosting from the J/ψ rest frame where $k^\mu \sim \Lambda$ to the proton-photon center of mass frame. In light-cone coordinates

$$k^\mu \sim \Lambda\left(\frac{\sqrt{s}}{M_\psi}, \frac{M_\psi}{\sqrt{s}}, 1\right), \tag{7}$$

where $z \sim 1$ has been used. Note the plus light-cone component is enhanced by $\sqrt{s}/M_\psi$, while the minus component is suppressed by the same amount. Including the residual momentum in Eq. (6) gives

$$p_X^2 = \frac{sM}{M_\psi}\left(\frac{M_\psi}{M} - z\right)\left(1 - \frac{MM_\perp^2}{szM_\psi}\right) + \frac{M^2}{M_\psi^2}p_\perp^2 - \sqrt{s}\left(1 - \frac{MM_\perp^2}{szM_\psi}\right)n\cdot k + \ldots \tag{8}$$

where suppressed pieces are dropped. The last term in the equation above scales as $s\Lambda/M_\psi$ so the final state becomes sensitive to nonperturbative physics when $z \gtrsim (M_\psi - \Lambda)/M$.

3. Soft Collinear Effective Theory

SCET describes the dynamics of highly energetic particles moving close to the light-cone interacting with a background field of soft quanta. The interaction of the collinear particle with the background introduces a small residual momentum component into the light-like collinear momentum so that collinear particles have momentum $p^\mu = Qn^\mu + k^\mu$, where $k^\mu \sim \Lambda$. However, collinear particles do not only interact with the soft background, they also couple to other collinear particles. As a consequence the SCET Lagrangian consists of two sectors: soft and collinear. The SCET Lagrangian was first derived in what is called the label formalism [16,17,18,19], and was subsequently formulated in position space [27,28]. To illustrate some important properties of SCET consider the quark sector of the Lagrangian. It can be split into two pieces: one coupling collinear to soft

$$\mathcal{L}_{cs} = \bar{\xi}_{n,p} in\cdot D \frac{\bar{n}\!\!\!/}{2}\xi_{n,p}\,, \tag{9}$$

where $\xi_{n,p}$ is the collinear quark field labelled by collinear momentum $p^\mu = n^\mu\, \bar{n}\cdot p/2 + p_\perp^\mu$, $\bar{n} = (1,0,0,1)$, and $in\cdot D = in\cdot\partial + gn\cdot A_s$, with A_s the soft gauge field. This expression looks very much like the HQET Lagrangian with the velocity v^μ replaced with the light-like vector n^μ, and was first derived in Ref. 29. The second piece of the collinear Lagrangian consists of interactions of only collinear particles among themselves

$$\mathcal{L}_c = \bar{\xi}_{n,p'}\left\{ gn \cdot A_{n,q} + i\not{D}_c^\perp \frac{1}{i\bar{n}\cdot D_c} i\not{D}_c^\perp \right\} \frac{\not{\bar{n}}}{2}\xi_{n,p}\,, \tag{10}$$

where $A_{n,q}$ is the collinear gauge field, and the collinear covariant derivative is $iD_c^\mu = \mathcal{P}^\mu + igA_{n,q}$. The operator $\mathcal{P}^\mu$ projects out collinear label momentum: $\mathcal{P}^\mu \xi_{n,p} = (n^\mu\, \bar{n}\cdot p/2 + p_\perp^\mu)\xi_{n,p}$ [17]. The SCET Lagrangian is invariant under separate collinear and soft gauge transformations which provide a powerful restriction on the operators allowed in the theory [19]. Furthermore the Lagrangian is invariant under a global $U(1)$ helicity spin symmetry, and must be invariant under certain types of reparameterizations of the collinear sector of the Lagrangian [30,31].

A remarkable consequence of the gauge symmetries of SCET is the factorization of soft and collinear effects. Towards this end the soft Wilson line is introduced

$$Y(x) = \mathrm{Pexp}\left(ig\int_{-\infty}^{x} ds\, n\cdot A_{us}(ns)\right)\,, \tag{11}$$

and the collinear fields are redefined as follows:

$$\xi_{n,p} = Y\xi_{n,p}^{(0)} \qquad A_{n,q}^\mu = YA_{n,q}^{(0)\mu}Y^\dagger\,. \tag{12}$$

After the field redefinitions the soft gluons decouple from the collinear fields, *i.e.* $\mathcal{L}_{cs} \to 0$ in Eq. (9), and the collinear Lagrangian becomes independent of soft physics. At higher orders in the SCET expansion this decoupling does not occur, and factorization is broken.

4. Factorization at the Endpoint

The soft-collinear factorization properties of SCET can be used to obtain a factored form for the J/ψ production cross section at the kinematic endpoint. An important condition for the derivation to be valid is that a sufficient range of $p_\perp$ is integrated over. To be precise we must smear over a range $p_\perp^2 \gtrsim \Lambda M$. If this is not done, or if only a range $p_\perp \sim \Lambda$ is integrated over the intrinsic transverse momentum of the partons in the proton must be taken into account [32,33,34]. This is an important point because an experimental cut of $p_\perp \geq 1\mathrm{GeV}$ is usually implemented in HERA data. This cut lies in the regime where non-perturbative momentum is important and may introduce a sensitivity to the transverse momentum of the partons in the proton.

There are two steps in factorizing the photo-production differential cross section at the endpoint. The first step is matching QCD onto SCET where the collinear particles have a typical off-shellness of order ΛM. This formulation of SCET is called $\mathrm{SCET_I}$. In the next step $\mathrm{SCET_I}$ is matched onto $\mathrm{SCET_{II}}$ where collinear fields have a typical off-shellness $\mathcal{O}(\Lambda^2)$ [35]. To be concrete one of the two leading color-octet contributions to J/ψ photo-production at the endpoint is considered: a $c\bar{c}$ pair produced in a color-octet ${}^1S_0$ configuration which hadronizes to J/ψ via a chromomagnetic spin-flip transition.

The derivation is easiest in the parton-photon CM frame where

$$p_\gamma^\mu = \frac{1}{2}\sqrt{\hat{s}}\bar{n}^\mu, p_g = x p_P^\mu = \frac{1}{2}\sqrt{\hat{s}} n^\mu, p_\psi^\mu = \frac{z\sqrt{\hat{s}}}{2}\bar{n}^\mu + p_\perp^\mu + \frac{M_\perp^2}{2z\sqrt{\hat{s}}} n^\mu , \quad (13)$$

with the J/ψ approximately at rest, $v^\mu = (1,0,0,0)$. First the QCD current is matched onto the $\mathrm{SCET_I}$ current

$$J^\mu(x) = \sum_{\boldsymbol{\omega}} e^{i(Mv-\boldsymbol{\omega})\cdot x} C_\alpha^\mu(\omega) J_{\boldsymbol{\omega}}^\alpha(x) ,$$
$$J_{\boldsymbol{\omega}}^\alpha(x) = \left[\psi_{\mathbf{p}}^\dagger \delta_{\vec{\mathcal{P}},\boldsymbol{\omega}}^{(3)} B_\perp^\alpha \chi_{-\mathbf{p}}\right](x) , \quad (14)$$

where $\boldsymbol{\omega} = \omega n^\mu/2 + \omega_\perp^\mu$, and $\vec{\mathcal{P}} = \mathcal{P} n^\mu/2 + \mathcal{P}_\perp$. At lowest order in $\alpha_s(M)$

$$C_{\mu_\alpha} = \frac{2ee_c g_s}{M}\epsilon_{\mu\alpha}^\perp . \quad (15)$$

The second step is to integrate out the scale ΛM by matching the differential cross section in $\mathrm{SCET_I}$ onto $\mathrm{SCET_{II}}$. To derive the cross section it is best to take a step back and consider the differential cross section in QCD:

$$2E_\psi \frac{d\sigma}{d^3p_\psi} = \frac{-g^{\mu\nu}}{16\pi^3 s}\sum_X \int d^4y\, e^{-ip_\gamma\cdot y}\langle p|J_\mu^\dagger(0)|J/\psi + X\rangle\langle J/\psi + X|J_\nu(y)|p\rangle \quad (16)$$

Inserting the expression in Eq. (14) gives the $\mathrm{SCET_I}$ cross section

$$2E_\psi \frac{d\sigma}{d^3p_\psi} = \frac{-g_{\mu\nu}}{16\pi^3 s}\sum_{\boldsymbol{\omega}_{1,2}} [C_\alpha^\mu(\boldsymbol{\omega}_1)]^\dagger C_\beta^\nu(\boldsymbol{\omega}_2)\int d^4y\, e^{-i(p_\gamma - Mv + \boldsymbol{\omega}_2)\cdot y}$$
$$\times\langle p|[\chi_{-\mathbf{p}}^\dagger \delta_{\vec{\mathcal{P}},\boldsymbol{\omega}}^{(3)} B_\perp^\alpha \psi_{\mathbf{p}}](0)|J/\psi + X\rangle\langle J/\psi + X|[\psi_{\mathbf{p}}^\dagger \delta_{\vec{\mathcal{P}},\boldsymbol{\omega}}^{(3)} B_\perp^\alpha \chi_{-\mathbf{p}}](y)|p\rangle , \quad (17)$$

and soft-collinear factorization properties of SCET outlined in the previous section can be used to obtain a factored form for this cross section:

$$\frac{d\sigma}{dz} = \frac{\alpha\alpha_s e_c^2}{\pi M z s}\langle \mathcal{O}_8^\psi({}^1S_0)\rangle \int dn\cdot k\, \hat{S}\left(\sqrt{\hat{s}}\left(1 - \frac{M}{M_\psi}z\right) + n\cdot k\right)\mathcal{J}_{\boldsymbol{\omega}_+}(nk) . \quad (18)$$

Here the momentum space functions $\mathcal{J}$ and $\hat{S}$ are Fourier transforms of collinear and soft matrix elements respectively:

$$\langle p_I|\mathrm{Tr}\big[B_\perp^{(0)}(0)\delta_{\vec{\mathcal{P}}_+,\boldsymbol{\omega}_+}B_\perp^{(0)}(y)\big]|p_I\rangle \\ = \delta(n\cdot y)\delta^{(2)}(y_\perp)\int\frac{dn\cdot k}{2\pi}e^{-\frac{i}{2}n\cdot k\bar{n}\cdot y}\mathcal{J}_{\boldsymbol{\omega}_+}(n\cdot k)\,. \quad (19)$$

and

$$S_\psi(\omega) \equiv 2M\langle\mathcal{O}_8^\psi(^1S_0)\rangle\hat{S}_\psi(\omega) \\ = \int\frac{d\bar{n}\cdot y}{4\pi}e^{-\frac{i}{2}\omega\bar{n}\cdot y}\langle 0|\chi^\dagger_{-\mathbf{p}}YT^AY^\dagger\psi_{\mathbf{p}}(0)\mathcal{P}_\psi\psi^\dagger_{\mathbf{p}'}YT^AY^\dagger\chi_{-\mathbf{p}'}(\bar{n}\cdot y)|0\rangle\,. \, (20)$$

By construction $\int d\omega\hat{S}(\omega) = 1$.

Next $\mathrm{SCET_I}$ is matched onto $\mathrm{SCET_{II}}$. This entails performing an operator product expansion (OPE) on $\mathcal{J}_{\boldsymbol{\omega}_+}(n\cdot k)$, where the large scale is set by the invariant mass of the collinear degrees of freedom in $\mathrm{SCET_I}$:

$$\mathcal{J}_{\boldsymbol{\omega}}(n\cdot k) \approx C_{II}\left(\frac{n\cdot k}{\omega_+}\right)\delta^{(2)}_{0,\omega_+^\perp} \\ \times\langle p_{II}|\Big[\mathrm{Tr}\big\{T^BB_\perp^{II}(0)\big\}\delta_{\bar{n}P_+,\omega_+}\mathrm{Tr}\big\{T^BB_\perp^{II}(0)\big\}\Big]|p_{II}\rangle\,, \quad (21)$$

where corrections are suppressed by $\mathcal{O}(\Lambda/M)$. The coefficient function is dimensionless, and therefore must be a function of the ratio of the invariant mass squared and the large momentum component squared. Furthermore, at leading order in $\alpha_s(\Lambda M)$ the perpendicular momentum of the J/ψ is $\mathcal{O}(\Lambda)$ so that the labels $\omega_+^\perp$ must be zero. The $\mathrm{SCET_{II}}$ operator above can be related to the familiar parton distribution function which gives the probability of finding a gluon in the proton [36]

$$\frac{1}{2}\sum_{\mathrm{spin}}\langle p_{II}|\Big[\mathrm{Tr}\big\{T^BB_\perp^{II}(0)\big\}\delta_{\bar{n}P_+,\omega_+}\mathrm{Tr}\big\{T^BB_\perp^{II}(0)\big\}\Big]|p_{II}\rangle \\ = -n\cdot p_p\int_0^1 dx\,\delta(\omega_+ - 2xn\cdot P_p)f_{g/p}(x)\,, \quad (22)$$

where $f_{g/p}(x)$ is the parton distribution function. Using this result the final factored form of the cross section is

$$\frac{d\sigma}{dz} = M\big|C^{\mu\alpha}(M)\big|^2\sum_{v_\perp}\delta^{(2)}_{0,v_\perp}\int d\xi\,S_\psi(\xi)\,C_{II}(\xi - 1 + z - \overline{\Lambda})\,f_{g/p}\left(\frac{M^2}{s}\right). \quad (23)$$

5. Summing Logarithms

Large logarithms are summed using the renormalization group equations (RGEs), which is a two step procedure in this case. First the effective theory currents in Eq. (14) are run from the scale $\mu_H = \sqrt{s}$ to the scale $\mu_{\rm I} = M\sqrt{1-z}$ using the SCET$_{\rm I}$ RGEs. Next the SCET$_{\rm II}$ operators in Eq. (23) are run to the scale $\mu \sim \Lambda$. Details of the running are left for a later publication [37]. Here preliminary results are presented in Fig. 3, which shows the differential cross section (in units of the total color-octet $^1S_0$ photo-production cross section). The solid line includes both perturbative resummation and non-perturbative resummation *i.e.* the shape function. The dashed line includes only perturbative resummation. This should be compared to Fig. 2 where only the shape function is included. In the case where either the shape function alone is used, or only perturbation theory is resummed the spectrum is too sharply peaked to be compatible with data. However, when both the shape function and perturbative resummation is included the spectrum softens considerably, which gives hope that theory will be compatible with data in a complete analysis.

Acknowledgments

A.L. was supported in part by the National Science Foundation under Grant No. PHY-0244599. T.M. was supported in part by the Department of Energy under grant numbers DE-FG02-96ER40945 and DE-AC05-84ER40150.

References

1. G. T. Bodwin, E. Braaten and G. P. Lepage, Phys. Rev. D **51**, 1125 (1995) [Erratum-ibid. D **55**, 5853 (1995)] [arXiv:hep-ph/9407339].
2. N. Brambilla, A. Pineda, J. Soto and A. Vairo, Nucl. Phys. B **566**, 275 (2000).
3. M. E. Luke, A. V. Manohar and I. Z. Rothstein, Phys. Rev. D **61**, 074025 (2000) [arXiv:hep-ph/9910209].
4. E. Braaten, K. m. Cheung and T. C. Yuan, Phys. Rev. D **48**, 4230 (1993) [arXiv:hep-ph/9302307].
5. E. Braaten and T. C. Yuan, Phys. Rev. Lett. **71**, 1673 (1993) [arXiv:hep-ph/9303205].
6. E. Braaten and S. Fleming, Phys. Rev. Lett. **74**, 3327 (1995) [arXiv:hep-ph/9411365].
7. M. Cacciari and M. Krämer, Phys. Rev. Lett. **76**, 4128 (1996) [arXiv:hep-ph/9601276].
8. J. Amundson, S. Fleming and I. Maksymyk, Phys. Rev. D **56**, 5844 (1997) [arXiv:hep-ph/9601298].

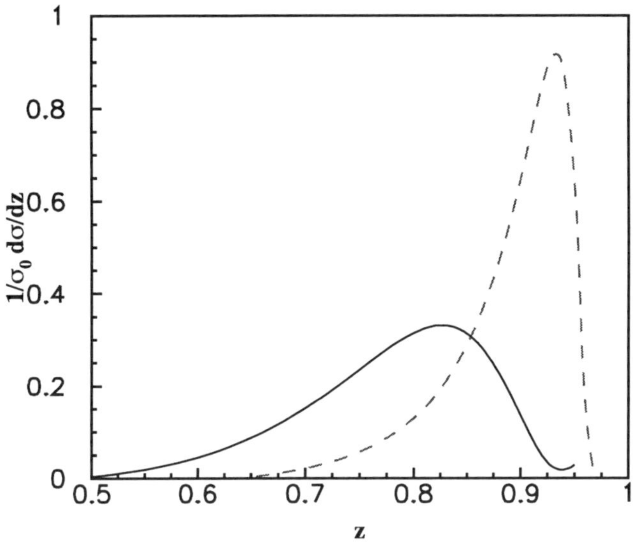

Figure 3. The differential cross section in units of the total color-octet $^1S_0$ cross section. The dashed curve only includes resummation of singular plus-distribution terms arising in perturbation theory. The solid line includes both non-perturbative resummation in the form of the shape function and perturbative resummation.

9. P. Ko, J. Lee and H. S. Song, Phys. Rev. D **54**, 4312 (1996) [Erratum-ibid. D **60**, 119902 (1999)] [arXiv:hep-ph/9602223].
10. Contributions to the Int. Europhys. Conf. on HEP, Brussels, 1995: H1 Coll., *Photoproduction of J/ψ Mesons at HERA*, EPS-0468.
11. I. Z. Rothstein and M. B. Wise, Phys. Lett. B **402**, 346 (1997) [arXiv:hep-ph/9701404].
12. M. Beneke, I. Z. Rothstein and M. B. Wise, Phys. Lett. B **408**, 373 (1997) [arXiv:hep-ph/9705286].
13. M. Beneke, G. A. Schuler and S. Wolf, Phys. Rev. D **62**, 034004 (2000) [arXiv:hep-ph/0001062].
14. S. Aid *et al.* (H1 Collaboration), Nucl. Phys. **B472**, 3 (1996);
    J. Breitweg *et al.* (ZEUS Collaboration), Z. Phys. **C76**, 599 (1997);

15. S. Fleming, A. K. Leibovich and T. Mehen, Phys. Rev. D **68**, 094011 (2003) [arXiv:hep-ph/0306139].
16. C. W. Bauer, S. Fleming and M. Luke, Phys. Rev. D **63**, 014006 (2001) [hep-ph/0005275].
17. C. W. Bauer, S. Fleming, D. Pirjol and I. W. Stewart, Phys. Rev. D **63**, 114020 (2001) [arXiv:hep-ph/0011336].
18. C. W. Bauer and I. W. Stewart, Phys. Lett. B **516**, 134 (2001) [arXiv:hep-ph/0107001].
19. C. W. Bauer, D. Pirjol and I. W. Stewart, Phys. Rev. D **65**, 054022 (2002) [arXiv:hep-ph/0109045].
20. C. W. Bauer, C. W. Chiang, S. Fleming, A. K. Leibovich and I. Low, Phys. Rev. D **64**, 114014 (2001) [arXiv:hep-ph/0106316].
21. S. Fleming and A. K. Leibovich, Phys. Rev. Lett. **90**, 032001 (2003) [arXiv:hep-ph/0211303].
22. S. Fleming and A. K. Leibovich, Phys. Rev. D **67**, 074035 (2003) [arXiv:hep-ph/0212094].
23. X. Garcia i Tormo and J. Soto, Phys. Rev. D **69**, 114006 (2004) [arXiv:hep-ph/0401233].
24. S. Fleming and A. K. Leibovich, Phys. Rev. D **70**, 094016 (2004) [arXiv:hep-ph/0407259].
25. S. Fleming, C. Lee and A. K. Leibovich, Phys. Rev. D **71**, 074002 (2005) [arXiv:hep-ph/0411180].
26. X. Garcia i Tormo and J. Soto, Phys. Rev. D **72**, 054014 (2005) [arXiv:hep-ph/0507107].
27. M. Beneke, A. P. Chapovsky, M. Diehl and T. Feldmann, Nucl. Phys. B **643**, 431 (2002) [arXiv:hep-ph/0206152]
28. M. Beneke and T. Feldmann, Phys. Lett. B **553**, 267 (2003) [arXiv:hep-ph/0211358]
29. M. J. Dugan and B. Grinstein, Phys. Lett. B **255**, 583 (1991).
30. A. V. Manohar, T. Mehen, D. Pirjol and I. W. Stewart, Phys. Lett. B **539**, 59 (2002) [arXiv:hep-ph/0204229].
31. J. g. Chay and C. Kim, arXiv:hep-ph/0205117.
32. K. Sridhar, A. D. Martin and W. J. Stirling, Phys. Lett. B **438**, 211 (1998) [arXiv:hep-ph/9806253].
33. S. P. Baranov and N. P. Zotov, J. Phys. G **29**, 1395 (2003) [arXiv:hep-ph/0302022].
34. A. V. Lipatov and N. P. Zotov, Eur. Phys. J. C **27**, 87 (2003) [arXiv:hep-ph/0210310].
35. C. W. Bauer, D. Pirjol and I. W. Stewart, Phys. Rev. D **67**, 071502 (2003) [arXiv:hep-ph/0211069].
36. C. W. Bauer, S. Fleming, D. Pirjol, I. Z. Rothstein and I. W. Stewart, Phys. Rev. D **66**, 014017 (2002) [arXiv:hep-ph/0202088].
37. S. Fleming, A. K. Leibovich, T. Mehen, work in progress.

5 Diffractive *ep* Scattering

EXCLUSIVE AND INCLUSIVE DIFFRACTION AT HERA

H. KOWALSKI

Deutsches Elektronen-Synchrotron, DESY
Notkestr. 85,
22607 Hamburg, Germany
E-mail: Henri.Kowalski@desy.de

This talk describes the measurement of F_2 and inclusive and exclusive diffractive cross sections in the low-x region by HERA experiments. The abundance of diffractive reactions observed at HERA indicates the presence of perturbative multi-ladder exchanges. The exclusive diffractive vector-meson and diffractive dijet production are discussed in terms of dipole models which connect the measurement of F_2 with diffractive processes and in which multiple exchanges and saturation processes are natural. The diffractive dijets are also discussed within the diffractive parton density approach. Good description of diffractive dijets in the dipole picture and in the diffractive parton density approach indicates that these two seemingly different views on diffraction are not really distinct.

1. F_2 and Diffraction at HERA

The HERA machine is a large electron-proton collider, in which electrons with energy of 27.5 GeV scatter on protons of 920 GeV. The collision products are recorded by the two large, multipurpose experiments ZEUS and H1. The detectors consist of inner tracking detectors surrounded by large calorimeters measuring the spatial energy distribution, event by event. The calorimeters are in addition surrounded by muon detector systems. Fig. 1 shows, as an example, a picture of a high Q^2 DIS event measured by the H1 and ZEUS detectors. From the amount and positions of energy deposited by the scattered electron and the hadronic debris, the total γ^*p CMS energy, W, and the virtuality of the exchanged photon, Q^2, are determined. Counting the events at given Q^2 and W^2 allows the determination of the total cross section for the collisions of the virtual photon with the proton, $\sigma_{\gamma^*p}(W^2, Q^2)$, and in turn the structure function,

$$F_2(x, Q^2) = \frac{Q^2}{4\pi^2\alpha_{em}}\sigma_{\gamma^*p}(W^2, Q^2)$$

with $x \approx Q^2/W^2$ when $Q^2 \ll W^2$.

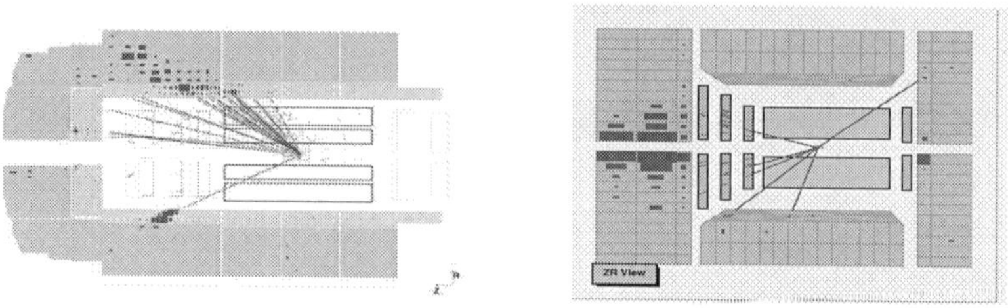

Figure 1. *Two examples of DIS events seen in the H1 (left) and ZEUS (right) detector.*

Deep inelastic scattering and the structure function F_2 have a simple and intuitive interpretation when viewed in the fast moving proton frame. The incoming electron scatters on the proton by emitting an intermediate photon with a virtuality Q^2. The incoming proton consists of a fluctuating cloud of quarks, antiquarks and gluons. Since the lifetime of the virtual photon is much shorter than the lifetime of the $q\bar{q}$-pair, the photon scans the "frozen" parton cloud and picks up quarks with longitudinal momentum x, see Fig. 2. F_2 measures then the density of partons with a size which is larger than the photon size, $1/Q$, at a given x. Fig. 3 shows the structure function F_2 as measured by H1, ZEUS and fixed target experiments for selected Q^2 values [1].

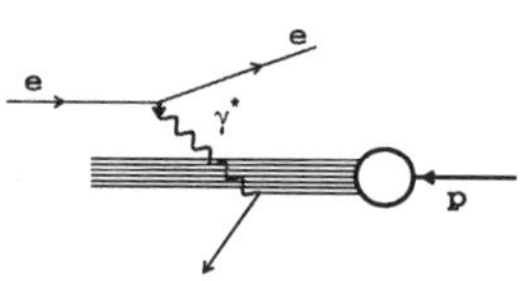

Figure 2. *Schematic view of deep inelastic scattering (DIS).*

In the low-x regime, F_2 measured at HERA exhibits a striking behavior. At low Q^2 values, $Q^2 < 1$ GeV^2, where the photon is large, F_2 rises only moderately with diminishing x, whereas as Q^2 increases, i.e. the photon becomes smaller, the rise of F_2 accelerates quickly. The rise of F_2 at low Q^2 values, i.e. when the photon is of similar size as a hadron, corresponds to the rise of the hadronic cross sections with energy. The fast rise at large Q^2 indicates the strong growth of the cloud of partons in the proton. The onset of the fast growth at Q^2 values larger than 1 GeV^2 indicates that these partons are of perturbative origin.

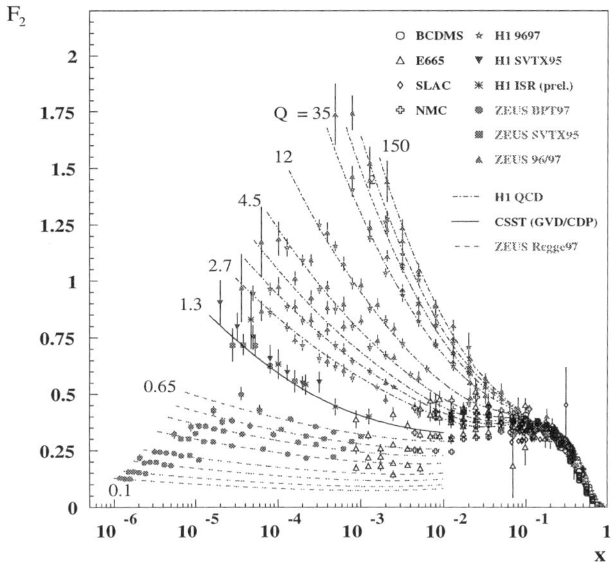

Figure 3. *The structure function F_2 as a function of x as measured by H1, ZEUS and fixed target experiments for selected Q^2 between 0.1 and 150 GeV^2.*

For sufficiently large Q^2 perturbative QCD provides a set of leading-twist linear evolution equations (DGLAP) which describe the variation of the cross section as a function of Q^2; see Fig. 4. Moreover, a closer look at the x-dependence of the parton splitting functions has led to the prediction that the gluon density, at small x, should rise with $1/x$. This rise should translate into a growth with energy of the total $\gamma^* p$ cross section or, equivalently, of F_2 with diminishing x. The data show that the growth of F_2 starts in the low-x regime which indicates that this is mainly due to the abundant gluon production. This is confirmed by all detailed theoretical

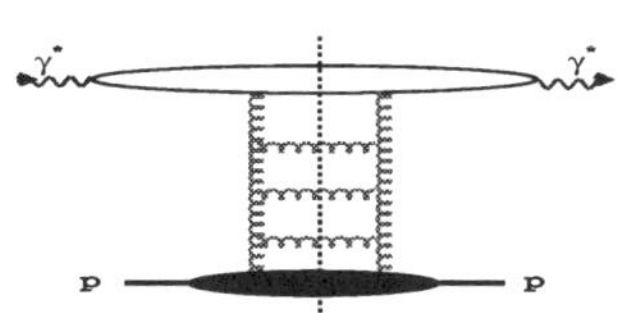

Figure 4. *Illustration of the pQCD description of the total cross section $\sigma_{tot}^{\gamma^* p}$. The gluon ladder represents the linear QCD evolution equations.*

investigations of HERA data. As an example Fig. 5 shows the results of the ZEUS and MRST analyses of parton densities. Both analyses show that in the low-x region the gluon density dwarfs all quark densities with exception of the sea quarks. The sea quarks, in perturbative QCD, are generated from the gluon density.

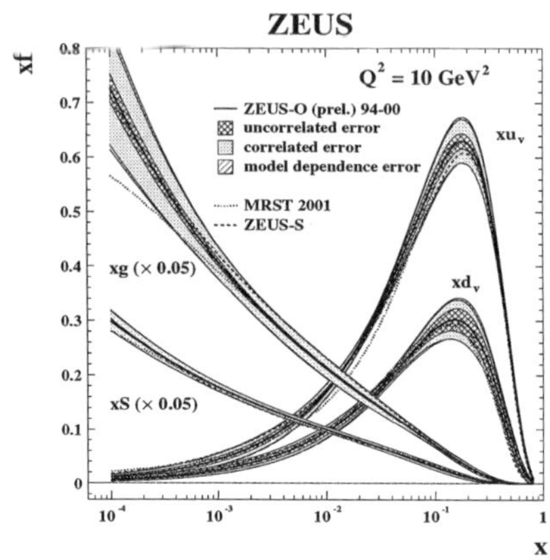

Figure 5. *Quark and gluon densities at $Q^2 = 10$ GeV^2 as determined from HERA data. Note that the gluon and sea-quark densities are displayed diminished by a factor 0.05.*

One of the most important observations of the HERA experiments is that, in addition to the usual DIS events, in which the struck proton is transformed into a swarm of particles, there are also events in which the proton remains intact after collision. Whereas the usual DIS events are characterized by large energy depositions in the forward (proton) direction, see Fig. 1, the events with intact protons show no activity in this region; see Fig. 6. By analogy to the absorption of light waves on a black disk,

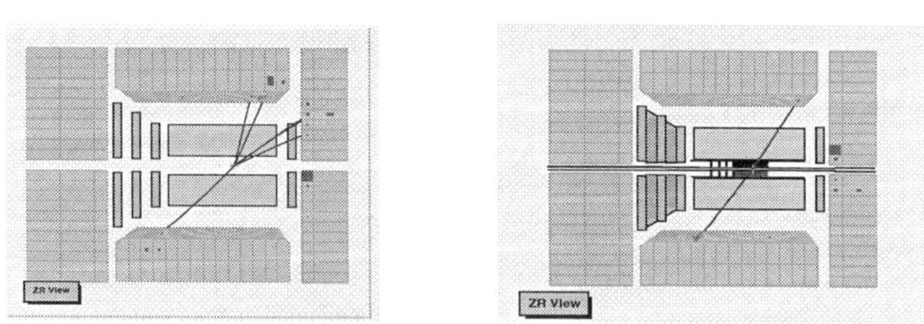

Figure 6. *Two examples of diffractive events seen in the ZEUS detector.*

the events of this type are called diffractive events and the process in which they are produced is called diffractive scattering. The intact forward proton corresponds in optics to the forward white spot observed in the center of the disc shadow.

The measurement of diffractive reactions requires the determination of two additional variables: the diffractive mass, M_X, and the square of the four-momentum transferred by the outgoing proton, t. The variable M_X, which is equal to the invariant mass of all particles emitted in the reaction with exception of the outgoing proton (or the proton dissociated system), is determined from energy depositions recorded by the central detectors of the H1 and ZEUS experiments. The variable t is determined by forward detectors, which measure the momentum of the outgoing diffractively scattered proton. In exclusive diffractive vector-meson production the t variable can also be determined from the precise measurement of the momenta of the vector-meson decay products measured in the tracking chamber systems of central detectors.

The analysis of the observed $\ln M_X^2$ distribution allows a separation of diffractive and non-diffractive events as indicated in Fig. 7. The plateau like structure, most notably seen at higher W values, is due to diffractive events since in diffraction $dN/d\ln M_X^2 \approx$ const. The high mass peaks in Fig. 7, which are due to non-diffractive events, have a steep exponential fall-off, $dN/d\ln M_X^2 \propto \exp(\lambda \ln M_X^2)$, towards smaller $\ln M_X^2$ values. This exponential fall-off is directly connected to the exponential suppression of large rapidity gaps in a single gluon ladder exchange diagram, Fig. 4, which represents the dominant QCD contribution.

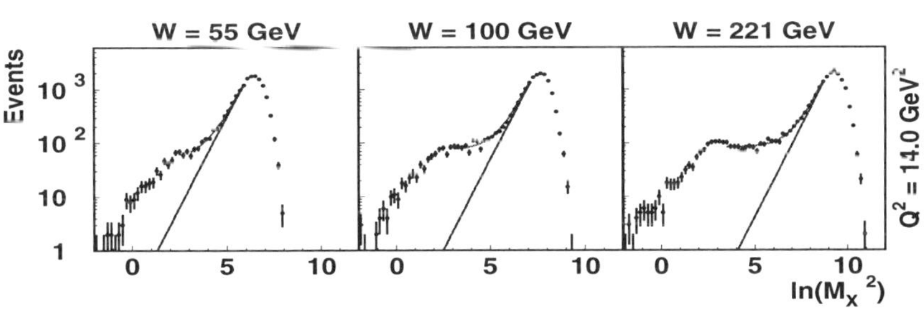

Figure 7. *Distribution of M_X in terms of ln M_X^2. The straight lines give the non-diffractive contribution as obtained from the fits. Note that the* $\ln M_X^2$ *distribution can be viewed as a rapidity gap distribution since* $\Delta Y = \ln(W^2/M_X^2)$ *for* $M_X^2 \gg Q^2$.

In the ZEUS investigation [2,3] the diffractive contribution was therefore identified as the excess of events at small M_X above the exponential fall-off of the non-diffractive contribution in $\ln M_X^2$. This selection procedure is called the M_X method. In the H1 investigation [4] the selection of diffractive events was performed by the requirement of a large rapidity gap in the event. The ZEUS M_X and the H1 rapidity gap methods allow only to measure the diffractive cross section integrated over the square of the four-momentum transfer t.

The measured diffractive cross sections show a clear rise with increasing energy W in all M_X regions. It is interesting to note that the increase of the differential diffractive cross sections with W is very similar to the increase of the total inclusive DIS cross sections, i.e. $\sigma_{diff}/\sigma_{\gamma^*p}^{tot}$ is approximately independent of energy in all Q^2 and M_X regions as seen in Fig. 8. The ratio of the diffractive to the total DIS cross section integrated over the whole accessible M_X range, $M_X < 35$ GeV, was evaluated at the highest energy of $W \approx 220$ GeV. At $Q^2 = 4$ GeV^2, $\sigma_{diff}/\sigma_{\gamma^*p}^{tot}$ reaches $\sim 16\%$. It decreases slowly with increasing Q^2, reaching $\sim 10\%$ at $Q^2 = 27$ GeV^2.

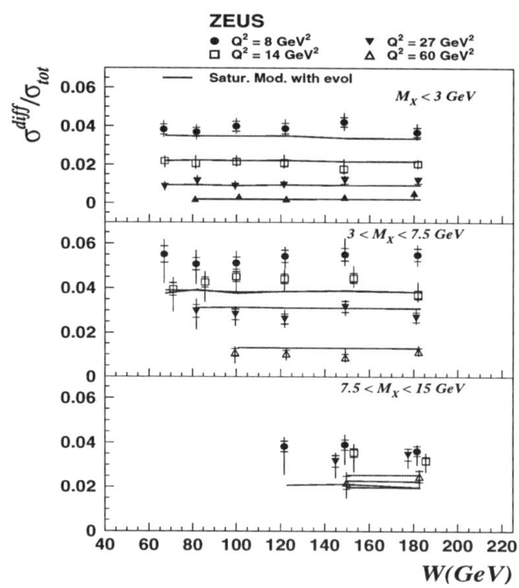

Figure 8. *The ratio of the inclusive diffractive and total DIS cross sections versus the $\gamma^* p$ energy W.*

The observation of such a large fraction of diffractive events was unexpected since according to the intuitive interpretation of DIS the incoming proton consists of a parton cloud and at least one of the partons is kicked out in the hard scattering process. In the language of QCD diagrams, at low-x and not so small Q^2, the total cross section or F_2 is dominated by the abundant gluon emission as described by the single ladder exchange shown in Fig. 4; the ladder structure also illustrates the linear DGLAP evolution equations that are used to describe the F_2 data. In the region of small x gluonic ladders are expected to dominate over quark ladders. The cut line in Fig. 4 marks the final states produced in a DIS event: a cut parton (gluon) hadronizes and leads to jets or particles seen in the detector. It is generally expected that partons produced from a single chain are unlikely to generate large rapidity gaps between them, since large gaps are exponentially suppressed as a function of the gap size. This is a general property of QCD evolution equations of the DGLAP, BFKL or other types.

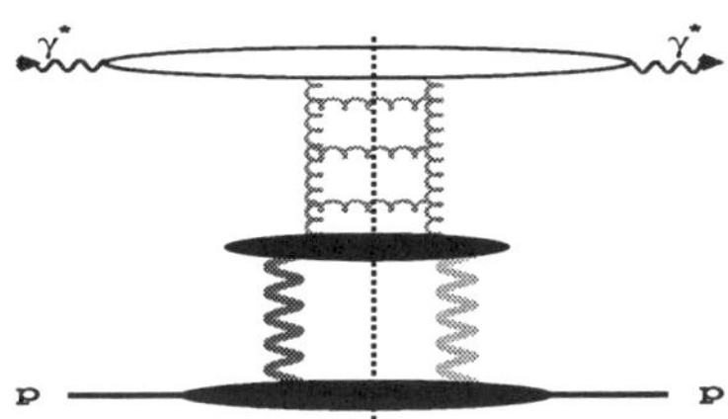

Figure 9. *Diffractive final states as part of the initial condition to the evolution equation in F_2. The thick vertical wavy lines denote the non-perturbative Pomeron exchanges which generate the rapidity gap in DIS diffractive states.*

In the single ladder contribution of Fig. 4, diffractive final states can, therefore, only reside inside the blob at the lower end, i.e. lie below the initial scale Q_0^2 which separates the parton description from the non-perturbative strong interaction, as shown in Fig. 9. The thick vertical wavy lines denote the non-perturbative Pomeron exchanges which generate the rapidity gap in DIS diffractive states [a]. The diagram of Fig. 9 exemplifies therefore the "Regge factorization" approach to diffractive parton densities as description of diffractive phenomena in DIS. In this approach the diffractive states are essentially of non-perturbative origin but they evolve accord-

[a]It is customary to call the exchange of a colourless system in scattering reactions a Pomeron. The simplest example of a (perturbative) Pomeron is given by the ladder diagram of Fig. 4.

ing to the perturbative QCD evolution equations. Note, however, that the effective Pomeron intercept, $\alpha_{I\!P}$, extracted from diffractive DIS data lies significantly above the 'soft' Pomeron intercept, indicating a substantial contribution to diffractive DIS from perturbative Pomeron exchange [3,5].

The properties of special diffractive reactions at HERA, like exclusive diffractive vector-meson and jets production, give clear indications that the diffractive processes could be hard and of perturbative origin. A significant contribution from perturbative multi-ladder exchanges should be present, in particular from the double ladder exchange of Fig. 10. This diagram provides a potential source for the harder diffractive states: the cut blob at the upper end may contain $q\bar{q}$ and $q\bar{q}g$ states which hadronize into harder jets or particles. The evidence for the presence of multi-ladder contributions is emerging mostly from the interconnections between the various DIS processes: inclusive γ^*p reaction, inclusive diffraction, exclusive diffractive vector-meson production and diffractive jet-jet production. These interconnections are naturally expressed in the dipole saturation models, which have been shown to successfully describe HERA F_2 data in the low-x region. These models are explicitly built on the idea of summing over multiple exchanges of single ladders. In the following we will discuss the exclusive and inclusive diffractive DIS processes and their connection with the total DIS cross section in terms of dipole models.

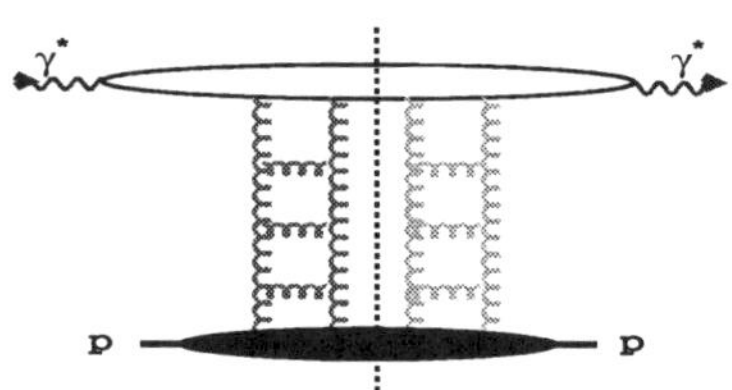

Figure 10. *The double gluon ladder contribution to the inclusive diffractive γ^*p cross section.*

2. Dipole Models

In the dipole model, deep inelastic scattering is viewed as interaction of a colour dipole, i.e. mostly a quark-antiquark pair, with the proton. The size of the pair is denoted by r and a quark carries a fraction z of the photon momentum. In the proton rest frame, the dipole life-time is much longer than the life-time of its interaction with the target proton. Therefore, the interaction is assumed to proceed in three stages: first the incoming virtual photon fluctuates into a quark-antiquark pair, then the $q\bar{q}$ pair elastically scatters on the proton, and finally the $q\bar{q}$ pair recombines to form a virtual photon. The amplitude for the complete process is simply the product of these three processes.

The amplitude of the incoming virtual photon to fluctuate into a quark-antiquark pair is given by the photon wave function ψ, which is determined from light cone perturbation theory to leading order in the fermionic charge (for simplicity, the indices of the quark and antiquark helicities are suppressed). Similarly the amplitude for the $q\bar{q}$ to recombine to a virtual photon is ψ^*. The cross section for elastic scattering of the $q\bar{q}$ pair with squared momentum transfer $\Delta^2 = -t$ is described by the elastic scattering amplitude, $A_{el}^{q\bar{q}}(x, r, \Delta)$, as

$$\frac{d\sigma_{q\bar{q}}}{dt} = \frac{1}{16\pi}|A_{el}^{q\bar{q}}(x, r, \Delta)|^2. \tag{1}$$

To evaluate the connections between the total cross section and various diffractive reactions it is convenient to work in coordinate space and define the S-matrix element at a particular impact parameter b

$$S(b) = 1 + \frac{1}{2}\int d^2\Delta \exp(i\vec{b}\cdot\vec{\Delta})A_{el}^{q\bar{q}}(x, r, \Delta). \tag{2}$$

This corresponds to the intuitive notion of impact parameter when the dipole size is small compared to the size of the proton. The Optical Theorem then connects the total cross section of the $q\bar{q}$ pair to the imaginary part of iA_{el}

$$\sigma_{q\bar{q}}(x, r) = \Im i A_{el}^{q\bar{q}}(x, r, 0) = \int d^2b\, 2[1 - \Re S(b)]. \tag{3}$$

The integration over the S-matrix element motivates the definition of the elastic $q\bar{q}$ differential cross section as

$$\frac{d\sigma_{q\bar{q}}}{d^2b} = 2[1 - \Re S(b)]. \tag{4}$$

The total cross section for $\gamma^* p$ scattering, or equivalently F_2, is obtained by averaging the dipole cross sections with the photon wave functions, $\psi(r, z)$:

$$\sigma^{\gamma^* p} = \int d^2r \int \frac{dz}{4\pi}\psi^*\, \sigma_{q\bar{q}}(x, r)\, \psi. \tag{5}$$

In the dipole picture the elastic vector-meson production appears in a similarly transparent way. The amplitude is given by

$$A_{\gamma^* p\to pV}(\Delta) = \int d^2r \int \frac{dz}{4\pi} \int d^2b\, \psi_V^* \psi \exp(-i\vec{b}\cdot\vec{\Delta}) 2[1 - S(b)]. \quad (6)$$

We denote the wave function for a vector meson to fluctuate into a $q\bar{q}$ pair by ψ_V. Assuming that the S-matrix element is predominantly real, we may substitute $2[1 - S(b)]$ with $d\sigma_{q\bar{q}}/d^2b$. Then, the elastic diffractive cross section is

$$\frac{d\sigma^{\gamma^* p\to Vp}}{dt} = \frac{1}{16\pi} \left| \int d^2r \int \frac{dz}{4\pi} \int d^2b\, \psi_V^* \psi \exp(-i\vec{b}\cdot\vec{\Delta}) \frac{d\sigma_{q\bar{q}}}{d^2b} \right|^2. \quad (7)$$

The equations (5) and (7) determine the inclusive and exclusive diffractive vector-meson production using the universal elastic differential cross section $d\sigma_{q\bar{q}}/d^2b$ which contains all the interaction dynamics.

The inclusive diffractive cross section can be obtained from the eq. (7) summing over all (generalized) vector-meson states as

$$\left. \frac{d\sigma_{dif}^{\gamma^* p}}{dt} \right|_{t=0} = \frac{1}{16\pi} \int d^2r \int \frac{dz}{4\pi} \psi^* \, \sigma_{q\bar{q}}^2 \, \psi. \quad (8)$$

Thus, properties of inclusive diffraction are also determined by the elastic cross section only and, contrary to vector-meson production, are not dependent on the wave function of the outgoing diffractive state.

2.1. *Dipole Cross Section and Saturation*

The dipole models became an important tool in investigations of deep inelastic scattering due to the initial observation of K. Golec-Biernat and M. Wüsthoff (GBW) [6] that a simple ansatz for the dipole cross section integrated over the impact parameter b, $\sigma_{q\bar{q}}$, is able to describe simultaneously the total inclusive and diffractive DIS cross sections:

$$\sigma_{q\bar{q}}^{GBW} = \sigma_0[1 - \exp(-r^2/4R_0^2)] \quad (9)$$

where σ_0 is a constant and R_0 denotes the x dependent saturation radius $R_0^2 = (x/x_0)^{\lambda_{GBW}} \cdot (1/GeV^2)$. The parameters $\sigma_0 = 23$ mb, λ_{GBW} and $x_0 = 3 \cdot 10^{-4}$ were determined from a fit to the data. Although the dipole model is theoretically well justified for small size dipoles only, the GBW model provides a good description of data from medium size Q^2 values ($\sim$30 GeV^2) down to low Q^2 ($\sim$0.1 GeV^2). The inverse of the saturation radius R_0 is analogous to the gluon density. The exponent λ_{GBW} determines therefore the growth of the total and diffractive cross sections with decreasing x. For dipole sizes which are large in comparison to R_0 the dipole cross section saturates by approaching a constant value σ_0, which becomes independent of λ_{GBW}. It is a characteristic of the model that a good description of data is due to large saturation effects, i.e. the strong growth due to the factor $(1/x)^{\lambda_{GBW}}$ is, for large dipoles, significantly flattened by the exponentiation in eq. (9).

The assumption of dipole saturation provided an attractive theoretical background for investigation of the transition from the perturbative to non-perturbative regime in the HERA data. Despite the appealing simplicity and success of the GBW model it suffers from clear shortcomings. In particular it does not include scaling violation, i.e. at large Q^2 it does not match with QCD evolution (DGLAP). Therefore, Bartels, Golec-Biernat and Kowalski (BGBK) [7] proposed a modification of the original ansatz of eq.(9) by replacing $1/R_0^2$ by a gluon density with explicit DGLAP evolution:

$$\sigma_{q\bar{q}}^{BGBK} = \sigma_0[1 - \exp(-\pi^2 r^2 \alpha_s(\mu^2) x g(x, \mu^2)/3\sigma_0)] \qquad (10)$$

The scale of the gluon density, μ^2, was assumed to be $\mu^2 = C/r^2 + \mu_0^2$, and the density was evolved according to DGLAP equations.

The BGBK form of the dipole cross section led to significantly better fits to the HERA F_2 data than the original GBW model, especially in the region of larger Q^2. The good agreement of the original model with the DIS diffractive HERA data was also preserved, as seen from the comparison of the predictions of the model with data for the ratio of the diffractive to the total cross section, Fig. 8.

The BGBK analysis found, surprisingly, that there exist two distinct solutions giving very good description of HERA data, depending on the quark mass in the photon wave function. The first solution is obtained assuming $m_q = 140$ MeV and leads to the initial gluon density distribution with the value of exponent $\lambda_g = 0.28$, which is very similar to the λ_{GBW}. As in the original model, the good agreement with data is due to substantial saturation effects. In the second solution, $m_q \approx 0$, and the value of the exponent is very different, $\lambda_g = -0.41$. The initial gluon density no longer rises at small x, it is valence-like, and QCD evolution plays a much more significant role than in the first solution.

The DGLAP evolution, which is generally used in the analysis of HERA data, may not be appropiate when x approaches the saturation region. Therefore, Iancu, Itakura and Munier (IIM) [8] proposed a new saturation model, the Colour Glass Condensate model, in which gluon saturation effects are incorporated via an approximate solution of the Balitsky-Kovchegov equation. Later, also Forshaw and Shaw (FS) [9] proposed a Regge type model with saturation effects. The IIM and FS models provide a description of HERA F_2 and diffractive data which is better than the original GBW model and comparable in quality to the BGBK analysis. Both models find strong saturation effects in HERA data comparable to the GBW model and the first solution of the BGBK model.

All approaches to dipole saturation discussed so far ignored a possible impact parameter (IP) dependence of the dipole cross section. This dependence was introduced by Kowalski and Teaney (KT) [10], who assumed that the dipole cross section is a function of the opacity Ω:

$$\frac{d\sigma_{qq}}{d^2b} = 2\left(1 - \exp(-\frac{\Omega}{2})\right). \tag{11}$$

At small x the opacity Ω can be directly related to the gluon density, $xg(x, \mu^2)$, and the transverse profile of the proton, $T(b)$:

$$\Omega = \frac{\pi^2}{N_C}\, r^2\, \alpha_s(\mu^2)\, xg(x, \mu^2)\, T(b). \tag{12}$$

The transverse profile is assumed to be of the form:

$$T(b) = \frac{1}{2\pi B_G} \exp(-b^2/2B_G), \tag{13}$$

since the Fourier transform of $T(b)$ has the exponential form:

$$\frac{d\sigma^{\gamma^* p}_{VM}}{dt} = \exp(-B_G|t|) \tag{14}$$

The formula of eq. (11) and (12) is called the Glauber-Mueller dipole cross section. The diffractive cross section of this type was used around 50 years ago to study the diffractive dissociation of the deuterons by Glauber and reintroduced by A. Mueller [11] to describe dipole scattering in deep inelastic processes.

The parameters of the gluon density are determined from the fit to the total inclusive DIS cross section, as shown in Fig. 11 [10]. The transverse profile was determined from the exclusive diffractive J/Ψ cross sections shown in the same figure. In this approach the charm quark was explicitly taken into account with the mass $m_c = 1.25$ GeV. For a small value of Ω the dipole cross section, eq. (11), is equal to Ω and therefore proportional to the gluon density. This allows one to identify the opacity with the single Pomeron exchange amplitude of Fig. 4.

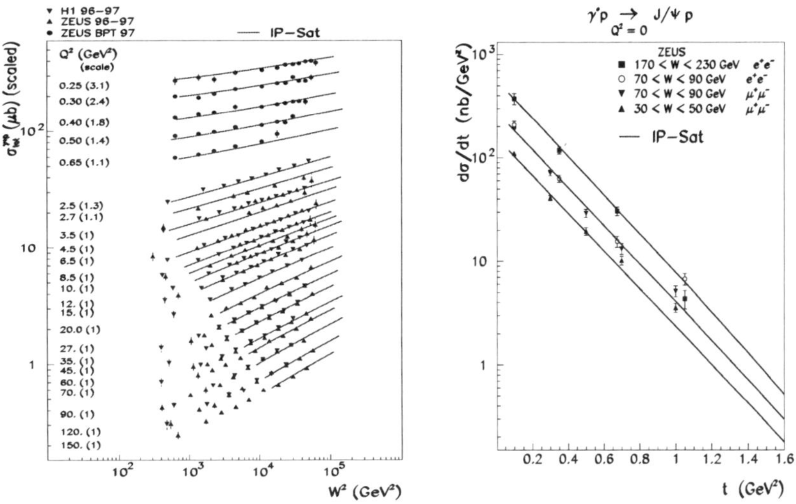

Figure 11. *LHS: The $\gamma^* p$ cross section as a function of W^2. RHS: The differential cross section for exclusive diffractive J/Ψ production as a function of the four-momentum transfer t. The solid line shows a fit by the IP saturation model (KT).*

The KT model with parameters determined in this way has predictive properties which go beyond the models discussed so far; it allows a description of the other measured reactions, e.g. the charm structure function [12] or elastic diffractive J/Ψ production [13] shown in Fig. 12. The initial gluon distribution determined in the model is valence-like, with $\lambda_g = -0.12$ and the fit pushes the quark mass to small values, $m_q \approx 50$ MeV. The resulting gluon distribution is therefore similar to the second solution of the BGBK model. The first solution of the BGBK model was disfavoured by the data. This behaviour is presumably due to the assumption of the Gaussian-like proton shape, eq. (13). In the tail of the Gaussian, the gluon density is low, but the relative contribution of the tail to the cross section is large. The saturation effects cannot therefore be as large as in the GBW-like models (i.e. BGBK-1, IIM, FS). In addition, as noted in the KT paper and also in the Thorne analysis [14], the introduction of charm in the analysis of HERA data lowers the gluon density and therefore diminishes the saturation effects. Nevertheless, the KT analysis shows that in the center of the proton ($b \approx 0$) the saturation effects are similar to the ones in the GBW-like models in which charm is properly taken into account. This can be seen from the evaluation of the saturation scale in the center of the proton in the KT

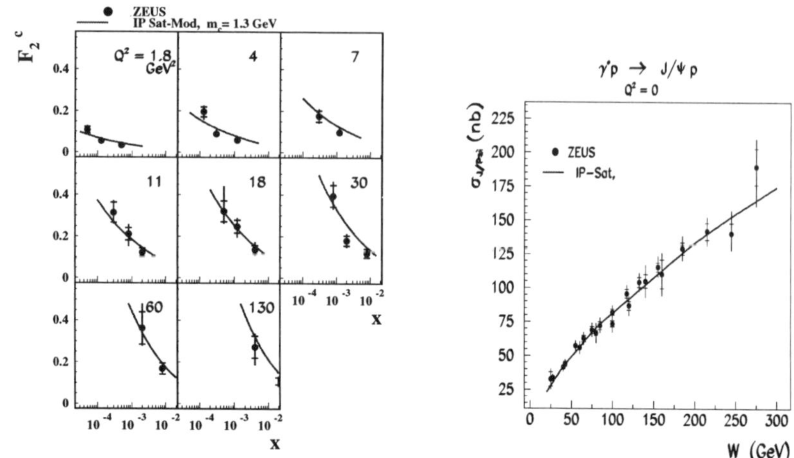

Figure 12. *LHS: Charm structure function, F_2^c. RHS: Total elastic J/Ψ cross section. The solid line shows the result of the IP saturation model (KT).*

paper and the comparison to the value of the saturation scale evaluated with charm in the original GBW paper.

3. Exclusive Diffractive Vector-Meson Production

The exclusive diffractive vector-meson production is very interesting because, in the low-x region, it is driven by the square of the gluon density. It was, therefore, investigated by many authors [15,16,17,10,18,19,20]. In addition, the information contained in the Q^2, W and t dependence of the cross sections allows to determine vector-meson wave functions together with the proton shape. The analysis can also be performed separately for the longitudinal and transverse photons.

The recent analysis of vector-meson production by Kowalski, Motyka and Watt (KMW) [21] shows that it is possible to describe the measured differential cross sections making simple assumptions about the vector-meson wave functions [15,19]. The analysis shows that using the gluon density determined from the total cross sections and the size of the interaction region determined from the t distribution of the J/Ψ meson at $Q^2 = 0$, it is possible to simultaneously describe not only the shape of various differential cross sections as a function of Q^2, W and t but also their absolute magnitude. In this analysis the assumption that vector-meson size should be much smaller than proton size was relaxed. Following the work of Bartels,

Golec-Biernat and Peters [22] the Fourier transform of eq. (7) was modified to take into account the finite size of the vector meson:

$$\exp(-i\vec{b}\cdot\vec{\Delta}) \to \exp(-i(\vec{b}+(1-z)\vec{r})\cdot\vec{\Delta}). \tag{15}$$

In this way, the information about the size of the vector meson, contained in the wave function, is contributing to the size of the interaction region B_D, together with the size of the proton. As an example of results obtained in this analysis Fig. 13 shows the comparison of KMW model predictions for the total exclusive diffractive vector-meson cross section and the size of the interaction region with data. Here, the profile function is assumed to have

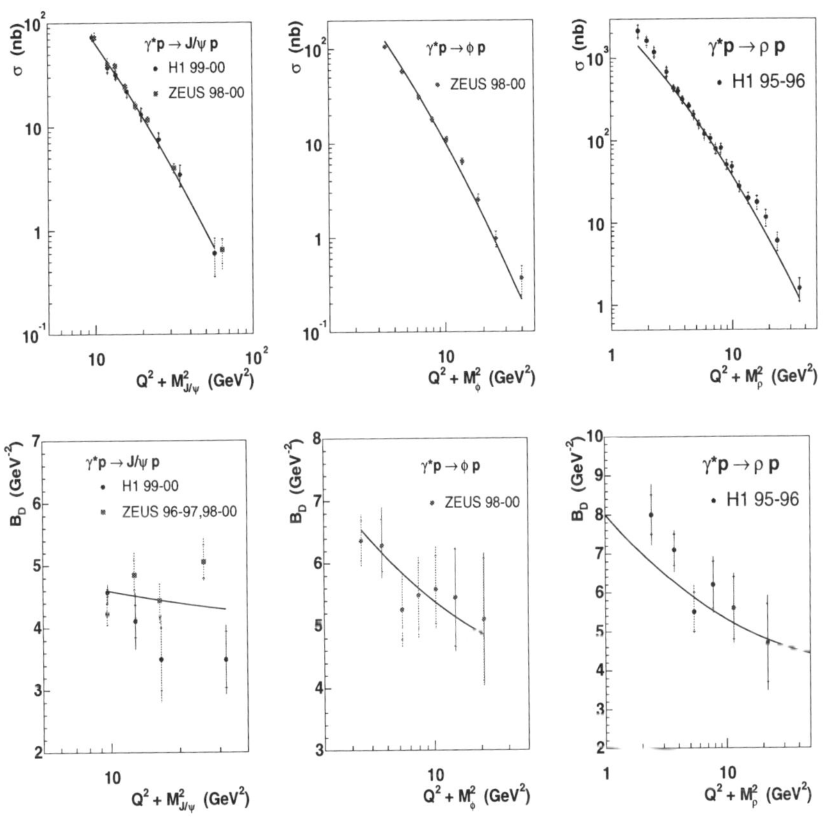

Figure 13. *(Top) The exclusive diffractive cross sections for J/Ψ, ϕ and ρ vector-meson production as a function of $Q^2 + M_V^2$. (Bottom) The interaction size B_D defined by $d\sigma/dt \propto \exp(B_D t)$, extracted from t distributions of J/Ψ, ϕ and ρ vector meson as a function of $Q^2 + M_V^2$. The solid line shows predictions of the KMW model. (Preliminary results)*

a Gaussian form (13), with the parameter $B_G = 4$ GeV^{-2}. The 'boosted Gaussian' vector-meson wave functions [19] are used. The light quark masses are $m_q = 140$ MeV, with $m_c = 1.4$ GeV.

4. Diffractive Dijets

The diffractive dijet production is analysed within QCD in two seemingly different ways: the diffractive parton density approach and the dipole picture. The diffractive structure function is proportional to the diffractive parton density which, in the simplest case, factorizes into the Pomeron flux factor and the Pomeron parton density. The Pomeron flux factor gives the probability to find a Pomeron within a proton with the momentum fraction $x_{I\!P}$. The Pomeron parton density gives the probability to find a parton within a Pomeron carrying a Pomeron momentum fraction $z_{I\!P}$[23]. This parton interacts then with the incoming photon in a boson-gluon, Fig. 14, or QCD-Compton process.

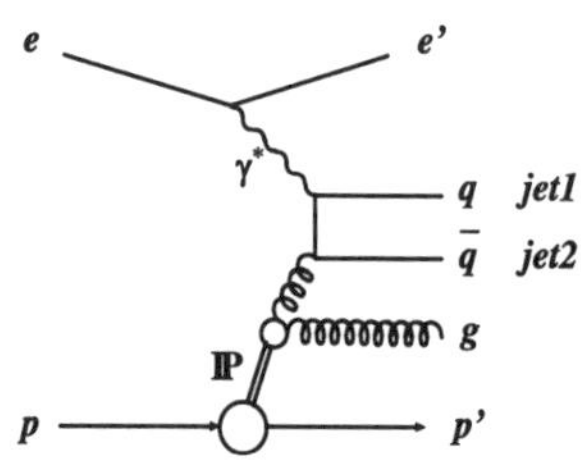

Figure 14. *Diagram describing diffractive production of a $q\bar{q}g$ state in the diffractive structure function approach.*

The diffractive parton distribution is extracted from the inclusive diffractive measurements at an initial scale, Q_0^2, and then evolved according to the DGLAP evolution to the scale Q^2 of jet production. The events are generated in the RAPGAP Monte Carlo in which the Q^2, x, $x_{I\!P}$ and $z_{I\!P}$ variables are obtained from the diffractive structure function and the jet variables, E_t, η_{jets} and β are obtained from the usual QCD matrix elements of the boson-gluon fusion and QCD-Compton processes. The results are compared to data in Fig. 16 [24].

In the dipole picture, diffractive dijet production can be directly determined from the unintegrated gluon density in the low-x region, $f(x, l^2)$, where $\vec{l}$ is the transverse momentum of the gluon coupled to the quark pair. The unintegrated gluon density is then obtained from the dipole cross section by [7]:

$$\frac{\alpha_s f(x,l^2)}{l^4} = \frac{3}{8\pi^2}\int_0^\infty dr\, r J_0(lr)[\sigma_{q\bar{q}}(x,\infty) - \sigma_{q\bar{q}}(x,r)]. \qquad (16)$$

The jet cross sections are then computed in momentum space from diagrams shown in Fig. 15 [6]. In addition to the contributions of the $q\bar{q}$ states it is important to include the contributions of the $q\bar{q}g$ final states [25]. The

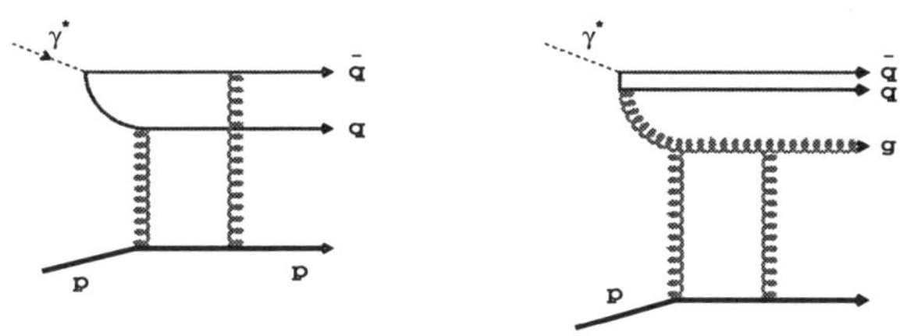

Figure 15. *Diagrams describing the diffractive production of a $q\bar{q}$ and $q\bar{q}g$ system in the dipole picture approach.*

results from the evaluation of the diagrams of Fig. 15, obtained within the SATRAP Monte Carlo, are compared to data[26,24] in Fig. 16. The figure shows that the distributions of typical jet variables are well described in both the diffractive parton densities (RAPGAP) and dipole picture (SATRAP) approaches. It is remarkable that in addition to the shapes also the absolute magnitudes of the diffractive jet cross sections are well described.

Finally, let us note for completeness that the diffractive parton density approach uses as input the inclusive diffractive data at low Q^2. It was observed that small systematic differences of input data tend to be amplified by the fitting procedure [24]. The dipole approach uses as input the dipole cross section determined from the F_2 measurements, which have smaller systematic errors than the F_2^D measurements.

5. Conclusions

One of the most important results of HERA measurements is the observation of the large amount of diffractive processes. Inclusive diffraction, diffractive jet process and exclusive diffractive vector-meson production are connected to inclusive deep inelastic scattering and, in the dipole picture,

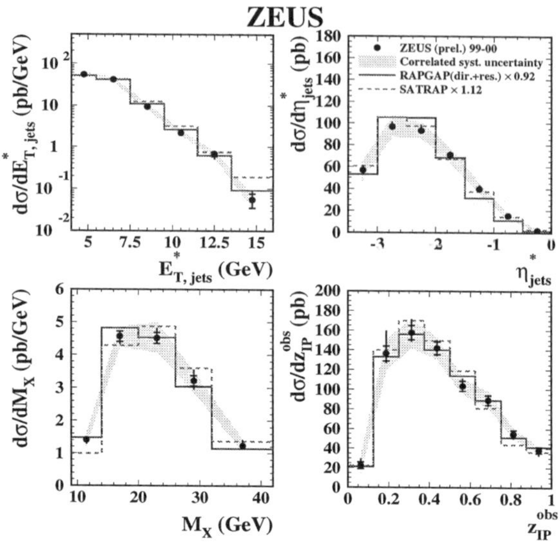

Figure 16. *The differential diffractive jet cross sections shown as a function of jet variables. The solid lines show the predictions from the LO RAPGAP MC normalised by a factor 0.92. SATRAP predictions normalised by a factor 1.12 are shown as dashed lines.*

can be successfully derived from the measured F_2. In the dipole approach, the Pomeron is essentially of the perturbative type, since the dipole models are explicitly built on the idea of summing over multiple exchanges of single ladders.

Inclusive diffraction and diffractive dijet production are also well described in the diffractive parton density approach, in which the Pomeron could be of non-perturbative origin. However, the effective Pomeron intercept extracted from diffractive DIS data lies significantly above the soft Pomeron intercept[3,5], indicating a substantial contribution to diffractive DIS from perturbative Pomeron exchange. In addition, the initial scale chosen for the analysis is relatively high, $Q_0^2 = 3$ GeV^2. At this scale F_2 exhibits a clear growth with diminishing x indicating that the exchanged Pomeron should be of perturbative type.

The good agreement between the diffractive parton density and dipole model analysis in the description of diffractive dijets indicates that both approaches, although seemingly different, are not really distinct. An attempt to combine these two approaches is recently discussed in Ref.[23].

Acknowledgments

I would like to thank Graeme Watt for discussions and help in preparing this manuscript.

References

1. G. Wolf, International Europhysics Conference on HEP, 2001
2. ZEUS Collab., J. Breitweg et al., *Eur. Phys. J.* C **6** 43 1999.
3. ZEUS Collab., S. Chekanov at al., *Nucl. Phys.* B **713** 3 2005.
4. H1 Collab., C. Adloff et al., *Z. Phys.* C **76** 613 1997.
5. H1 Collab., paper 089, EPS03, Aachen.
6. K. Golec-Biernat and M. Wüsthoff, *Phys. Rev.* D **60** 014023 1999.
7. J. Bartels, K. Golec-Biernat and H. Kowalski, *Phys. Rev.* D **66** 014001 2002.
8. E. Iancu, K. Itakura and S. Munier, *Phys. Lett.* B **590** 199 2004.
9. J. R. Forshaw and G. Shaw, JHEP **0412** 052 2004.
10. H. Kowalski and D. Teaney, *Phys. Rev.* D **68** 114005 2003.
11. A. H. Mueller, *Nucl. Phys.* B **335** 115 1990.
12. ZEUS Collab., J. Breitweg et al., *Eur. Phys. J.* C **12** 35 2000.
13. ZEUS Collab., S. Chekanov et al., *Eur. Phys. J.* C **24** 345 2002.
14. R.S. Thorne, *Phys. Rev.* D **71** 054024 2005.
15. J. Nemchik, N.N. Nikolaev, E. Predazzi and B.G. Zakharov, *Z. Phys.* C **75** 71 1997.
16. S. Munier, A. Mueller and A. Stasto, *Nucl. Phys.* B **603** 427 2001.
17. A. Caldwell and M. Soares, *Nucl. Phys.* A **696** 125 2001.
18. H.G. Dosch, T. Gousset, G. Kulzinger and H.J. Pirner, *Phys. Rev.* D **55** 2602 1997.
19. J. R. Forshaw, R. Sandapen and G. Shaw, *Phys. Rev.* D **69** 094013 2004.
20. A. Levy, *Nucl. Phys.* B **92** 146 2005.
21. H. Kowalski, L. Motyka, G. Watt, in preparation.
22. J. Bartels, K. Golec-Biernat and K. Peters, *Acta Phys. Polon.* B **34** 3051 2003.
23. G. Watt, these proceedings.
24. ZEUS Collab., XXII Int. Symp. on Lepton-Photon Interactions, Uppsala 2005, Abstract 295 and Addendum.
25. J. Bartels, J. Ellis, H. Kowalski and M. Wüsthoff, *Eur. Phys. J.* C **7** 443 1999.
26. H1 Collab., Int. Conf. on HEP ICHEP, Beijing 2004, Abstract 6-0177.

DIFFRACTIVE PRODUCTION OF VECTOR MESONS AND THE GLUON AT SMALL x

THOMAS TEUBNER

The University of Liverpool
Liverpool L69 3BX, England, U.K.
E-mail: thomas.teubner@liv.ac.uk

Perturbative QCD calculations of diffractive production of vector mesons are discussed. The MRT approach which uses k_T-factorization and Parton Hadron Duality is explained in some detail, and the uncertainties of the predictions are scrutinized. Skewing corrections turn out to be important. With recent data from HERA the strong sensitivity on the input gluon distribution at small x and small to intermediate scales will allow to better constrain the gluon.

1. Introduction

Diffractive production of vector mesons (VMs) has traditionally attracted a lot of attention for more than one reason. Experimentally several VMs are now accessible at HERA in an increasing range of the γp energy W and Q^2 of the photon, making them a very interesting laboratory to study the dynamics of QCD. Theoretically, the appearance of hard scales allows to make predictions within perturbative QCD by factorizing the perturbative photon splitting $\gamma \to q\bar{q}$ from the non-perturbative parton distribution functions (PDFs) of the proton. The diffractive cross section depends then quadratically on the (non-diffractive) PDFs and is hence an extremely sensitive probe. It will be shown that the good (and still growing) statistics together with refined theoretical predictions involving generalized parton distribution functions will allow for a better determination of the gluon, in a regime where its precise knowledge will be crucial at the LHC, e.g. for exclusive Higgs production or the modelling of underlying events.

In the following we will review QCD calculations within the approach of Martin, Ryskin and Teubner (MRT) and discuss uncertainties and possible improvements. We will then confront the predictions with recent data in the case of J/ψ production and briefly show the applicability for the case of the diffractive structure function F_2^D in the region of large β.

2. Theoretical Description in QCD

In the presence of a large scale, (semi-) hard diffractive scattering can be calculated in perturbative QCD, similar to heavy quark or jet production in DIS.[a] In leading order QCD, diffractive production of VMs is then described by colourless two-gluon exchange, and in the forward limit[b] is given by [1]

$$\frac{\mathrm{d}\sigma}{\mathrm{d}t}(\gamma^* p \to Vp)\bigg|_{t=0} = \frac{\Gamma^V_{ee} M_V^3 \pi^3}{48\alpha} \frac{\alpha_s(\overline{Q}^2)^2}{\overline{Q}^8} \left[x\, g(x, \overline{Q}^2)\right]^2 \left(1 + \frac{Q^2}{M_V^2}\right) . \quad (1)$$

Here M_V and Γ^V_{ee} are the mass and electronic width of the VM, $\overline{Q}^2 = (Q^2 + M_V^2)/4$ is an effective scale, and $x = (Q^2 + M_V^2)/(Q^2 + W^2)$ (W the c.m.s. energy of $\gamma^* p$). Equation (1) is valid in the high energy (small x) regime, where quark contributions are suppressed. Those contributions are usually neglected but could, in principle, be added on the same footing as the gluon. Equation (1) assumes the leading $\log Q^2$, $\log 1/x$ approximation, resulting in the collinear factorization of the *integrated, forward* gluon distribution. In addition the non-relativistic limit for the VM wave function is taken. These approximations only give a very crude leading order prediction which has to be improved for a realistic description of the data. In the following we will discuss such improvements in the framework of the MRT model [2].

2.1. *Parton Hadron Duality.*

Equation (1) assumes that the q and $\bar{q}$ forming the VM are produced with zero transverse momentum (non-relativistic limit). This can be overcome by introducing the quark's transverse momentum k_T (see Fig. 1) and convoluting the k_T dependent amplitude with a suitable VM wave function, thus accounting for Fermi motion inside the VM, see e.g. [3,4]. This introduces considerable uncertainties due to the limited knowledge of the form of the wave function. An alternative approach is to assume Parton Hadron Duality and to integrate *open* $q\bar{q}$ production over a given mass interval, $M^2 = M_{q\bar{q}}^2 \sim M_V^2$. The size of the interval is a somewhat free parameter (though chosen universally for different VMs in the MRT predictions) which results in an uncertainty of the overall normalization. To assure that only $q\bar{q}$ pairs with the quantum numbers allowing hadronization to the VM

[a] We will quantify below to which extent diffusion into the infrared regime persists even if a hard scale governs the production process.

[b] In the following, the cross section is always taken in the forward limit $t \sim 0$, assuming an exponential t dependence $\sigma \sim \exp(-bt)$ with a slope b determined either experimentally or by phenomenological models.

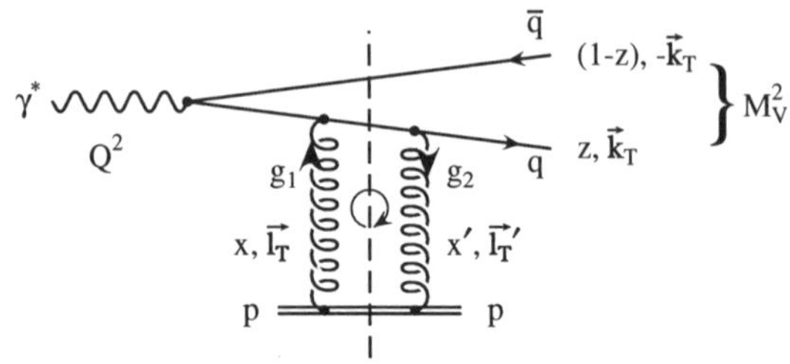

Figure 1. One of four leading order amplitudes for the diffractive process $\gamma^* p \to q\bar{q}p$.

are taken into account, a projection of the $\gamma^*_{L,T}p \to q\bar{q}p$ amplitudes on the quantum numbers $J^P = 1^-$ is performed, see [2] for details. Note that in the case of light quarks this is crucial as it leads to the suppression of endpoint-singularities present for production through transverse (T) photons. Without this projection, we would rely on some ad-hoc regularization, like introducing a constituent mass for the light quarks as done in the prediction of the diffractive structure function F_2^D at $\beta = Q^2/(Q^2+M^2) \to 1$.[5]

2.2. k_T-factorization; skewing corrections.

The identification of the two gluons with the integrated, forward gluon is only valid in the limit $l_T^2 \ll \overline{Q}^2 + k_T^2$ and $x \simeq x'$, respectively. (Here l_T is the transverse momentum of the gluons and x, x' are the longitudinal momentum fractions of the two gluons g_1, g_2, see Fig. 1.) However, beyond the leading $\log Q^2$ and $\log(1/x)$ approximations the amplitudes depend on generalized (unintegrated, non-forward) parton distributions (GPDs). Employing k_T-factorization[c] the basic amplitudes can be written as

$$\mathcal{A}(\gamma^*_{L,T}\, p \to q\bar{q}\, p) = \int_0^\infty \frac{\mathrm{d}l_T^2}{l_T^4}\, \alpha_s(l_T^2)\, f(x, x', l_T^2)\, \phi^{L,T}(Q^2, m^2, k_T^2, M^2, l_T^2)\,. \tag{2}$$

The (semi-) hard matrix-elements for longitudinal and transverse photons are given by $\phi^{L,T}$ whereas f is the non-forward, unintegrated gluon distribution. It contains the non-perturbative information about the proton, and not much is known about it a priori. In the predictions from MRT, normal (integrated, forward) gluons $g(x,\mu^2)$ are used as input to derive unintegrated distributions via the relation

$$f(x, l_T^2) = \left.\frac{\partial[xg(x, q_0^2)T(q_0^2, \mu^2)]}{\partial \ln q_0^2}\right|_{q_0^2 = l_T^2}\,. \tag{3}$$

[c] As the factorization scale is l_T one may prefer the term l_T-factorization.

The Sudakov factor $T = \exp[\frac{-C_A\alpha_s(\mu^2)}{4\pi}\ln^2\frac{\mu^2}{q_0^2}]$ resums virtual corrections, assuring that no gluon emission is allowed in the interval $q_0^2\ldots\mu^2 \sim (Q^2+M^2)/4$, which would destroy the rapidity gap. As known from other applications, the use of k_T-factorization leads to a significant enhancement of the cross section. This is expected as the k_T- (or high-energy-) factorization includes terms which in collinear factorization would only be generated at higher orders.

For VM production at small x and $t \sim 0$ the momentum fractions x, x' satisfy

$$1 \gg x \simeq \frac{Q^2+M^2}{W^2+Q^2} \gg x' \simeq \frac{l_T^2}{W^2+Q^2}\,. \tag{4}$$

In this regime the effect from skewing is mainly coming from the off-diagonal evolution and no new non-perturbative input is required. The enhancement factor R_g of the skewed (integrated) gluon $H_g(x,x')$ over the normal forward gluon $H_g(x,x)$ is approximated by [6] (see also [7] for a different approach)

$$R_g = \frac{H_g(x,x'\ll x)}{H_g(x,x)} = \frac{2^{2\lambda+3}}{\sqrt{\pi}}\frac{\Gamma(\lambda+\frac{5}{2})}{\Gamma(\lambda+4)}\,. \tag{5}$$

The effective power $\lambda(Q^2)$ of the gluon can be calculated numerically for the amplitudes $\mathcal{A}^{L,T}$ through $\lambda = \partial\log\mathcal{A}^{L,T}/\partial\log(1/x)$. This description assumes a behaviour of the gluon $xg \sim x^{-\lambda}$ as expected at small x. We have checked that this approximation of the enhancement factor R_g due to skewing is sufficient in the cases discussed below. Note that skewing is a leading $\ln Q^2$ effect and can be sizeable when the gluon is sampled at 'large' scales Q^2 or M^2, e.g. $R_g^2 \sim 2$ for Υ photoproduction at HERA.

2.3. *Soft contributions.*

It is clear from Eqs. (2, 3) that k_T-factorization for the two-gluon exchange requires the knowledge of the unintegrated gluon for all scales l_T^2. Parametrizations of the parton distributions are typically given only down to scales ~ 2 GeV^2. For smaller scales we have to extrapolate into the infrared (IR) regime.[d] The simplest ansatz is the 'linear extrapolation'

$$\alpha_s(l_T^2)g(x,l_T^2) = (l_T^2/l_0^2)\alpha_s(l_0^2)g(x,l_0^2) \tag{6}$$

[d]Contributions from large momenta are suppressed through the $1/l_T^4$ term in (2), so there is no problem with the upper integration limit.

for scales $l_T^2 < l_0^2$. This gives a relatively smooth continuation of the phenomenological gluon distributions and corresponds to a constant gluon-proton cross section at small scales. Of course other, more complicated, descriptions at low scales are possible. We will discuss the ambiguity introduced through the modelling of the IR regime below.

2.4. *Real part and higher order corrections.*

When the basic amplitudes $\mathcal{A}^{L,T}$ are derived from Feynman diagrams in the high energy (leading $\log 1/x$) limit, typically only their (leading) imaginary part is obtained analytically. Under the assumption of crossing symmetry and a power behaviour $\text{Im}\,\mathcal{A} \sim s^\lambda$ the real part is then calculated (numerically) through $\text{Re}\,\mathcal{A} = \tan(\pi\lambda/2)\,\text{Im}\,\mathcal{A}$, where $\lambda(\mu^2) = \partial \log \mathcal{A}/\partial \log(1/x)$. It is found that the inclusion of the real part enhances the cross sections for diffractive VM production at HERA by about 15 to 30%, where the effect is bigger for a larger VM mass or Q^2.

Although the framework as described above goes well beyond the naive leading order formula (1), a complete prediction at next-to-leading order (NLO) has not been achieved yet. First steps in this direction have been undertaken, see [8], but full NLO corrections to the $(q\bar{q})(2g)$-vertex within k_T-factorization and for general scale assignments and photon polarizations are still not available. If one estimates the full corrections from the π^2 enhanced terms analogous to the well known corrections in the Drell-Yan process, the cross section enhancement would be $\sigma = \sigma^0 \exp[\pi^2 C_F \alpha_s/\pi]$.[9] Of course other, unknown terms may well compensate or even enhance this large correction. However, as the NLO corrections are expected to be approximately scale independent, we may parametrize them in form of a K factor. The (constant) K factor can be obtained from fits to experimental data by sacrificing e.g. one bin in a Q^2 or W distribution. Thus the overall normalization, which is anyway uncertain due to the unknown mass interval integrated over in the Parton Hadron Duality approach, can be fixed.

3. Results

Before confronting the theoretical predictions with data, let us further discuss the uncertainties from the IR regime and the effects from skewing. Figure 2 displays MRT predictions for the W dependence of the diffractive J/ψ production cross section for three different gluon distributions used as input. The dash-dotted lines are the complete predictions including all corrections as discussed above. The solid lines are obtained neglecting the

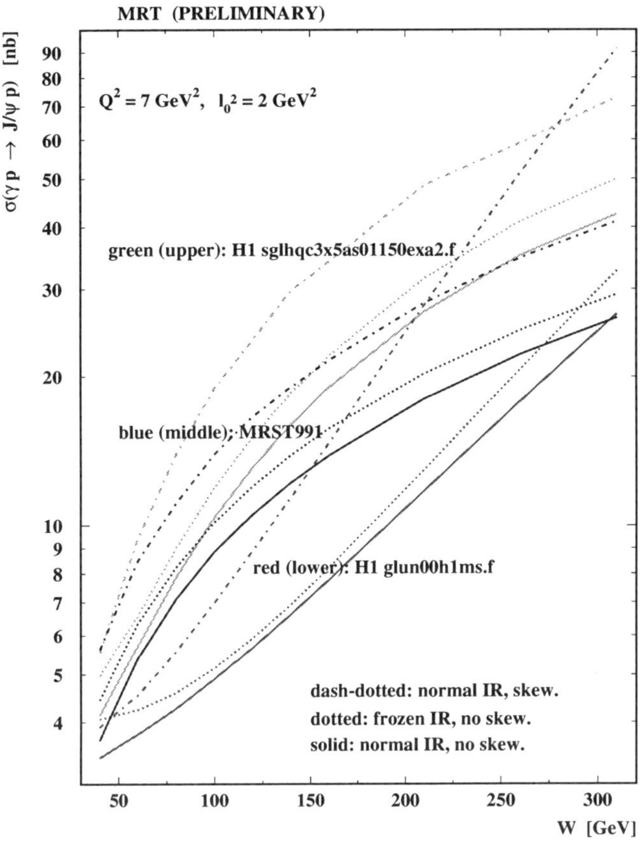

Figure 2. MRT predictions for the W dependence of diffractive J/ψ production for different input gluon distributions. Solid (and dotted) lines are without skewing effects, dash-dotted ones with. The dotted lines are obtained with a different treatment of the IR regime, see the text for more details.

corrections from skewing. Clearly the skewing enhances the cross section considerably. The different W behaviour of the predictions for different gluon inputs is a direct consequence of the large variation of the x dependence at small to intermediate scales of the various fits used. This is demonstrated in Fig. 3 where we display $xg(x, \mu^2)$ as a function of x at three different scales μ^2 for the three gluons used in Fig. 2. The stronger curvature with a steep rise at small x of the H1 2000 fit (routine 'glun00h1ms.f') leads to the markedly different W behaviour of the cross section. This effect is further enhanced through the skewing corrections.

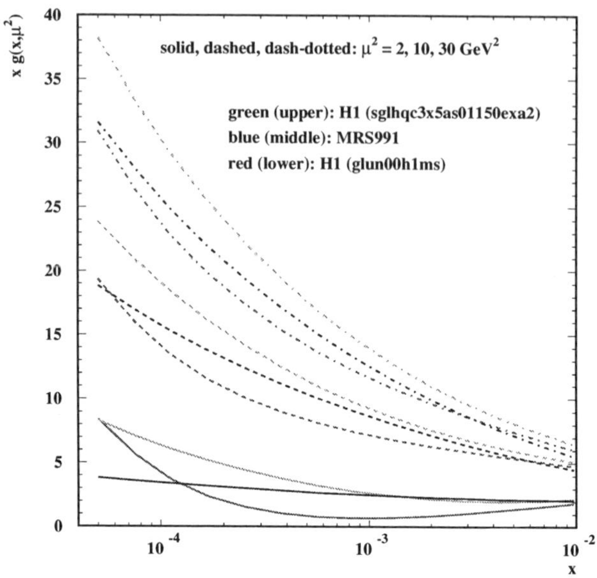

Figure 3. The three different gluon distributions $xg(x, \mu^2)$ as used in Fig. 2 in the relevant x range for three values of the scale μ^2.

In contrast, the predictions depend only mildly on the way the IR regime is modelled. This can be seen by comparing the solid with the dotted curves in Fig. 2. For the former the 'linear extrapolation' of the gluon, Eq. 6, is used in the IR regime. The latter are obtained with the IR approximation where only the leading term $\sim l_T^2$ in the expansion of the $\gamma \to q\bar{q}$ subamplitude ϕ is used in (2) for $l_T^2 < l_0^2$. In this approximation (labelled 'frozen IR' on Fig. 2) the IR part of the integral in (2) is proportional to $xg(x, l_0^2)$. As the neglected subleading terms are negative, this approximation constitutes an upper bound for the contribution from the IR regime and should overestimate the true contribution. However, we see from Fig. 2 that the effect is small compared to the skewing corrections and that the W dependence is nearly unaffected; only for the H1 2000 gluon with its strong x and scale dependence the shape of the cross section is altered towards very small or large values of W.

Thus the IR regime is fairly well controlled in the MRT approach using k_T-factorization.[e] Nevertheless, through interference effects, the total

[e]This is plausible as both the average scale sampled in the l_T^2 integral in (2) and the factorization scale in collinear factorization are clearly in the perturbative regime for

contributions of the IR regime to the total cross section are sizeable and a better understanding of the gluon at small scales is desirable.

It should be noted that the variation of the cross section predictions w.r.t. the choice of the input gluon distribution is by far larger than the uncertainty of the prediction which could be derived by scanning over the 'error band' of individual gluon fits. Again, this just reflects the fact that at small x and small to intermediate scales both normalization and form of the gluon are not well constrained by the data used in the fits. Quite different functional forms can lead to good fits, and the uncertainty of the PDF will be underestimated if these systematic uncertainties are not taken into account.

It is now most interesting to confront the predictions using different input gluon distributions with HERA data. The two panels in Fig. 4 display the cross section data as a function of W for diffractive J/ψ photo- and electroproduction from H1 [10] and ZEUS [11] compared to MRT predictions. As indicated the curves are multiplied with a K factor which is fitted in photo- and kept for electroproduction. Although this factor is quite different for different gluons, the data at large W (small x) are not well fitted by and clearly disfavour gluon parametrizations which are very steep or flat (or even turning negative) at small x like the H1 2000 or MRST02 fits, respectively.

4. Diffractive structure function F_2^D for $\beta \to 1$

A similar approach can also be used to predict inclusive diffraction in the region of large $\beta = Q^2/(Q^2 + M^2) \to 1$.[5] In this limit the longitudinal diffractive structure function F_L^D dominates the one from transverse photons. While the former is infrared safe the latter has to be regularized by hand, e.g. by a light quark constituent mass m_q, which indroduces additional uncertainties. Figure 5 shows the predictions together with data (see Ref. [5] for more details). Again, the enhancement due to skewing is very relevant and roughly a factor of two (lower panel of Fig. 5), and the sensitivity to the input gluon is high (upper panel).

J/ψ (or ρ electro-) production.

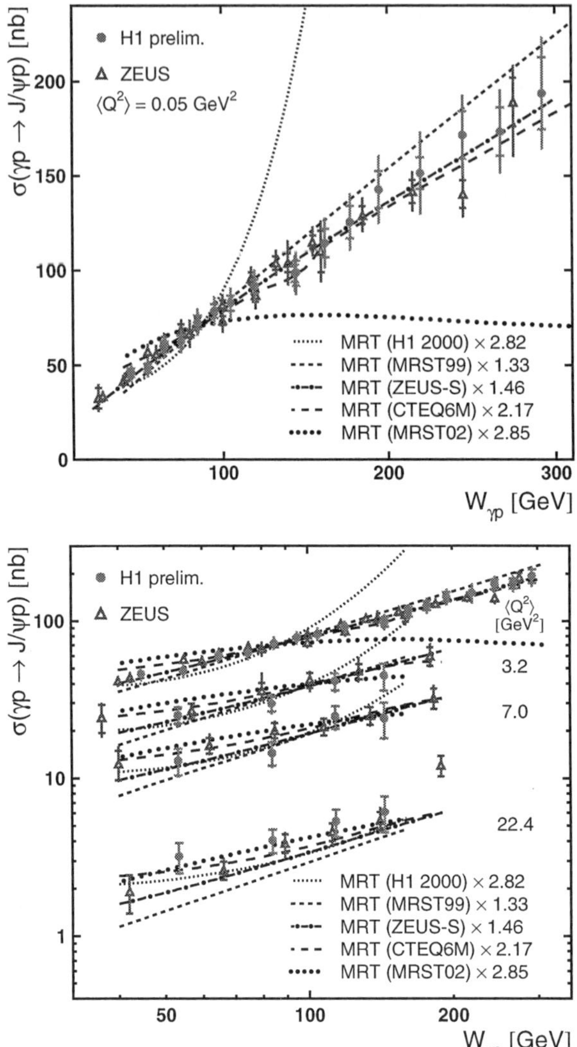

Figure 4. H1 [10] and ZEUS [11] data for diffractive J/ψ production compared to MRT predictions. Figure provided by P. Fleischmann, H1 Collaboration.

5. Conclusions

Diffractive production of vector mesons at HERA can be calculated in QCD. Using Parton Hadron Duality the energy and Q^2 behaviour of the cross section (as well as the ratio L/T not discussed here) is predicted

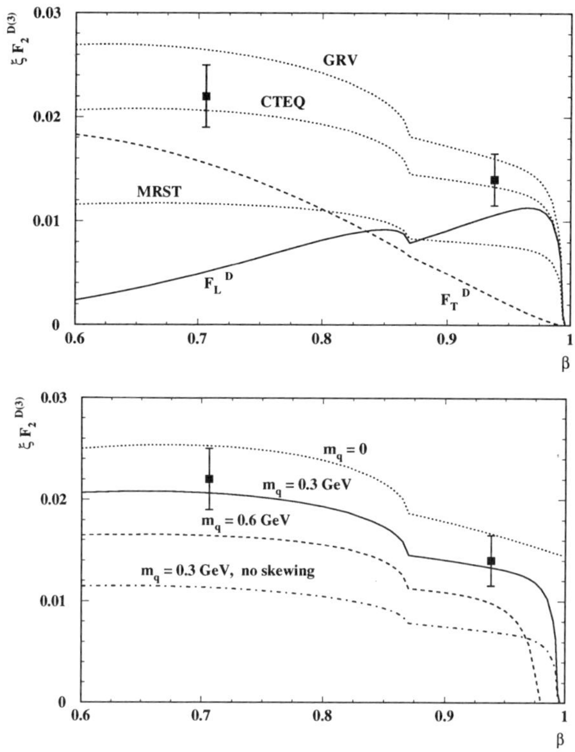

Figure 5. Diffractive structure function F_2^D for large β in perturbative QCD. Upper panel: Contributions F_L^D, F_T^D separately and the dependence of F_2^D on the input gluon. Lower panel: Effect of skewing corrections and IR-regularization via a light quark constituent mass. Figure taken from Ref. [5].

without dependence on the VM wave function. Non-perturbative input is parametrized in form of (unintegrated) generalized parton distribution functions where effects from skewing turn out to be sizeable. The cross sections have a strong sensitivity on the input gluon in the region of small x and small to intermediate scales where the standard PDF fits are not well constrained. The complicated relation between cross sections and input PDF makes a direct determination of the gluon difficult. However, with recent data from HERA certain fits are already strongly disfavoured. Data for different VMs ($\rho, \omega, \phi, J/\psi, \Upsilon$) in an as large as possible range of W and Q^2 (and possibly F_2^D at large β) will allow to better constrain the gluon at small x. On the theoretical side the complete calculation of the next-to-leading order corrections to the $q\bar{q}(gg)$ vertex is most important.

Acknowledgments

It is my pleasure to thank all the organizers for their efforts in making this workshop such a productive and enjoyable meeting.

References

1. M. G. Ryskin, *Z. Phys.* **C57**, 89 (1993).
2. A. D. Martin, M. G. Ryskin and T. Teubner, *Phys. Rev.* **D55**, 4329 (1997); *Phys. Lett.* **B454**, 339 (1999); *Phys. Rev.* **D62**, 014022 (2000).
3. M. G. Ryskin, R. G. Roberts, A. D. Martin and E. M. Levin, *Z. Phys.* **C76**, 231 (1997).
4. L. Frankfurt, W. Koepf and M. Strikman, *Phys. Rev.* **D54**, 3194 (1996); *Phys. Rev.* **D57**, 512 (1998).
5. A. Hebecker and T. Teubner, *Phys. Lett.* **B498**, 16 (2001).
6. A. G. Shuvaev, K. J. Golec-Biernat, A. D. Martin and M. G. Ryskin, *Phys. Rev.* **D60**, 014015 (1999).
7. S. V. Goloskokov and P. Kroll, *Eur. Phys. J.* **C42**, 281 (2005).
8. D. Yu. Ivanov, M. I. Kotsky and A. Papa, *Eur. Phys. J.* **C38**, 195 (2004), and references therein.
9. E. M. Levin, A. D. Martin, M. G. Ryskin and T. Teubner, *Z. Phys.* **C74**, 671 (1997).
10. H1 Collaboration, A. Aktas *et al.*, `arXiv:hep-ex/0510016`.
11. ZEUS Collaboration, S. Chekanov *et al.*, *Nucl. Phys.* **B695**, 3 (2004); *Eur. Phys. J.* **C24**, 345 (2002). ZEUS Collaboration, J. Breitweg *et al.*, *Eur. Phys. J.* **C6**, 603 (1999); *Z. Phys.* **C75**, 215 (1997).

INCLUSIVE DIFFRACTION

L. FAVART
I.I.H.E., CP-230
Université Libre de Bruxelles,
1050 Brussels
Belgium
lfavart@ulb.ac.be

Results are given on the measurements of the hard diffractive interactions at HERA *ep* collider. The structure of the diffractive exchange in terms of partons and the factorisation properties are discussed, in particular by comparing the QCD predictions for dijets and D^* with measurements both in the photo and electroproduction regimes.

1. Introduction

The high energies of the HERA *ep* collider and of the Tevatron $p\bar{p}$ collider allow us for the first time to study diffraction in terms of perturbative QCD (pQCD), i.e. in the presence of a hard scale, which is called hard diffraction. At HERA the diffractive interaction takes place between the hadronic behaviour of the exchanged virtual photon and the proton (see Fig. 1a), and between the two protons at Tevatron (see Fig. 1b). The possible hard scales are large Q^2, the negative of the four-momentum squared of the exchanged virtual photon, large transverse energy, E_T, in jet production, heavy-quark mass or large momentum transfer squared at the proton vertex, t. The hard scale presence of large Q^2 at HERA and large E_T at Tevatron, in particular, gives the possibility to probe the partonic structure of the diffractive exchange.

This article concentrates on the factorisation properties using a structure function approach giving insight into the understanding of the nature of diffraction in terms of partons. Before reviewing recent measurements, a short discussion on the parton densities of the diffractive exchange and their relation to the diffractive cross section is given.

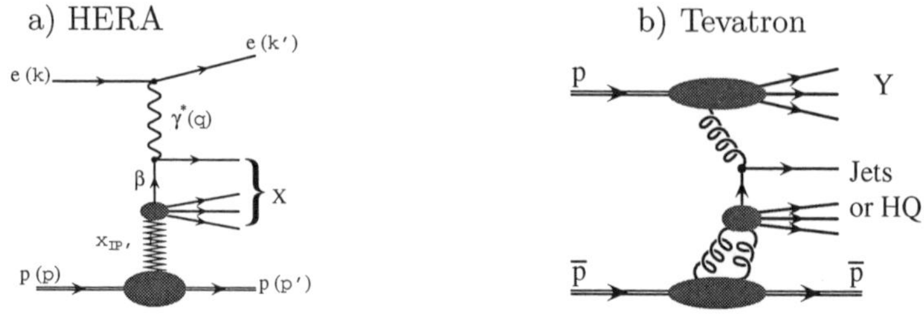

Figure 1. Basic diagrams for diffraction in presence of a hard scale at HERA and at Tevatron

Although it will not be covered here, the study of diffraction has several other topics of interest on top of the diffractive exchange measurement in terms of partons. It gives access to the partons correlation through the exclusive final state measurements (generalised parton distribution formalism). It allows to test the region of validity of the different asymptotic dynamical approaches of QCD that are DGLAP and BFKL from the measurement of vector meson or photon exclusive production at large values of $|t|$. Finally, the alternative approach of colour Dipole models allows us to study the transition between non-perturbative and perturbative regimes and to test the presence of a gluon density saturation in the proton.

2. Partonic structure of diffractive exchanges and factorisation properties

The inclusive diffractive cross section at HERA, $ep \to eXp$, can be defined with the help of four kinematic variables conveniently chosen as Q^2, $x_{I\!P}$, β and t, where $x_{I\!P}$ and β are defined as

$$x_{I\!P} \simeq \frac{Q^2 + M_X^2}{Q^2 + W^2}, \qquad \beta \simeq \frac{Q^2}{Q^2 + M_X^2};$$

M_X being the invariant mass of the X system, and W the $\gamma^* - p$ center of mass energy. $x_{I\!P}$ can be interpreted as the fraction of the proton momentum carried by the exchanged Pomeron and β is the fraction of the exchanged momentum carried by the quark struck by the photon, or in other terms, the fraction of the exchanged momentum reaching the photon. These variables are related to the Bjorken x_{Bj} scaling variable by the relation $x_{Bj} = \beta \cdot x_{I\!P}$. The presence of the hard scale, Q^2, ensures that the virtual photon is point-like and that the photon probes the partonic structure of the diffractive exchange (Fig. 1a), in analogy with the inclusive DIS processes.

Factorisation properties in hard diffraction

For hard QCD processes in general, like high-E_T jet production or DIS, the cross section can be factorized into two terms: the parton density and the hard parton-parton cross section. The cross section can be written as

$$\sigma = \sum_i f_i(\xi, \mu^2)\, \hat{\sigma}_{i\gamma}(\xi, \mu^2)$$

where i runs over all parton types, f_i is the parton density function for the i-th parton with longitudinal momentum fraction ξ, which is probed at the factorisation scale μ. $\hat{\sigma}_{i\gamma}$ denotes the cross section for the interaction of the i-th parton and the virtual photon. Such an expression, often referred to as the QCD factorisation theorem, is well supported by the data. If the theorem holds, only one parton per hadron is coupled to the hard scattering vertex.

The theorem is proven to be applicable at all orders in the strong force coupling constant α_s for the leading log Q^2 for hard inclusive diffraction [1] in ep collisions at large Q^2, namely

$$\frac{d\sigma(x, Q^2, x_{I\!P}, t)}{dx_{I\!P}dt} = \sum_i \int_x^{x_{I\!P}} dz\, \hat{\sigma}_{i\gamma}(z, Q^2, x_{I\!P})\, f_i^D(z, Q^2, x_{I\!P}, t)$$

where z is the longitudinal momentum fraction of the parton in the proton, $\hat{\sigma}_{i\gamma}$ is again the hard scattering parton-photon cross section for hard diffraction and f_i^D is the diffractive parton density for the i-th parton. f_i^D can be regarded as the parton density of the diffractive exchange occuring at a given $(x_{I\!P}, t)$. If such a theorem holds, f_i^D should be universal for all hard processes, e.g. inclusive diffraction, jet or heavy-quark production etc.

If the scattered proton is not detected in Roman Pots, the t variable is not measured with enough accuracy and is integrated over. In analogy with non-diffractive DIS scattering, the measured cross section is expressed, in the neutral current case, in the form of a three-fold diffractive structure function $F_2^{D(3)}(Q^2, x_{I\!P}, \beta)$ (neglecting the longitudinal contribution and Z exchange),

$$\frac{\mathrm{d}^3\sigma\ (ep \to eXp)}{\mathrm{d}Q^2\ \mathrm{d}x_{I\!P}\ \mathrm{d}\beta} = \frac{4\pi\alpha^2}{\beta Q^4}\ (1 - y + \frac{y^2}{2})\ F_2^{D(3)}(Q^2, x_{I\!P}, \beta),$$

where y is the usual scaling variable, with $y \simeq W^2/s$.

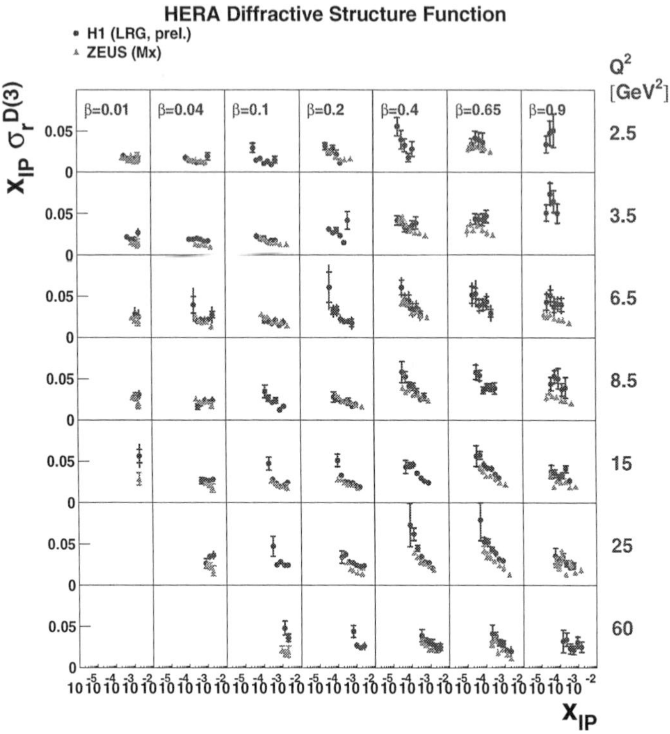

Figure 2. H1 and ZEUS reduced cross section (equal to $F_2^{D(3)}$ if F_L^D and F_3^D can be neglected) measurements as a function $x_{I\!P}$ for fixed values of Q^2 and β.

Conveniently, the Regge factorisation is applied where $F_2^{D(3)}$ is written in the form

$$F_2^{D(3)}(Q^2, x_{I\!P}, \beta) = f_{I\!P/p}(x_{I\!P}) \cdot F_2^D(Q^2, \beta),$$

using the approximation that the pomeron ($I\!P$) flux $f_{I\!P/p}(x_{I\!P})$ is independent of the $I\!P$ structure $F_2^{D(2)}$, by analogy with the hadron structure functions, β playing the role of Bjorken x_{Bj}. The $I\!P$ flux is parameterised using a Regge inspired form $f_{I\!P/p} = e^{bt}/x_{I\!P}^{2\alpha_{I\!P}(t)-1}$.

H1 [2] and ZEUS [3] measurements of $F_2^{D(3)}$ diffractive structure function are presented in Fig. 2 as a function of $x_{I\!P}$ for fixed values of Q^2 and β. The reduced cross section σ_r^D presented in the figure is equal to $F_2^{D(3)}$ if F_L^D and F_3^D can be neglected.

Both measurements are in good general agreement, although the Q^2

dependence at fixed β is stronger in the H1 measurement and the last bin in β exhibits higher values for the H1 measurement for $Q^2 < 25$ GeV^2.

To fit the $F_2^{D(3)}(Q^2, x_{I\!P}, \beta)$ points H1 includes a sub-leading Reggeon ($I\!R$) trajectory in addition to the Pomeron which is not included for the ZEUS measurement as it does not improve the quality of the fit in that case. Corresponding value of the Pomeron intercept is $\alpha_{I\!P}(0) = 1.173$ ($\alpha_{I\!P}(0) = 1.132 \pm 0.006$) from H1 (ZEUS) fit.

As shown by H1 [2], the Regge factorisation ansatz holds within the present precision of the measurement as the $F_2^{D(2)}(Q^2, \beta)$ measurement is not sensitive to the $x_{I\!P}$ value. The Q^2 dependence exhibits a more important scaling violation than in the F_2 structure function measured in non-diffractive deep inelastic scattering indicating that the exchanged object in diffraction has an important gluon content.

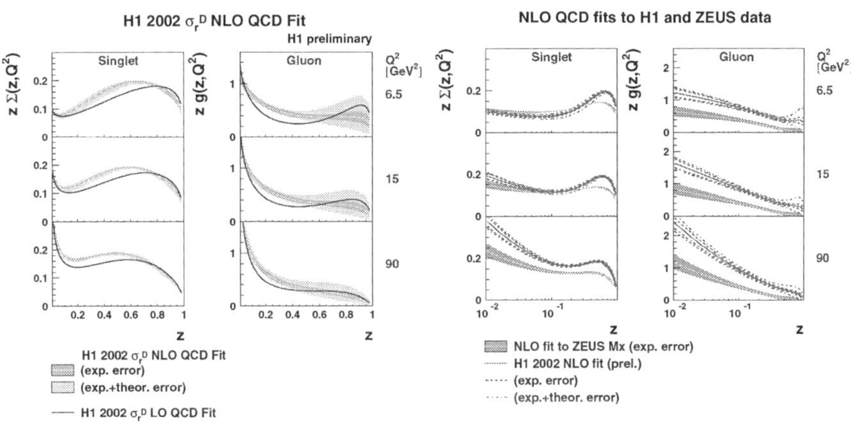

Figure 3. Quark (singlet) and gluon densities in the $I\!P$ extracted from the QCD fit of $F_2^{D(2)}(Q^2, \beta)$ as a function of z. **left)** From H1 measurement. **right)** From ZEUS measurement.

By analogy to the QCD evolution of the proton structure function F_2, one can attempt to extract the partonic structure of the Pomeron from the Q^2 evolution of $F_2^{D(2)}$. Starting the QCD evolution at $Q_0^2 = 6.5$ GeV^2, the extracted partonic distributions [4,5] are shown in Fig. 3 separately for the gluon and the singlet quark components as a function of z, the Pomeron momentum fraction carried by the parton entering the hard interaction. The left 2 columns of the plots correspond to the singlet quark and the gluon components extracted from the H1 measurement; the right 2 columns from the ZEUS measurement. As a consequence of the differences between the H1 and ZEUS $F_2^{D(2)}$ a weaker gluon component is found from the ZEUS fit.

The QCD factorisation is expected to break down at large β values where higher twist terms may become important. A part of them corresponds to the contribution of the hard scale integration in the Pomeron [6]. Such a QCD Pomeron corresponds to the dominant term for exclusive Vector meson contribution like J/Ψ and for the Deeply Virtual Compton Scattering [7].

Dijet and charm productions in diffractive electroproduction To test QCD factorisation for diffractive dijet production in the electroproduction regime ($Q^2 >> 1$ GeV^2), the H1 dijet cross section [8] in the kinematic range $Q^2 > 4$ GeV^2 and $x_{I\!P} < 0.03$ is compared to the NLO QCD prediction in Fig. 4, using the extracted diffractive parton densities as obtained by H1. The cross sections were corrected to asymmetric cuts on the jet transverse momentum $p_{T,1(2)} > 5(4)$ GeV, to facilitate comparisons with NLO calculations. The inner error band of the NLO calculations represents the renormalisation scale uncertainty, whereas the outer band includes the uncertainty in the hadronisation corrections. Within the uncertainties, the data are well described in both shape and normalisation by the NLO calculations, in agreement with QCD factorisation. ZEUS measures the same process in an equivalent kinematic range and comparing it to different diffractive parton densities concludes: "the differences observed between the sets of predictions may be interpreted as an estimate of the uncertainty associated with the dPDFs... A better understanding of the dPDFs and their uncertainties is required before a firm statement about the validity of QCD factorisation can be made." [9].

The D^* meson production measurements in diffractive electroproduction were achieved by H1 [10] and ZEUS [3] for the kinematic range $Q^2 > 2$ GeV^2, $x_{I\!P} < 0.03$ and $p^*_{T,D^*} > 2$ GeV, where the latter variable corresponds to the transverse momentum of the D^* meson in the photon-proton centre-of-mass frame. NLO QCD calculations were performed interfacing the H1 diffractive parton distributions. The renormalisation and factorisation scales were set to $\mu^2 = Q^2 + 4m_c^2$. A comparison of the calculations with the D^* H1 and ZEUS data is shown in Fig. 4-right. The inner error

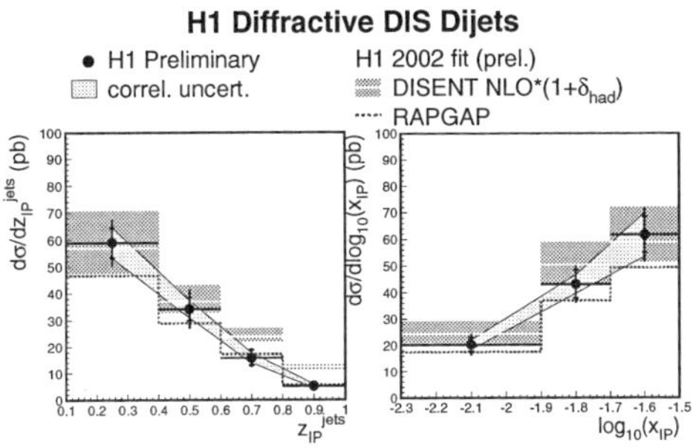

(a) (b)

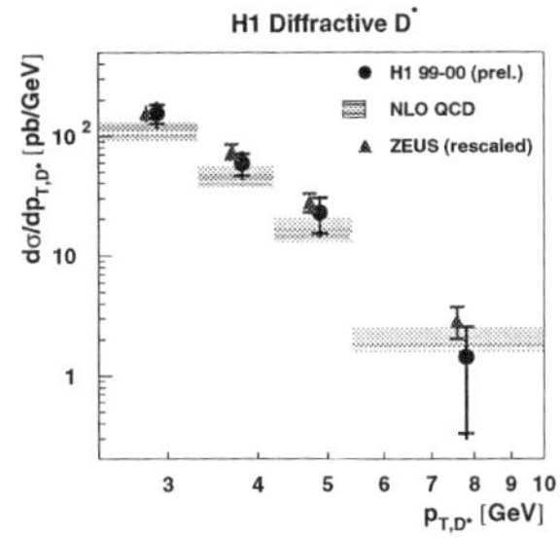

Figure 4. **left**: H1 measurement of diffractive dijet cross section in electroproduction as a function of $z_{I\!P}^{jets}$, an estimator for the parton momentum fraction of the diffractive exchange entering the hard sub-process, and as a function of $x_{I\!P}$ as measured by H1 and ZEUS. **right**: Diffractive D^* meson cross sections in electroproduction differential in p^*_{T,D^*}.

band of the NLO calculation represents the renormalisation scale uncertainty, whereas the outer error band includes variations of the charm mass and of the Peterson fragmentation function. Within the uncertainties, the data are well described in both shape and normalisation by the NLO calculations, supporting the idea of QCD factorisation.

3. Comparison with hard diffraction at the TEVATRON

One of the most striking features of the hard diffractive process measured at the Tevatron is a large suppression of the cross section with respect to the prediction based on the diffractive parton densities obtained from the HERA F_2^D data. Figure 5-left shows the comparison of the dijet cross section in the single-diffractive process measured by CDF at the Tevatron [11] to the prediction using the diffractive parton densities discussed in the previous section based on the H1 measurement. Although the prediction reproduces the shape of the data in the low-β region, the magnitude of the cross section is smaller by a factor 5 to 10. This indicates a strong factorisation breaking between HERA and the Tevatron: the diffractive parton densities are not universal between these two environments. Additionally, CDF has measured the ratio of double-diffractive over single-diffractive processes (shown in the right plot of Fig. 5) to be 0.19 ± 0.07. This indicates that the formation of a second gap is not (or only slightly) suppressed.

The reason for the breaking is not yet clearly known. It is usually attributed to re-scattering between spectator partons in the two beam remnants where one or more colour-octet partons are exchanged, which destroys the already formed colour-singlet state.

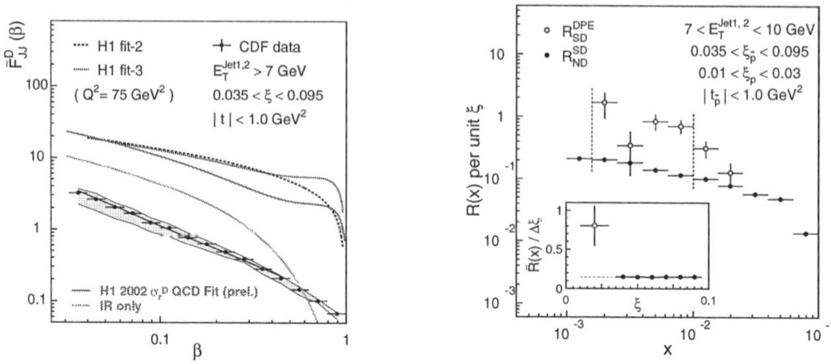

Figure 5. **left** : measured dijet cross section in single-diffractive processes in comparison with the prediction using the diffractive PDF's obtained by H1. **right** : measured dijet cross section in single-diffractive processes compared to double-diffractive processes by CDF.

Factorisation test with hard photoproduced diffraction at HERA

QCD factorisation can be further investigated within HERA looking at the diffractive dijet photoproduction ($Q^2 \sim 0$), where the hard scale is provided by the E_T of the jets. Factorisation is expected not to hold in photoproduction events, where the resolved process has a photon remnant, allowing re-scattering. On the other hand, the direct process does not have a beam remnant and the suppression of diffractive events is expected to be much smaller than in the resolved process [12].

Figure 6-left shows the dijet cross section in diffractive photoproduction measured by ZEUS as a function of x_γ^{jets} [13], the longitudinal momentum fraction of the parton that participated in the hard scattering (see Fig. 6-right diagram), reconstructed from the dijet momenta. Resolved events dominate in the low-x_γ^{jets} region while the direct process is concentrated at x_γ^{jets} close to one. The cross sections compared to the NLO QCD prediction exhibit a factorisation breaking both in the direct and in the resolved parts. The NLO QCD prediction required a global factor of 0.5 to be able to describe the data. Similar results have been measured by H1 [8].

Recently D^* meson photoproduction cross section in diffraction has been measured by ZEUS [14] for the kinematic range $P_T^{D^*} > 1.9$ GeV, $\eta_{D^*} < 1.6$, $130 < W < 300$ GeV and $0.001 < x_{I\!P} < 0.035$. The measurement is found

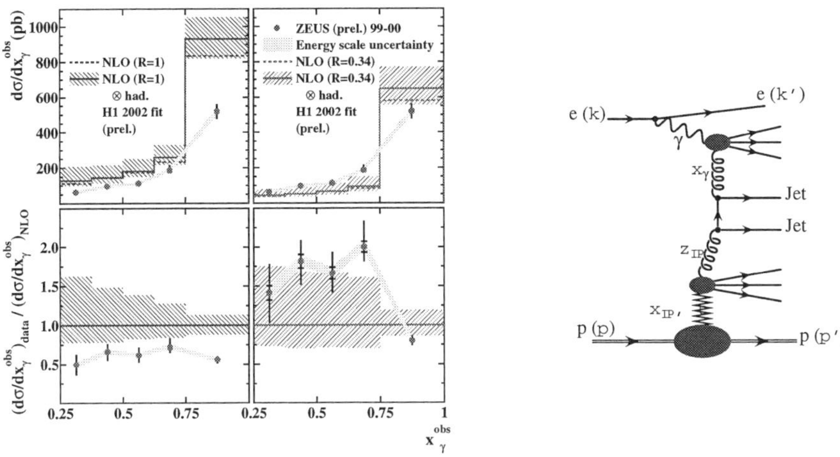

Figure 6. **left**: Dijet cross section in photoproduction at HERA measured by ZEUS as a function of $x_\gamma^{\rm jets}$. **right**: Diagram for dijet in photoproduction at HERA.

to be in good agreement in shape and in normalisation with the NLO QCD prediction as shown in Fig. 7 presenting the data ratio to the theory.

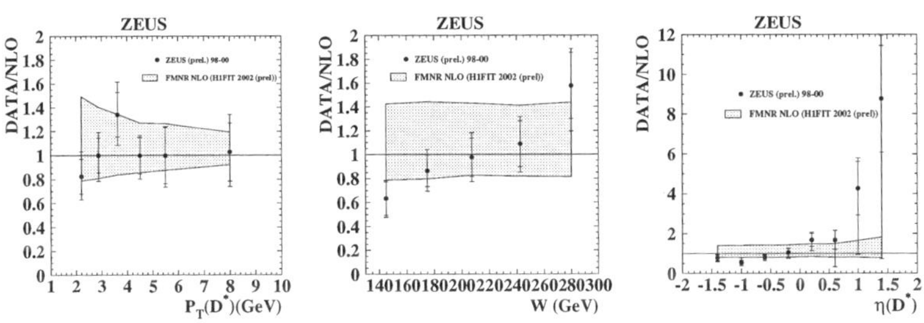

Figure 7. **left**: Ratio of the D^* meson cross section to the NLO QCD prediction in photoproduction measured by ZEUS as a function of $P_T^{D^*}$, W and η.

4. Concluding remarks

After almost ten years of research at HERA we are reaching a perturbative QCD understanding of hard diffraction in *ep* interactions. The partonic structure of the diffractive exchange has been measured and is found to be dominated by gluons. In its validity domain ($Q^2 >> 1$ GeV^2 in *ep* collisions), the QCD factorisation holds, as confirmed by charm and dijet productions. In $p\bar{p}$ collisions, at Tevatron, a breakdown of the QCD factorisation of a factor 5 to 10 is observed for one gap formation and no (or weak) additional gap suppression is observed for a second gap formation. Rescattering corrections seem to be important in $p\bar{p}$ and in ep collisions in dijet photoproduction. The D^* photoproduction is in agreement with the NLO QCD prediction within the present precision.

A global understanding of inclusive and exclusive hard diffractions has progressed . Many more results and a deeper understanding are needed and expected with the coming data at HERA II, Run II at Tevatron, Compass and in a further future at LHC.

5. Acknowledgments

It is a pleasure to thank the organizers of the Ringberg workshop on *New Trends in HERA Physics 2005* for their kind invitation and the perfect organization of this very interesting workshop. This work is supported by the Fonds National de la Recherche Scientifique Belge (FNRS).

References

1. J.C. Collins, *Phys. Rev.* D **57**, 3051 (1998), [erratum-ibid. **D 61**, 019902 (2000)].
2. H1 Coll., paper 980 subm. to ICHEP 2002, Amsterdam.
3. S. Chekanov *et al.* [ZEUS Collaboration], Eur. Phys. J. C **38** 43, (2004) [hep-ex/0408009].
4. H1 Coll., paper 113 subm. to EPS 2003, Aachen.
5. P. Newman and F. P. Schilling, Proceedings of the Workshop on the *Implications of HERA for LHC Physics*, Hamburg, 2005, [hep-ex/0511032].
6. G. Watt, These proceedings, [hep-ph/0511333].
7. L. Favart, Proceedings of the International Conference on Elastic and Diffractive Scattering, Rencontres de Blois, EDS05, [hep-ex/0510031].
8. H1 Coll., paper 6-0177 subm. to ICHEP 2004, Beijing.
9. ZEUS Coll., paper 295 subm. to Lepton-Photon 2005, Uppsala and Addendum on: `http://www-zeus.desy.de/physics/phch/conf/lp05_eps05/295/addendum_295.pdf`
10. H1 Coll., paper 5-0165 subm. to ICHEP 2004, Beijing.
11. CDF Coll., T. Affolder *et al.*, *Phys. Rev. Lett.* **84**, 5043 (2000).
12. M. Klasen, These proceedings, [hep-ph/0512255].
13. ZEUS Coll., paper 6-0249 subm. to ICHEP 2004, Beijing.
14. ZEUS Coll., paper 268 subm. to Lepton-Photon 2005, Uppsala.

FROM FACTORIZATION TO ITS BREAKING IN DIFFRACTIVE DIJET PRODUCTION

M. KLASEN
Laboratoire de Physique Subatomique et de Cosmologie, Université Joseph Fourier/CNRS-IN2P3, 53 Avenue des Martyrs, 38026 Grenoble, France
E-mail: klasen@lpsc.in2p3.fr

When comparing recent experimental data from the H1 and ZEUS Collaborations at HERA for diffractive dijet production in deep-inelastic scattering (DIS) and photoproduction with next-to-leading order (NLO) QCD predictions using diffractive parton densities, good agreement is found for DIS. However, the dijet photoproduction data are overestimated by the NLO theory, showing that factorization breaking occurs at this order. While this is expected theoretically for resolved photoproduction, the fact that the data are better described by a global suppression of direct *and* resolved contribution by about a factor of two comes as a surprise. We therefore discuss in some detail the factorization scheme and scale dependence between direct and resolved contributions and propose a new factorization scheme for diffractive dijet photoproduction.

1. Diffractive *ep* Scattering

It is well known that in high-energy deep-inelastic *ep*-collisions a large fraction of the observed events are diffractive. These events are defined experimentally by the presence of a forward-going system Y with four-momentum p_Y, low mass M_Y (in most cases a single proton and/or low-lying nucleon resonances), small momentum transfer squared $t = (p - p_Y)^2$, and small longitudinal momentum transfer fraction $x_{I\!P} = q(p - p_Y)/qp$ from the incoming proton with four-momentum p to the system X (see Fig. 1). The presence of a hard scale, as for example the photon virtuality $Q^2 = -q^2$ in deep-inelastic scattering (DIS) or the large transverse jet momentum p_T^* in the photon-proton centre-of-momentum frame, should then allow for calculations of the production cross section for the central system X with the known methods of perturbative QCD. Under this assumption, the cross section for the inclusive production of two jets, $e+p \to e+2\,\text{jets}+X'+Y$, can be predicted from the well-known formulæ for jet production in non-diffractive *ep* collisions, where in the convolution of the partonic cross section with the

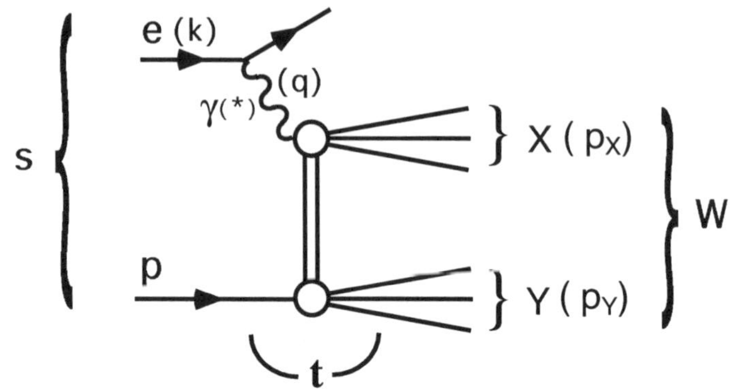

Figure 1. Diffractive scattering process $ep \to eXY$, where the hadronic systems X and Y are separated by the largest rapidity gap in the final state.

parton distribution functions (PDFs) of the proton the latter ones are replaced by the diffractive PDFs. In the simplest approximation, they are described by the exchange of a single, factorizable pomeron/Regge-pole.

2. Diffractive Parton Distribution Functions

The diffractive PDFs have been determined by the H1 Collaboration at HERA from high-precision inclusive measurements of the DIS process $ep \to eXY$ using the usual DGLAP evolution equations in leading order (LO) and next-to-leading order (NLO) and the well-known formula for the inclusive cross section as a convolution of the inclusive parton-level cross section with the diffractive PDFs [1]. A similar analysis of inclusive measurements has been published by the ZEUS Collaboration [2,3]. A longer discussion of the extraction of diffractive PDFs can be found elsewhere [4,5].

3. QCD Factorization in Hard Diffraction

For inclusive diffractive DIS it has been proven by Collins that the formula referred to above is applicable without additional corrections and that the inclusive jet production cross section for large Q^2 can be calculated in terms of the same diffractive PDFs [6]. The proof of this factorization formula, usually referred to as the validity of QCD factorization in hard diffraction, may be expected to hold for the direct part of photoproduction ($Q^2 \simeq 0$) or low-Q^2 electroproduction of jets [6]. However, factorization does not hold for hard processes in diffractive hadron-hadron scattering. The problem is that soft interactions between the ingoing two hadrons and their remnants occur in both the initial and final state. This agrees with experimental measurements at the Tevatron [7]. Predictions of diffractive dijet cross sections for $p\bar{p}$ collisions as measured by CDF using the same PDFs as determined by H1 [1] overestimate the measured cross section by up to an order of magnitude [7]. This suppression of the CDF cross section can be explained by considering the rescattering of the two incoming hadron beams which, by creating additional hadrons, destroy the rapidity gap [8].

4. Factorization Breaking in Diffractive Photoproduction

Processes with real photons ($Q^2 \simeq 0$) or virtual photons with fixed, but low Q^2 involve direct interactions of the photon with quarks from the proton as well as resolved photon contributions, leading to parton-parton interactions and an additional remnant jet coming from the photon (for a review see [9]). As already said, factorization should be valid for direct interactions as in the case of DIS, whereas it is expected to fail for the resolved process similar as in the hadron-hadron scattering process. In a two-channel eikonal model similar to the one used to calculate the suppression factor in hadron-hadron processes [8], introducing vector-meson dominated photon fluctuations, a suppression by about a factor of three for resolved photoproduction at HERA is predicted [10]. Such a suppression factor has recently been applied to diffractive dijet photoproduction [11,12] and compared to preliminary data from H1 [13] and ZEUS [14]. While at LO no suppression of the resolved contribution seemed to be necessary, the NLO corrections increase the cross section significantly, showing that factorization breaking occurs at this order at least for resolved photoproduction and that a suppression factor R must be applied to give a reasonable description of the experimental data.

5. Factorization Scale Dependence for Real Photons

As already mentioned elsewhere [11,12], describing the factorization breaking in hard photoproduction as well as in electroproduction at very low Q^2 [15] by suppressing the resolved contribution only may be problematic. An indication for this is the fact that the separation between the direct and the resolved process is uniquely defined only in LO. In NLO these two processes are related. The separation depends on the factorization scheme and the factorization scale M_γ. The sum of both cross sections is the only physically relevant cross section, which is approximately independent of the factorization scheme and scale [16]. As demonstrated in Fig. 2, multiplying

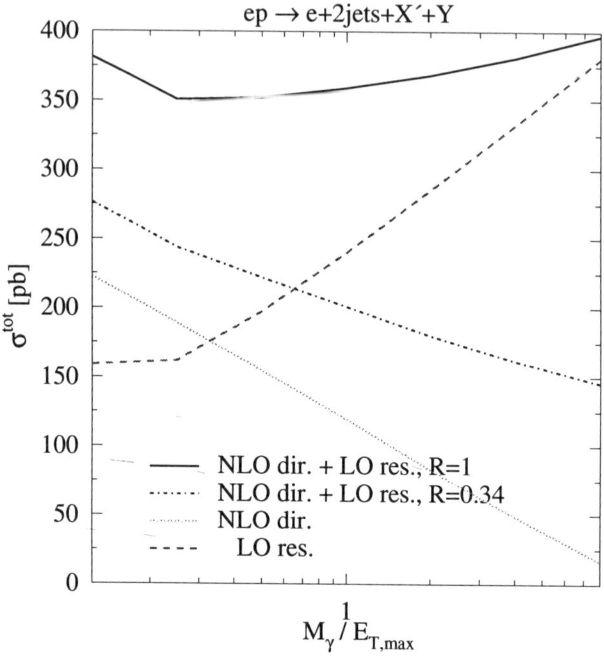

Figure 2. Photon factorization scale dependence of resolved (dashed) and direct (dotted) contributions to the diffractive dijet photoproduction cross section (full curve). Also shown is the sum of the direct and suppressed resolved contribution (dot-dashed curve).

the resolved cross section with the suppression factor $R = 0.34$ destroys the correlation of the M_γ-dependence between the direct and resolved part [11,12], and the sum of both parts has a stronger M_γ-dependence than for the unsuppressed case ($R = 1$), where the M_γ-dependence of the NLO direct cross section is compensated to a high degree by the M_γ-dependence of the LO resolved part.

The introduction of the resolved cross section is dictated by perturbation theory. At NLO, collinear singularities arise from the photon initial state, which are absorbed at the factorization scale into the photon PDFs. This way the photon PDFs become M_γ-dependent. The equivalent M_γ-

dependence, just with the opposite sign, is left in the NLO corrections to the direct contribution. With this knowledge, it is obvious that we can obtain a physical cross section at NLO, *i.e.* the superposition of the NLO direct and LO resolved cross section, with a suppression factor $R < 1$ and no M_γ-dependence left, if we also multiply the $\ln M_\gamma$-dependent term of the NLO correction to the direct contribution with the same suppression factor as the resolved cross section. We are thus led to the theoretical conclusion that, contrary to what one may expect, not *all* parts of the direct contribution factorize. Instead, the *initial state* singular part appearing beyond LO breaks factorization even in direct photoproduction, presumably through soft gluon attachments between the proton and the collinear quark-antiquark pair emerging from the photon splitting. This would be in agreement with the non-cancellation of initial state singularities in diffractive hadron-hadron scattering [6].

6. The Transition Region of Virtual Photoproduction

We now present the special form of the $\ln M_\gamma$-term in the NLO direct contribution and demonstrate that the M_γ-dependence of the physical cross section cancels to a large extent in the same way as in the unsuppressed case ($R = 1$). These studies can be done for photoproduction ($Q^2 \simeq 0$) as well as for electroproduction with fixed, small Q^2. Since in electroproduction the initial-state singularity in the limit $Q^2 \to 0$ is more directly apparent than for the photoproduction case, we shall consider in this contribution the low-Q^2 electroproduction case just for demonstration. This diffractive dijet cross section has been calculated recently [15].

A consistent factorization scheme for low-Q^2 virtual photoproduction has been defined and the full (direct and resolved) NLO corrections for inclusive dijet production have been calculated in [17]. In this work we adapt this inclusive NLO calculational framework to diffractive dijet production at low-Q^2 in the same way as in [15], except that we multiply the $\ln M_\gamma$-dependent terms as well as the resolved contributions with the same suppression factor $R = 0.34$, as an example, as in our earlier work [11,12,15]. The exact value of this suppression factor may change in the future, when better data for photoproduction and low-Q^2 electroproduction have been analyzed. We present the $\ln M_\gamma$-dependence of the partly suppressed NLO direct and the fully suppressed NLO resolved cross section $d\sigma/dQ^2$ and their sum for the lowest Q^2 bin.

The NLO corrections for virtual jet photoproduction have been implemented in the NLO Monte Carlo program JETVIP [18] and adapted to diffractive dijet production in [15]. The subtraction term, which is absorbed into the PDFs of the virtual photon $f_{a/\gamma}(x_\gamma, M_\gamma)$, can be found in [19]. The main term is proportional to $\ln(M_\gamma^2/Q^2)$ times the splitting function

$$P_{q_i \leftarrow \gamma}(z) = 2N_c Q_i^2 \frac{z^2 + (1-z)^2}{2}, \tag{1}$$

where $z = p_1 p_2 / p_0 q \in [x; 1]$ and Q_i is the fractional charge of the quark q_i. p_1 and p_2 are the momenta of the two outgoing jets, and p_0 and q are the momenta of the ingoing parton and virtual photon, respectively. Since $Q^2 = -q^2 \ll M_\gamma^2$, the subtraction term is large and is therefore resummed by the DGLAP evolution equations for the virtual photon PDFs. After this subtraction, the finite term $M(Q^2)_{\overline{\mathrm{MS}}}$, which remains in the matrix element for the NLO correction to the direct process [17], has the same M_γ-dependence as the subtraction term, *i.e.* $\ln M_\gamma$ is multiplied with the same factor. As already mentioned, this yields the M_γ-dependence before the evolution is turned on. In the usual non-diffractive dijet photoproduction these two M_γ-dependences cancel, when the NLO correction to the direct part is added to the LO resolved cross section [16]. Then it is obvious that the approximate M_γ-independence is destroyed, if the resolved cross section is multiplied by a suppression factor R to account for the factorization breaking in the experimental data. To remedy this deficiency, we propose to multiply the $\ln M_\gamma$-dependent term in $M(Q^2)_{\overline{\mathrm{MS}}}$ with the same suppression factor as the resolved cross section. This is done in the following way: we split $M(Q^2)_{\overline{\mathrm{MS}}}$ into two terms using the scale p_T^* in such a way that the term containing the slicing parameter y_s, which was used to separate the initial-state singular contribution, remains unsuppressed. In particular, we replace the finite term after the subtraction by

$$\begin{aligned} M(Q^2, R)_{\overline{\mathrm{MS}}} = & \left[-\frac{1}{2N_c} P_{q_i \leftarrow \gamma}(z) \ln \left(\frac{M_\gamma^2 z}{p_T^{*2}(1-z)} \right) + \frac{Q_i^2}{2} \right] R \\ & - \frac{1}{2N_c} P_{q_i \leftarrow \gamma}(z) \ln \left(\frac{p_T^{*2}}{zQ^2 + y_s s} \right), \end{aligned} \tag{2}$$

where R is the suppression factor. This expression coincides with the finite term after subtraction (see Ref. [19]) for $R = 1$, as it should, and leaves the second term in Eq. (2) unsuppressed. In Eq. (2) we have suppressed in addition to $\ln(M_\gamma^2/p_T^{*2})$ also the z-dependent term $\ln(z/(1-z))$, which is specific to the $\overline{\mathrm{MS}}$ subtraction scheme as defined in [17]. The second term in

Eq. (2) must be left in its original form, *i.e.* being unsuppressed, in order to achieve the cancellation of the slicing parameter (y_s) dependence of the complete NLO correction in the limit of very small Q^2 or equivalently very large s. It is clear that the suppression of this part of the NLO correction to the direct cross section will change the full cross section only very little as long as we choose $M_\gamma \simeq p_T^*$. The first term in Eq. (2), which has the suppression factor R, will be denoted by $\mathrm{DIR_{IS}}$ in the following.

To study the left-over M_γ-dependence of the physical cross section, we have calculated the diffractive dijet cross section with the same kinematic constraints as in the H1 experiment [20]. Jets are defined by the CDF cone algorithm with jet radius equal to one and asymmetric cuts for the transverse momenta of the two jets required for infrared stable comparisons with the NLO calculations [21]. The original H1 analysis actually used a symmetric cut of 4 GeV on the transverse momenta of both jets [22]. The data have, however, been reanalyzed for asymmetric cuts [20].

For the NLO resolved virtual photon predictions, we have used the PDFs SaS1D [23] and transformed them from the DIS_γ to the $\overline{\mathrm{MS}}$ scheme as in Ref. [17]. If not stated otherwise, the renormalization and factorization scales at the pomeron and the photon vertex are equal and fixed to $p_T^* = p_{T,jet1}^*$. We include four flavors, *i.e.* $n_f = 4$ in the formula for α_s and in the PDFs of the pomeron and the photon. With these assumptions we have calculated the same cross section as in our previous work [15]. First we investigated how the cross section $\mathrm{d}\sigma/\mathrm{d}Q^2$ depends on the factorization scheme of the PDFs for the virtual photon, *i.e.* $\mathrm{d}\sigma/\mathrm{d}Q^2$ is calculated for the choice SaS1D and SaS1M. Here $\mathrm{d}\sigma/\mathrm{d}Q^2$ is the full cross section (sum of direct and resolved) integrated over the momentum and rapidity ranges as in the H1 analysis. The results, shown in Fig. 2 of Ref. [19], demonstrate that the choice of the factorization scheme of the virtual photon PDFs has negligible influence on $\mathrm{d}\sigma/\mathrm{d}Q^2$ for all considered Q^2. The predictions agree reasonably well with the preliminary H1 data [20].

We now turn to the M_γ-dependence of the cross section with a suppression factor for $\mathrm{DIR_{IS}}$. To show this dependence for the two suppression mechanisms, (i) suppression of the resolved cross section only and (ii) additional suppression of the $\mathrm{DIR_{IS}}$ term as defined in Eq. (2) in the NLO correction of the direct cross section, we consider $\mathrm{d}\sigma/\mathrm{d}Q^2$ for the lowest Q^2-bin, $Q^2 \in [4, 6]$ GeV^2. In Fig. 3, this cross section is plotted as a function of $\xi = M_\gamma/p_T^*$ in the range $\xi \in [0.25; 4]$ for the cases (i) (light full curve) and (ii) (full curve). We see that the cross section for case (i) has an appreciable ξ-dependence in the considered ξ range of the order of 40%,

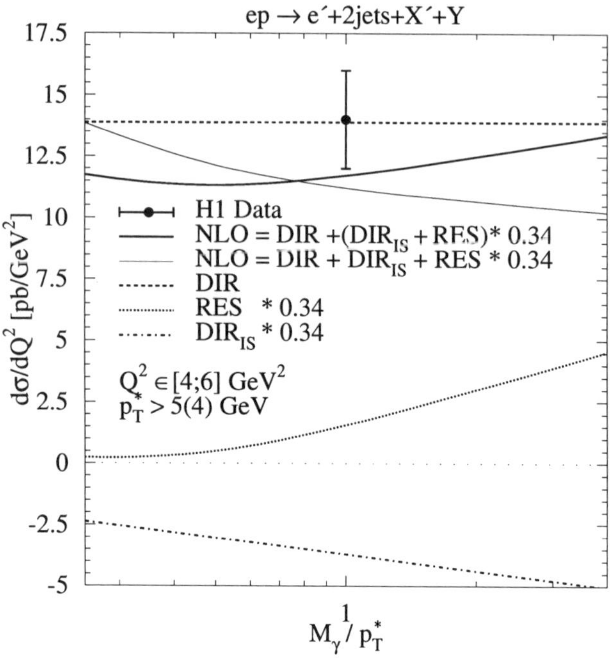

Figure 3. Photon factorization scale dependence of resolved and direct contributions to $d\sigma/dQ^2$ together with their weighted sums for (i) suppression of the resolved cross section and for (ii) additional suppression of $\mathrm{DIR_{IS}}$, using SaS1D virtual photon PDFs.

which is caused by the suppression of the resolved contribution only. With the additional suppression of the $\mathrm{DIR_{IS}}$ term in the direct NLO correction, the ξ-dependence of $d\sigma/dQ^2$ is reduced to approximately less than 20%, if we compare the maximal and the minimal value of $d\sigma/dQ^2$ in the considered ξ range. The remaining ξ-dependence is caused by the NLO corrections to the suppressed resolved cross section and the evolution of the virtual photon PDFs. How the compensation of the M_γ-dependence between the suppressed resolved contribution and the suppressed direct NLO term works in detail is exhibited by the dotted and dashed-dotted curves in Fig. 3. The suppressed resolved term increases and the suppressed direct NLO term decreases by approximately the same amount with increasing ξ. In addition we show also $d\sigma/dQ^2$ in the DIS theory, *i.e.* without subtraction of any $\ln Q^2$ terms (dashed line). Of course, this cross section must be independent of ξ. This prediction agrees very well with the experimental point, whereas the result for the subtracted and suppressed theory (full curve) lies slightly below. We notice, that for $M_\gamma = p_T^*$ the additional suppression of $\mathrm{DIR_{IS}}$ has only a small effect. It increases $d\sigma/dQ^2$ by 5% only [24,25].

7. Conclusion

When comparing experimental data from the H1 and ZEUS Collaborations at HERA for diffractive dijet production in DIS and photoproduction with NLO QCD predictions using diffractive parton densities from H1 and ZEUS, good agreement is found for DIS assuming the H1 diffractive PDFs. However, the dijet photoproduction data are overestimated by the NLO theory, showing that factorization breaking occurs at this order. While this is expected theoretically for resolved photoproduction, the fact that the data are better described by a global suppression of direct *and* resolved contribution by about a factor of two comes as a surprise. We have therefore discussed in some detail the factorization scheme and scale dependence between direct and resolved contributions and proposed a new factorization scheme for diffractive dijet photoproduction.

Acknowledgments

The author thanks the organizers of the Ringberg workshop on *New Trends in HERA Physics 2005* for the kind invitation, G. Kramer for his continuing collaboration, and the *Comité de Financement des Projets de Physique Théorique de l'IN2P3* for financial support.

References

1. H1 Collaboration, Abstract 980, contributed to the 31st *International Conference on High Energy Physics* (ICHEP 2002), Amsterdam, July 2002.
2. S. Chekanov *et al.* [ZEUS Collaboration], Eur. Phys. J. C **38** (2004) 43.
3. H. Abramowicz, M. Groys and A. Levy, Proceedings of the 13th International Workshop on *Deep Inelastic Scattering* (DIS 2005), Melville NY, 2005, p. 461.
4. P. Newman and F. P. Schilling, Proceedings of the Workshop on the *Implications of HERA for LHC Physics*, Hamburg, 2005, hep-ex/0511032.
5. A. D. Martin, M. G. Ryskin and G. Watt, Eur. Phys. J. C **37** (2004) 285 and ibid. **44** (2005) 69.
6. J. C. Collins, Phys. Rev. D **57** (1998) 3051 [Erratum-ibid. D **61** (2000) 019902].
7. T. Affolder *et al.* [CDF Collaboration], Phys. Rev. Lett. **84** (2000) 5043.
8. A. B. Kaidalov, V. A. Khoze, A. D. Martin and M. G. Ryskin, Eur. Phys. J. C **21** (2001) 521.

9. M. Klasen, Rev. Mod. Phys. **74** (2002) 1221.
10. A. B. Kaidalov, V. A. Khoze, A. D. Martin and M. G. Ryskin, Phys. Lett. B **567** (2003) 61.
11. M. Klasen and G. Kramer, hep-ph/0401202, Proceedings of the 12th International Workshop on *Deep Inelastic Scattering* (DIS 2004), Kosice, 2004, p. 492.
12. M. Klasen and G. Kramer, Eur. Phys. J. C **38** (2004) 93.
13. H1 Collaboration, Abstract 6-0177, contributed to the 32nd *International Conference on High Energy Physics* (ICHEP 2004), Beijing, August 2004.
14. ZEUS Collaboration, Abstract 6-0249, contributed to the 32nd *International Conference on High Energy Physics* (ICHEP 2004), Beijing, August 2004.
15. M. Klasen and G. Kramer, Phys. Rev. Lett. **93** (2004) 232002.
16. D. Bödeker, G. Kramer and S. G. Salesch, Z. Phys. C **63**, 471 (1994).
17. M. Klasen, G. Kramer and B. Pötter, Eur. Phys. J. C **1** (1998) 261.
18. B. Pötter, Comput. Phys. Commun. **133** (2000) 105.
19. M. Klasen and G. Kramer, J. Phys. G **31** (2005) 1391.
20. S. Schätzel, Proceedings of the 12th International Workshop on *Deep Inelastic Scattering* (DIS 2004), Kosice, 2004, p. 529; H1 Collaboration, Abstract 6-0176, contributed to the 32nd *International Conference on High Energy Physics* (ICHEP 2004), Beijing, August 2004.
21. M. Klasen and G. Kramer, Phys. Lett. B **366** (1996) 385.
22. C. Adloff *et al.* [H1 Collaboration], Eur. Phys. J. C **20** (2001) 29.
23. G. A. Schuler and T. Sjöstrand, Phys. Lett. B **376** (1996) 193.
24. M. Klasen and G. Kramer, Proceedings of the 13th International Workshop on *Deep Inelastic Scattering* (DIS 2005), Melville NY, 2005, p. 444.
25. A. Bruni, M. Klasen, G. Kramer and S. Schätzel, Proceedings of the Workshop on the *Implications of HERA for LHC Physics*, Hamburg, 2005.

DIFFRACTIVE PARTON DENSITY FUNCTIONS

G. WATT*

Deutsches Elektronen-Synchrotron DESY,
Notkestrasse 85, 22607 Hamburg, Germany
E-mail: graeme.watt@desy.de

We discuss the perturbative QCD description of diffractive deep-inelastic scattering and extract diffractive parton distributions from recent HERA data. The asymptotic collinear factorisation theorem has important modifications in the sub-asymptotic HERA regime. In addition to the usual *resolved* Pomeron contribution, the *direct* interaction of the Pomeron must also be accounted for. The diffractive parton distributions are shown to satisfy an *inhomogeneous* evolution equation, analogous to the parton distributions of the photon.

1. Introduction

A notable feature of deep-inelastic scattering is the existence of diffractive events, $\gamma^* p \to X + p$, in which the slightly deflected proton and the cluster X of outgoing hadrons are well-separated in rapidity. At high energies, the large rapidity gap is believed to be associated with 'Pomeron', or vacuum quantum number, exchange. The diffractive events make up an appreciable fraction of all (inclusive) deep-inelastic events, $\gamma^* p \to X$. We will refer to the diffractive and inclusive processes as DDIS and DIS, respectively. The recent improvement in the precision of the DDIS data [1–3] allows improved analyses to be performed and more reliable diffractive parton density functions (DPDFs) to be extracted. In this article we criticise the conventional extraction of DPDFs based on 'Regge factorisation' in which the exchanged proton is treated as a hadron-like object. We show using perturbative QCD (pQCD) that the treatment of *diffractive* PDFs has more in common with the *photon* PDFs than with the *proton* PDFs.

*In collaboration with A.D. Martin and M.G. Ryskin.

2. Diffractive parton distributions from Regge factorisation

Let the momenta of the incoming proton, the outgoing proton, and the photon be labelled p, p', and q respectively; see Fig. 1(a). Then the basic kinematic variables in DDIS are the photon virtuality, $Q^2 = -q^2$, the Bjorken-x variable, $x_B = Q^2/(2p \cdot q)$, the squared momentum transfer, $t = (p - p')^2$, the fraction of the proton's light-cone momentum transferred through the rapidity gap, $x_{\mathbb{P}} = 1 - p'^+/p^+$, and the fraction of the Pomeron's light-cone momentum carried by the struck quark, $\beta = x_B/x_{\mathbb{P}}$.

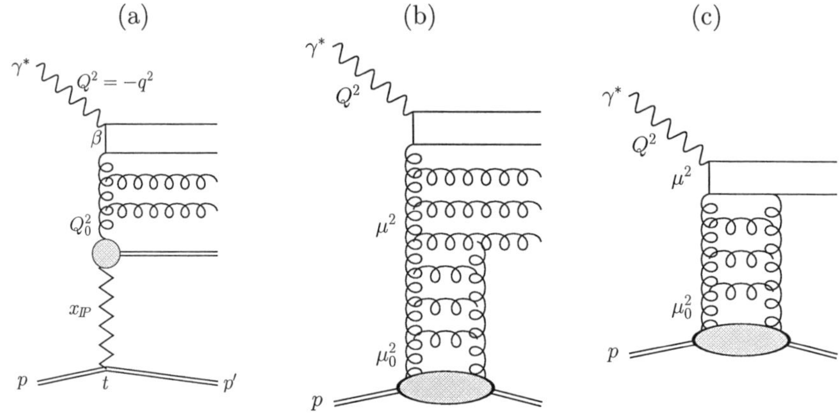

Figure 1. (a) Resolved Pomeron contribution in the 'Regge factorisation' approach. (b) Resolved Pomeron contribution in the 'perturbative QCD' approach. (c) Direct Pomeron contribution in the 'perturbative QCD' approach.

It is conventional to extract DPDFs from DDIS data using two levels of factorisation. Firstly, collinear factorisation means that the diffractive structure function can be written as [4]

$$F_2^{\mathrm{D}(3)}(x_{\mathbb{P}}, \beta, Q^2) = \sum_{a=q,g} C_{2,a} \otimes a^{\mathrm{D}}, \tag{1}$$

where the DPDFs $a^{\mathrm{D}} = zq^{\mathrm{D}}$ or zg^{D}, with $z \in [\beta, 1]$, satisfy DGLAP evolution:

$$\frac{\partial a^{\mathrm{D}}}{\partial \ln Q^2} = \sum_{a'=q,g} P_{aa'} \otimes a'^{\mathrm{D}}, \tag{2}$$

and where $C_{2,a}$ and $P_{aa'}$ are the *same* hard-scattering coefficients and splitting functions as in inclusive DIS. The factorisation theorem (1) applies when Q is made large, therefore it is correct up to power-suppressed corrections. It says nothing about the mechanism for diffraction, which is assumed to reside entirely in the input DPDFs fitted to data at a starting scale Q_0^2; see Fig. 1(a).

In a second stage [5] Regge factorisation is usually assumed, such that

$$a^{\mathrm{D}}(x_{\mathbb{P}}, z, Q^2) = f_{\mathbb{P}}(x_{\mathbb{P}})\, a^{\mathbb{P}}(z, Q^2), \tag{3}$$

where the Pomeron PDFs $a^{\mathbb{P}} = zq^{\mathbb{P}}$ or $zg^{\mathbb{P}}$. The Pomeron flux factor $f_{\mathbb{P}}$ is taken from Regge phenomenology,

$$f_{\mathbb{P}}(x_{\mathbb{P}}) = \int_{t_{\mathrm{cut}}}^{t_{\mathrm{min}}} \mathrm{d}t \quad \mathrm{e}^{B_{\mathbb{P}}\, t}\, x_{\mathbb{P}}^{1-2\alpha_{\mathbb{P}}(t)}. \tag{4}$$

Here, $\alpha_{\mathbb{P}}(t) = \alpha_{\mathbb{P}}(0) + \alpha'_{\mathbb{P}}\, t$, and the parameters $B_{\mathbb{P}}$, $\alpha_{\mathbb{P}}(0)$, and $\alpha'_{\mathbb{P}}$ should be taken from fits to soft hadron data. Although the first fits to use this approach assumed a 'soft' Pomeron, $\alpha_{\mathbb{P}}(0) \simeq 1.08$ [6], all recent fits require a substantially higher value to describe the data. In addition, a secondary Reggeon contribution is needed to describe the data for $x_{\mathbb{P}} \gtrsim 0.01$. This approach is illustrated in Fig. 1(a), where the virtualities of the t-channel partons are strongly ordered as required by DGLAP evolution. The Pomeron PDFs $a^{\mathbb{P}}$ are parameterised at some arbitrary low scale Q_0^2, then evolved up to the factorisation scale, usually taken to be the photon virtuality Q^2.

Although this approach has been found to give a good description of the DDIS data [1, 2, 7, 8], it has little theoretical justification. The 'Regge factorisation' of (3) is merely a simple way of parameterising the $x_{\mathbb{P}}$ dependence of the DPDFs. Note, however, that the effective Pomeron intercept $\alpha_{\mathbb{P}}(0)$ has been observed to depend on Q^2 [3], contrary to the 'Regge factorisation' of (3). The fact that the required $\alpha_{\mathbb{P}}(0)$ is greater than the 'soft' value indicates that there is a significant perturbative QCD (pQCD) contribution to DDIS.

3. Diffractive parton distributions from perturbative QCD

In pQCD, Pomeron exchange can be described by two-gluon exchange, two gluons being the minimum number needed to reproduce the quantum numbers of the vacuum. Two-gluon exchange calculations are the basis for the colour dipole model description of DDIS, in which the photon dissociates into $q\bar{q}$ or $q\bar{q}g$ final states. Such calculations have successfully been used to describe HERA data. The crucial question, therefore, is how to reconcile two-gluon exchange with collinear factorisation as given by (1) and (2). Are these two approaches compatible?

Generalising the $q\bar{q}$ or $q\bar{q}g$ final states to an arbitrary number of parton emissions from the photon dissociation, and replacing two-gluon exchange by exchange of a parton ladder, we have diagrams like that shown in Fig. 1(b) [9–12]. Again, the virtualities of the t-channel partons are strongly ordered: $\mu_0^2 \ll \ldots \ll \mu^2 \ll \ldots \ll Q^2$. The scale μ^2 at which the Pomeron-to-parton splitting occurs can vary between $\mu_0^2 \sim 1$ GeV^2 and the factorisation scale Q^2. Therefore, to calculate the inclusive diffractive structure function, $F_2^{\mathrm{D}(3)}$, we need to integrate over μ^2:

$$F_2^{\mathrm{D}(3)}(x_{\mathbb{P}}, \beta, Q^2) = \int_{\mu_0^2}^{Q^2} \frac{\mathrm{d}\mu^2}{\mu^2}\, f_{\mathbb{P}}(x_{\mathbb{P}}; \mu^2)\, F_2^{\mathbb{P}}(\beta, Q^2; \mu^2). \tag{5}$$

Here, the perturbative Pomeron flux factor can be shown to be [12]

$$f_{\mathbb{P}}(x_{\mathbb{P}}; \mu^2) = \frac{1}{x_{\mathbb{P}} B_D} \left[R_g \frac{\alpha_S(\mu^2)}{\mu}\, x_{\mathbb{P}} g(x_{\mathbb{P}}, \mu^2) \right]^2. \tag{6}$$

The diffractive slope parameter B_D comes from the t-integration, while the factor R_g accounts for the skewedness of the proton gluon distribution [13]. There is a similar contribution from sea quarks, where $g(x_{\mathbb{P}}, \mu^2)$ in (6) is replaced by $S(x_{\mathbb{P}}, \mu^2)$, together with an interference term. In the fits presented here, we use the MRST2001 NLO gluon and sea-quark distributions of the proton [14]. The Pomeron structure function in (5), $F_2^{\mathbb{P}}(\beta, Q^2; \mu^2)$, is calculated from Pomeron PDFs, $a^{\mathbb{P}}(z, Q^2; \mu^2)$, evolved using NLO DGLAP from a starting scale μ^2 up to Q^2, taking the input distributions to be LO Pomeron-to-parton splitting functions, $a^{\mathbb{P}}(z, \mu^2; \mu^2) = P_{a\mathbb{P}}(z)$ [11, 12]. At first glance, it would appear that the perturbative Pomeron flux factor (6) behaves as $f_{\mathbb{P}}(x_{\mathbb{P}}; \mu^2) \sim 1/\mu^2$, so that contributions from large μ^2 are strongly suppressed. However, at large μ^2, the gluon distribution of the proton behaves as $x_{\mathbb{P}} g(x_{\mathbb{P}}, \mu^2) \sim (\mu^2)^\gamma$, where γ is the anomalous dimension. In the BFKL limit of $x_{\mathbb{P}} \to 0$, $\gamma \simeq 0.5$, so $f_{\mathbb{P}}(x_{\mathbb{P}}; \mu^2)$ would be approximately independent of μ^2. The HERA domain is in an intermediate region: γ is not small, but is less than 0.5. In Fig. 2(a) we plot (6) multiplied by μ^2 to show that (6) does not behave as $1/\mu^2$ at small $x_{\mathbb{P}}$. It is also interesting to plot the integrand of (5) as a function of μ^2, as shown in Fig. 2(b). Notice that there is a large contribution from $\mu^2 > 2$–3 GeV^2, which is the value of the input scale Q_0^2 typically used in the 'Regge factorisation' fits of Sect. 2. Recall that fits using 'Regge factorisation' include contributions from $\mu^2 \leq Q_0^2$ in the input distributions, but neglect all contributions from $\mu^2 > Q_0^2$; from Fig. 2(b) this is clearly an unreasonable assumption.

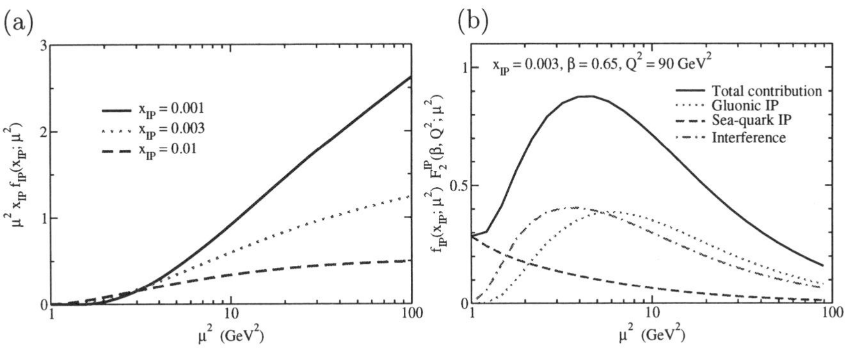

Figure 2. (a) The perturbative Pomeron flux factor (6) multiplied by μ^2. (b) Contributions to $F_2^{\mathrm{D}(3)}$, given by (5), as a function of μ^2.

As well as the *resolved* Pomeron contribution of Fig. 1(b), we must also account for the *direct* interaction of the Pomeron in the hard subprocess, Fig. 1(c), where there is no DGLAP evolution in the upper part of the diagram. Therefore, the diffractive structure function can be written as

$$F_2^{\mathrm{D}(3)} = \underbrace{\sum_{a=q,g} C_{2,a} \otimes a^{\mathrm{D}}}_{\text{Resolved Pomeron}} + \underbrace{C_{2,\mathbb{P}}}_{\text{Direct Pomeron}} \quad ; \tag{7}$$

cf. (1) where there is no direct Pomeron contribution. The direct Pomeron term, $C_{2,\mathbb{P}}$, calculated from Fig. 1(c), will again depend on $f_{\mathbb{P}}(x_{\mathbb{P}}; \mu^2)$ given by (6). Therefore, it is formally suppressed by a factor $1/\mu^2$, but in practice does not behave as such; see Fig. 2(a).

The contribution to the DPDFs from scales $\mu > \mu_0$ is

$$a^{\mathrm{D}}(x_{\mathbb{P}}, z, Q^2) = \int_{\mu_0^2}^{Q^2} \frac{\mathrm{d}\mu^2}{\mu^2}\, f_{\mathbb{P}}(x_{\mathbb{P}}; \mu^2)\, a^{\mathbb{P}}(z, Q^2; \mu^2). \tag{8}$$

Differentiating (8), we see that the evolution equations for the DPDFs are [12]

$$\frac{\partial a^{\mathrm{D}}}{\partial \ln Q^2} = \sum_{a'=q,g} P_{aa'} \otimes a'^{\mathrm{D}} + P_{a\mathbb{P}}(z)\, f_{\mathbb{P}}(x_{\mathbb{P}}; Q^2); \tag{9}$$

cf. (2) where the second term of (9) is absent. That is, the DPDFs satisfy an *inhomogeneous* evolution equation [10, 12], with the extra inhomogeneous term in (9) leading to more rapid evolution than in the 'Regge factorisation' fits described in Sect. 2. Note that the inhomogeneous term will change the $x_{\mathbb{P}}$ dependence evolving upwards in Q^2, in accordance with the data, and unlike the 'Regge factorisation' assumption (3). Again, the inhomogeneous term in (9) is formally suppressed by a factor $1/Q^2$, but in practice does not behave as such; see Fig. 2(a).

Therefore, the diffractive structure function is analogous to the photon structure function, where there are both resolved and direct components and the photon PDFs satisfy an inhomogeneous evolution equation, where at LO the inhomogeneous term accounts for the splitting of the point-like photon into a $q\bar{q}$ pair. If we consider, for example, diffractive dijet photoproduction, there are four classes of contributions; see Fig. 3. The relative importance of each contribution will depend on the values of x_γ, the fraction of the photon's momentum carried by the parton entering the hard subprocess, and $z_{\mathbb{P}}$, the fraction of the Pomeron's momentum carried by the parton entering the hard subprocess.

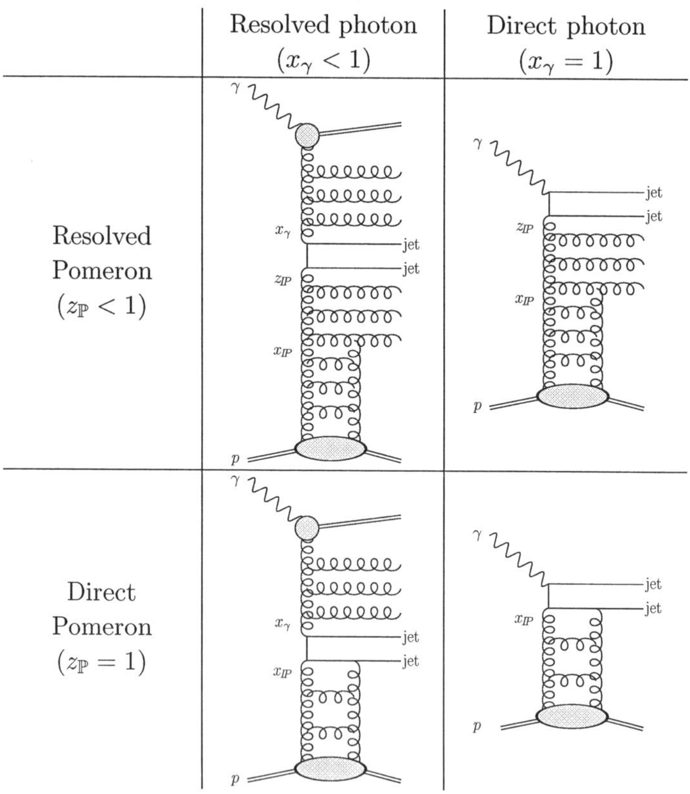

Figure 3. The four classes of contributions to diffractive dijet photoproduction at LO. Both the photon and the Pomeron can be either 'resolved' or 'direct'.

4. Description of DDIS data

A NLO analysis of DDIS data is not yet possible. The direct Pomeron terms, $C_{2,\mathbb{P}}$, and Pomeron-to-parton splitting functions, $P_{a\mathbb{P}}$, need to be calculated at NLO within a given factorisation scheme (for example, $\overline{\text{MS}}$). Here, we perform a simplified analysis where the usual coefficient functions $C_{2,a}$ and splitting functions $P_{aa'}$ $(a, a' = q, g)$ are taken at NLO, but $C_{2,\mathbb{P}}$ and $P_{a\mathbb{P}}$ are taken at LO [12]. We work in the fixed flavour number scheme, where there is no charm DPDF. Charm quarks are produced via $\gamma^* g^{\mathbb{P}} \to c\bar{c}$ at NLO [15] and $\gamma^*\mathbb{P} \to c\bar{c}$ at LO [16]. For light quarks, we include the direct Pomeron process $\gamma_L^*\mathbb{P} \to q\bar{q}$ at LO [12], which is higher-twist and known to be important at large β.

To see the effect of the direct Pomeron contribution and the inhomogeneous evolution, we make two types of fits:

"Regge" : The 'Regge factorisation' approach discussed in Sect. 2, where there is no direct Pomeron contribution and no inhomogeneous term in the evolution equation.

"pQCD" : The 'perturbative QCD' approach discussed in Sect. 3, where these effects are included.

We make separate fits to the recent H1 LRG (prel.) [1] and ZEUS M_X [3] $\sigma_r^{\mathrm{D}(3)}$ data, applying cuts $Q^2 \geq 3$ GeV^2 and $M_X \geq 2$ GeV, and allowing for overall normalisation factors of 1.10 and 1.43 to account for proton dissociation up to masses of 1.6 GeV and 2.3 GeV respectively. Statistical and systematic experimental errors are added in quadrature. The strong coupling is set via $\alpha_S(M_Z) = 0.1190$. We take the input forms of the DPDFs at a scale $Q_0^2 = 3$ GeV^2 to be

$$z\Sigma^{\mathrm{D}}(x_{\mathbb{P}}, z, Q_0^2) = f_{\mathbb{P}}(x_{\mathbb{P}})\, C_q\, z^{A_q}(1-z)^{B_q}, \tag{10}$$

$$zg^{\mathrm{D}}(x_{\mathbb{P}}, z, Q_0^2) = f_{\mathbb{P}}(x_{\mathbb{P}})\, C_g\, z^{A_g}(1-z)^{B_g}, \tag{11}$$

where $f_{\mathbb{P}}(x_{\mathbb{P}})$ is given by (4), and where $\alpha_{\mathbb{P}}(0)$, C_a, A_a, and B_a $(a = q, g)$ are free parameters. The secondary Reggeon contribution to the H1 data is treated in a similar way as in the H1 2002 fit [1], using the GRV pionic parton distributions [17]. Good fits are obtained in all cases, with $\chi^2/\text{d.o.f.} = 0.75$, 0.71, 0.76, and 0.84 for the "Regge" fit to H1 data, "pQCD" fit to H1 data, "Regge" fit to ZEUS M_X data, and "pQCD" fit to ZEUS M_X data respectively. The "pQCD" fits are shown in Fig. 4, including a breakdown of the different contributions. The DPDFs are shown in Fig. 5. Note that the "pQCD" DPDFs are smaller than the corresponding

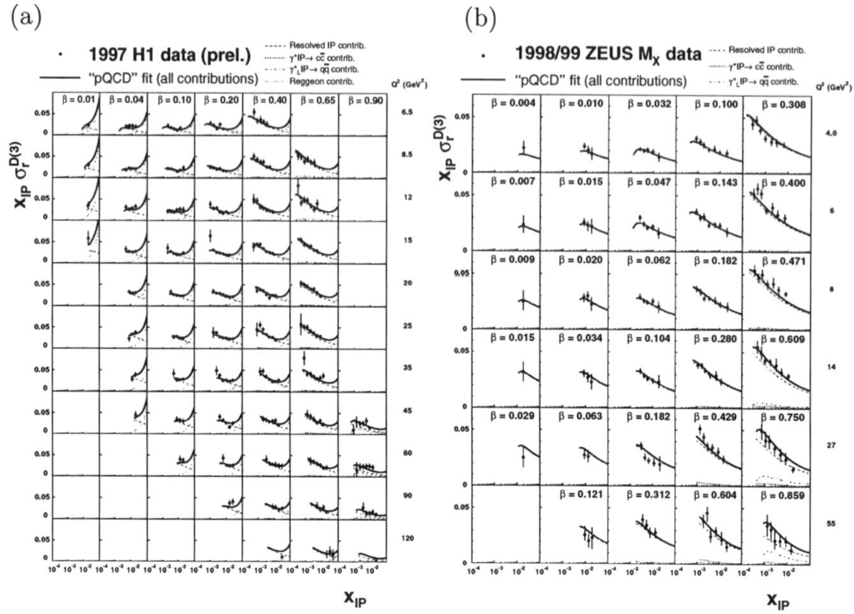

Figure 4. "pQCD" fits to (a) H1 LRG and (b) ZEUS M_X data.

"Regge" DPDFs at large z due to the inclusion of the higher-twist $\gamma_L^* \mathbb{P} \to q\bar{q}$ contribution. Also note that the "pQCD" DPDFs have slightly more rapid evolution than the "Regge" DPDFs due to the extra inhomogeneous term in the evolution equation (9). There is a large difference between the DPDFs obtained from the H1 LRG and ZEUS M_X data due to the different Q^2 dependence of these data sets; see also [7, 8].

The predictions from the two "pQCD" fits for the charm contribution to the diffractive structure function as measured by ZEUS using the LRG method [18] are shown in Fig. 6. Our H1 LRG fit gives a good description, while our ZEUS M_X fit is too small at low β. Note that the direct Pomeron contribution is significant at moderate β. These charm data points were included in the determination of DPDFs from ZEUS LPS data [2], but only the resolved Pomeron ($\gamma^* g^{\mathbb{P}} \to c\bar{c}$) contribution was included and not the direct Pomeron ($\gamma^* \mathbb{P} \to c\bar{c}$) contribution. Therefore, the diffractive gluon distribution from the ZEUS LPS fit [2] needed to be artificially large to fit the charm data at moderate β.

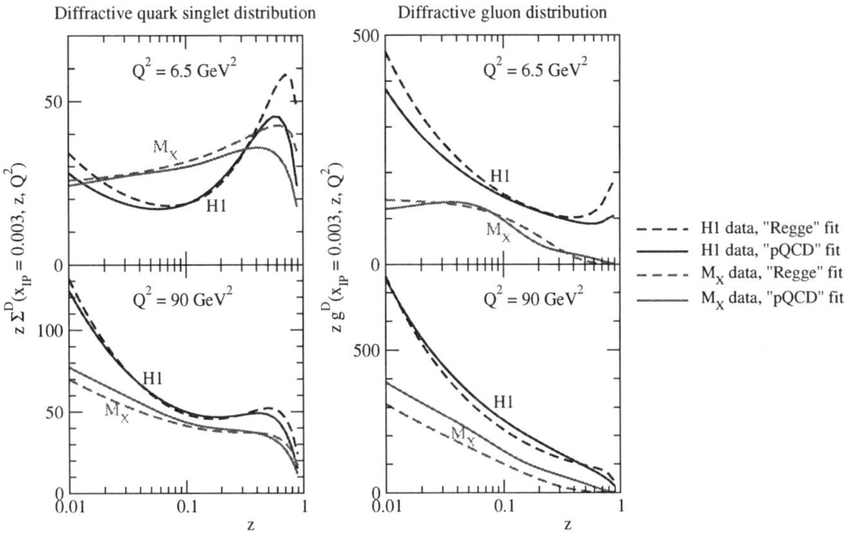

Figure 5. DPDFs obtained from separate fits to H1 LRG and ZEUS M_X data using the "Regge" and "pQCD" approaches.

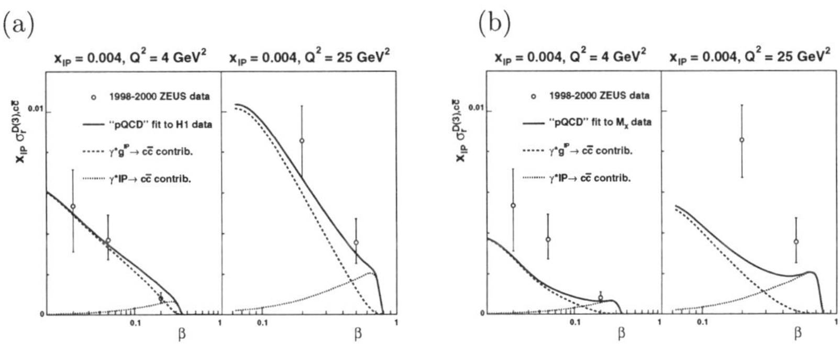

Figure 6. Predictions for ZEUS LRG diffractive charm production data using DPDFs from the "pQCD" fits to (a) H1 LRG and (b) ZEUS M_X data. Note the large direct Pomeron ($\gamma^* \mathbb{P} \rightarrow c\bar{c}$) contribution at moderate β.

5. Conclusions and outlook

To summarise, diffractive DIS is more complicated to analyse than inclusive DIS. Collinear factorisation holds, but we need to account for the direct

Pomeron coupling, leading to an inhomogeneous evolution equation (9).[a] Therefore, the treatment of DPDFs has more in common with photon PDFs than with proton PDFs. The H1 LRG and ZEUS M_X data have a different Q^2 dependence, leading to different DPDFs. This issue needs further attention. For a NLO analysis of DDIS data, the direct Pomeron terms, $C_{2,\mathbb{P}}$, and Pomeron-to-parton splitting functions, $P_{a\mathbb{P}}$, need to be calculated at NLO. There are indications [16] that there are large π^2-enhanced virtual loop corrections ('K-factors') similar to those found in the Drell–Yan process. As with all PDF determinations, the sensitivity to the form of the input parameterisation, (10) and (11), and input scale Q_0^2 needs to be studied. The inclusion of jet and heavy quark DDIS data, and possibly $F_L^{D(3)}$ if it is measured [22], would help to constrain the DPDFs further. The extraction of DPDFs from HERA data will provide an important input for predictions of diffractive processes at the LHC.

References

1. H1 Collaboration, paper 089 submitted to EPS 2003, Aachen.
2. S. Chekanov *et al.* [ZEUS Collaboration], Eur. Phys. J. C **38** (2004) 43.
3. S. Chekanov *et al.* [ZEUS Collaboration], Nucl. Phys. B **713** (2005) 3.
4. J. C. Collins, Phys. Rev. D **57** (1998) 3051.
5. G. Ingelman and P. E. Schlein, Phys. Lett. B **152** (1985) 256.
6. A. Donnachie and P. V. Landshoff, Phys. Lett. B **296** (1992) 227.
7. H. Abramowicz, M. Groys and A. Levy, arXiv:hep-ph/0507090.
8. P. Newman and F.-P. Schilling, arXiv:hep-ex/0511032.
9. M. G. Ryskin, Sov. J. Nucl. Phys. **52** (1990) 529.
10. E. Levin and M. Wusthoff, Phys. Rev. D **50** (1994) 4306.
11. A. D. Martin, M. G. Ryskin and G. Watt, Eur. Phys. J. C **37** (2004) 285.
12. A. D. Martin, M. G. Ryskin and G. Watt, Eur. Phys. J. C **44** (2005) 69.
13. A. G. Shuvaev *et al.*, Phys. Rev. D **60** (1999) 014015.
14. A. D. Martin *et al.*, Eur. Phys. J. C **23** (2002) 73.
15. S. Riemersma, J. Smith and W. van Neerven, Phys. Lett. B **347** (1995) 143.
16. E. M. Levin *et al.*, Z. Phys. C **74** (1997) 671.
17. M. Glück, E. Reya and A. Vogt, Z. Phys. C **53** (1992) 651.
18. S. Chekanov *et al.* [ZEUS Collaboration], Nucl. Phys. B **672** (2003) 3.
19. V. A. Abramovsky, V. N. Gribov and O. V. Kancheli, Sov. J. Nucl. Phys. **18** (1974) 308.
20. A. D. Martin, M. G. Ryskin and G. Watt, Phys. Rev. D **70** (2004) 091502.
21. G. Watt, A. D. Martin and M. G. Ryskin, Phys. Lett. B **627** (2005) 97.
22. P. Newman, arXiv:hep-ex/0511047.

[a]The inhomogeneous evolution of DPDFs leads, via the AGK cutting rules [19], to non-linear evolution of the inclusive PDFs [20, 21].

6 Beyond the Standard Model

BEYOND THE STANDARD MODEL AT HERA: STATUS AND PROSPECTS

E. PEREZ
CE-Saclay, DSM/DAPNIA/Spp, F-91191 Gif-sur-Yvette and DESY, Notkestrasse 85, D-22607 Hamburg.
E-mail: eperez@hep.saclay.cea.fr

An overview of experimental results on searches for new phenomena at HERA is presented. The complementarity with searches performed at other experiments is discussed and the prospects for a discovery, using the full HERA data to be delivered until mid-2007, are presented.

1. Introduction

Although remarkably confirmed so far by low and high energy experiments, the Standard Model (SM) of strong, weak and electromagnetic interactions remains unsatisfactory. Many models of "new physics" have been proposed to address the questions which are unexplained by the SM. Experimentally, the observation of deviations with respect to the SM predictions is a key part of the existing and future high energy programmes. At HERA, searches for new phenomena have been carried out using a luminosity of up to 280 pb^{-1}. Besides an excess of atypical W-like events observed at H1 all measurements are so far in good agreement with the SM expectation and constraints on models for new physics have been obtained.

2. Searches for new phenomena in inclusive DIS

Neutral Current Deep Inelastic Scattering (NC DIS) is measured at HERA for values of the photon virtuality Q^2 up to about 40000 GeV^2. Although the precision is still statistically limited at the highest Q^2, the good agreement of the measurements with the SM expectation allows stringent constraints on new physics to be set [1]. For example, a finite quark radius would reduce the high Q^2 DIS cross section with respect to the SM predictions, such that the current data rule out quark radii larger than $0.85 \cdot 10^{-18}$ m, assuming that the electron is point-like. Similarly, the effective scale of

$eeqq$ contact interactions is constrained to be larger than typically 5 TeV - a similar sensitivity being achieved from the preliminary Run II Drell-Yan data of the Tevatron experiments [2]. The effective mass scale associated to the exchange of Kaluza-Klein gravitons in models assuming additional large space dimensions is constrained to be larger than about 0.8 TeV, slightly below the LEP and Tevatron bounds.

The longitudinal polarisation of the lepton beam in the HERA II data has been exploited to measure the polarisation dependence of the Charged Current (CC) DIS cross section [3]. Although these data constrain in principle the existence of a right-handed W boson, the sensitivity is below the W_R mass bounds obtained at the Tevatron. However these measurements confirm the left-handed nature of the weak interaction in the t-channel.

3. Model-dependent searches

As HERA is not an annihilation machine, the pair-production of new heavy particles, which could occur in e^+e^- or pp collisions via their coupling to gauge bosons, has a very low cross-section at HERA. Instead, searches for the *single* production of new particles are performed at the H1 and ZEUS experiments. The cross-section for such processes depends on the unkown coupling of the new particle to the SM fields. Hence these searches do not provide *absolute* constraints on the mass of new particles. Conversely, the observation of a signal for the single production of a new particle would provide information not only on its mass, but also on this unknown coupling. In the following example constraints obtained on leptoquarks, squarks in R-parity violating supersymmetry and excited fermions are presented.

3.1. *Lepton-quark resonances: Leptoquarks*

An intriguing characteristic of the Standard Model is the observed symmetry between the lepton and the quark sectors, which is manifest in the representation of the fermion fields under the SM gauge groups, and in their replication over three family generations. This could be a possible indication of a new symmetry between the lepton and quark sectors, leading to "lepto-quark" interactions. Leptoquarks (LQs) are new scalar or vector colour-triplet bosons carrying both a baryon and a lepton number. Several types of LQs can be predicted, differing in their quantum numbers. The interaction of the LQ with a lepton-quark pair is parameterized by a coupling λ. Depending on its quantum numbers a LQ couples to eq, νq or both. The branching ratio β for a LQ to decay into an electron and a quark

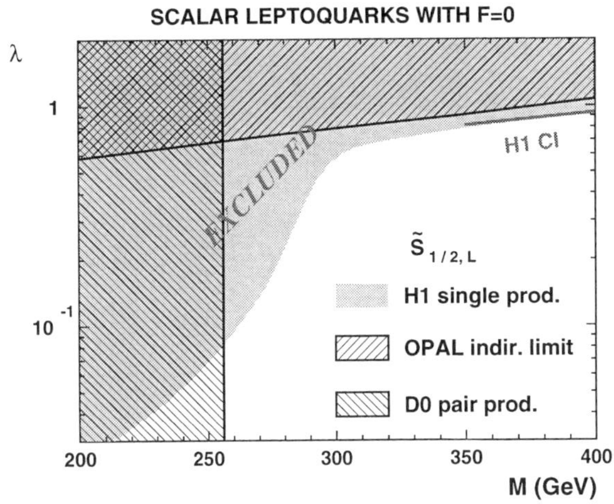

Figure 1. Example mass-dependent upper bounds on the Yukawa coupling λ of a first generation scalar leptoquark to the electron-quark pair. These are shown for a LQ of $F = 0$ coupling to an e^+ and a d quark, which decays exclusively into ed.

can be fixed by model assumptions, or can be treated as a free parameter. Decay modes other than eq and νq are usually neglected.

In electron-proton collisions, first generation LQs might be singly produced via the fusion of the incoming lepton with a quark coming from the proton. The production cross-section roughly scales with λ^2 times the parton density function of the relevant parton evaluated at $x_{Bj} = M_{LQ}^2/s$, $\sqrt{s}$ being the centre-of-mass energy and M_{LQ} the LQ mass. Hence, e^+p (e^-p) collisions provide a larger sensitivity to LQs with fermion number $F = 0$ ($F = 2$) since the u and d parton density is larger than that of antiquarks at large x_{Bj}. LQs might be observed as a resonant peak in the lepton-jet mass spectrum of NC or CC DIS events. No such signal has been observed by the H1 and ZEUS experiments [4].

Existing constraints on a scalar leptoquark which decays solely into an electron and a quark ($\beta = 1$) are summarised in Fig. 1. For an electromagnetic strength of the coupling λ ($\lambda^2/4\pi = \alpha_{em}$, i.e. $\lambda \simeq 0.3$), the HERA experiments rule out LQ masses below $\sim$ 290 GeV. Constraints derived from the search for pair produced LQs at the Tevatron do not depend on the coupling λ and a lower mass bound of 256 GeV is set by the D0 experiment [5]. Example constraints on more general LQ models where $\beta \neq 1$ are shown in Fig. 2 assuming that the LQ decays exclusively into

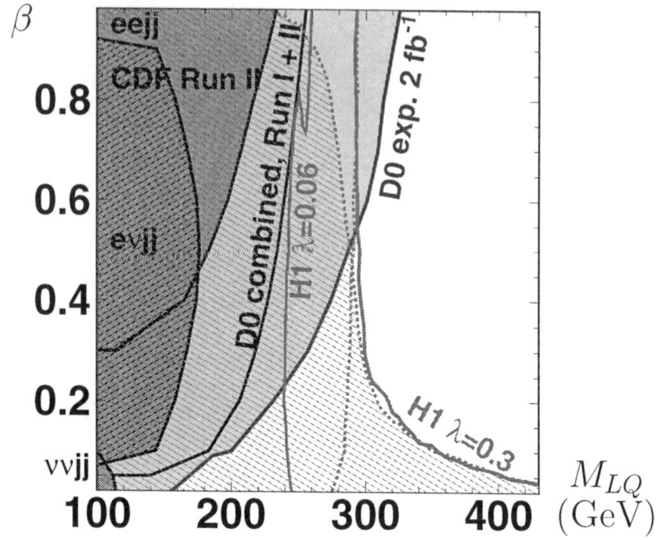

Figure 2. Example constraints on first generation scalar leptoquarks decaying exclusively into eq and νq. For $\lambda = 0.3$, the H1 constraints obtained from the ej and νj analyses are indicated by the dotted curves, and the combined bound is shown for two λ values. The future sensitivity of the Tevatron for a luminosity of 2 fb^{-1} is also shown.

eq and νq. Leptoquarks decaying with a large branching ratio into νq are not easily probed at the Tevatron due to the large background for final states containing only jets and missing transverse momentum. In contrast, with a similarly good signal to background ratio for the eq and νq final states, HERA experiments can be sensitive to LQs which decay with a large branching ratio into νq, provided that the coupling λ is not too small.

The current HERA bounds are more stringent for $F = 0$ LQs due to the much larger HERA I luminosity in e^+p collisions (about 110 pb^{-1} in e^+p and 15 pb^{-1} in e^-p). Assuming the existence of a $F = 0$ LQ in the HERA kinematic range, and that its coupling to the eq pair is equal to the HERA I upper limit, a significance of $\sim 4\,\sigma$ could be reached in each experiment with an e^+p luminosity of 350 pb^{-1} and combining the H1 and ZEUS data could allow the $5\,\sigma$ threshold to be reached. Assuming a LQ production cross-section equal to half the existing limit the discovery potential is limited. In contrast a much larger discovery potential remains for $F = 2$ leptoquarks, which will be probed with the e^-p data to be delivered until mid-2007.

It should be noted that an ep collider would be the ideal machine to study a leptoquark signal. The fermion number of the leptoquark could easily be determined by comparing the signal rates in e^+p and e^-p collisions; the polarisation of the lepton beam provides a handle to determine the chiral couplings of the LQ; and the good signal to background ratio in the νq channel allows the LQ coupling to the neutrino to be studied.

3.2. *R-parity violating supersymmetry*

In supersymmetric (SUSY) models where the so-called R-parity (R_p) is not conserved, squarks could be resonantly produced at HERA similarly to leptoquarks. In addition to the "LQ-like" decays into eq and possibly νq the squarks also undergo decays into gauginos (the supersymmetric partners of gauge bosons) and an exhaustive search requires a large number of final states to be analysed. This has been pioneered by the H1 collaboration in [6] where the full HERA I dataset has been used to set constraints on supersymmetric models. A similar preliminary analysis looking for the SUSY partner of the top quark has been performed by the ZEUS experiment. Example constraints are shown in Fig. 3. For a squark R_p-violating ($\not{R}_p$)

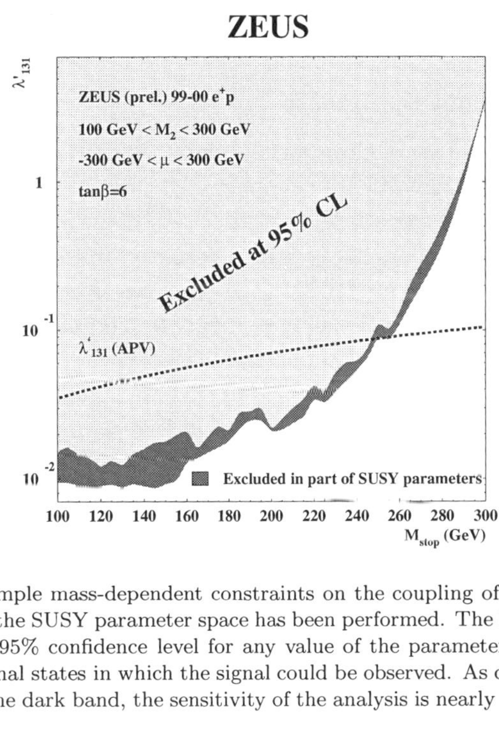

Figure 3. Example mass-dependent constraints on the coupling of the stop to an e^+d pair. A scan of the SUSY parameter space has been performed. The light shaded domain is ruled out at 95% confidence level for any value of the parameters, which determine the dominant final states in which the signal could be observed. As can be seen from the narrowness of the dark band, the sensitivity of the analysis is nearly model-independent.

coupling of electromagnetic strength, lower mass bounds of 270 – 280 GeV can be set. Within constrained SUSY models where a few parameters determine the full Higgs and supersymmetric spectrum, stringent bounds were derived on the squark mass from searches for Higgs, sfermions and gauginos at LEP. However, part of the SUSY parameter space remains open for a discovery at HERA II, for a reasonably large $R\!\!\!/_p$ coupling [6]. The case of a light stop or sbottom is of high interest for HERA II since the bounds coming from Tevatron are less stringent than those obtained assuming five degenerate squarks. In particular, the sensitivity to the sbottom, which has a larger production cross-section in e^-p than in e^+p collisions ($e^-u \rightarrow \tilde{b}$), will considerably increase with the HERA II e^-p data.

In case the squarks are too heavy to be produced at HERA, the t-channel exchange of a selectron or sneutrino could allow for single gaugino production. This process has been considered in two classes of SUSY models, differing in the dominant decay mode of the produced gaugino [7]. In both cases the analyses slightly improve the previous bounds if the relevant R-parity violating coupling is quite large. These are the first SUSY constraints set at HERA which are independent of the squark sector.

3.3. *Fermion-boson resonances*

The observed replication of three fermion families motivates the possibility of a yet unobserved new scale of matter. An unambiguous signature for a new scale of matter would be the direct observation of excited states of fermions (f^*), via their decay into a fermion and a gauge boson. In the most commonly used model [8], the interaction of an f^* with a fermion and a gauge boson is described by a magnetic coupling proportional to $1/\Lambda$ where Λ is a new scale. Proportionality constants f, f' and f_s result in different couplings to $U(1)$, $SU(2)$ and $SU(3)$ gauge bosons. Existing constraints on excited electrons are shown in Fig. 4, under the assumption that $f = f'$. Searches for pair produced e^* at LEP allowed to rule out masses below about 103 GeV, independently of the value of the coupling f/Λ. In contrast, searches for single e^* production at LEP, HERA [9] and Tevatron [10] set mass bounds which depend on f/Λ. The future HERA and Tevatron sensitivities, also depicted in Fig. 4, show the discovery potential of HERA II for excited electrons. The case of excited neutrinos is also very interesting for HERA II, since their production cross-section is larger in e^-p than in e^+p collisions by typically one order of magnitude.

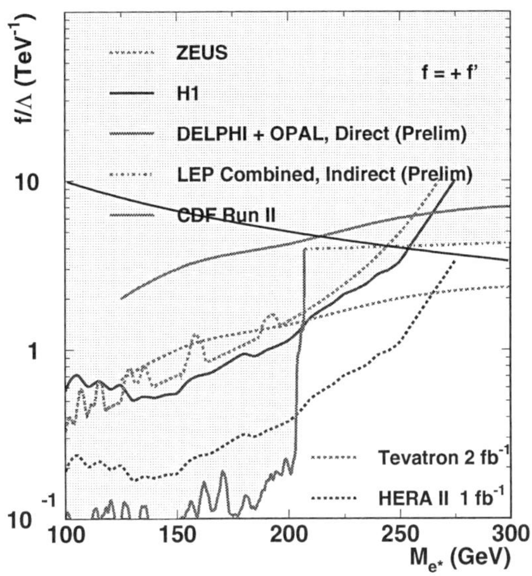

Figure 4. Existing constraints on excited electron masses and couplings, assuming that $f = f'$. The decreasing curve shows the hyperbola $f/\Lambda = 1/M_{e^*}$. The future HERA and Tevatron sensitivities are also shown as dotted curves.

4. Searches for deviations from the SM in rare processes

4.1. *The "isolated lepton events"*

Within the Standard Model, W production at HERA has a cross-section of about 1 pb. When the W decays leptonically, the final state contains an isolated lepton, missing transverse momentum, and a usually soft hadronic system. This process has been measured using the HERA I data [11] in the "electron" ($W \to e\nu_e$) and "muon" ($W \to \mu\nu_\mu$) channels, and a general agreement with the SM prediction was observed. However, for large values of the transverse momentum of the hadronic system, P_T^X, an excess of events was reported by the H1 Collaboration [11]. This excess was not confirmed by a ZEUS analysis [12], differing from the H1 analysis in terms of background rejection[a].

An abnormally large rate of high P_T^X events is also observed by the H1 experiment [13] in the HERA II data. Combining the e and μ channels and the HERA I and HERA II datasets, which amount to a total luminosity

[a]The non W contribution to the expected background is about 50% at large P_T^X in the ZEUS analysis, while it amounts to 15% only in the H1 analysis.

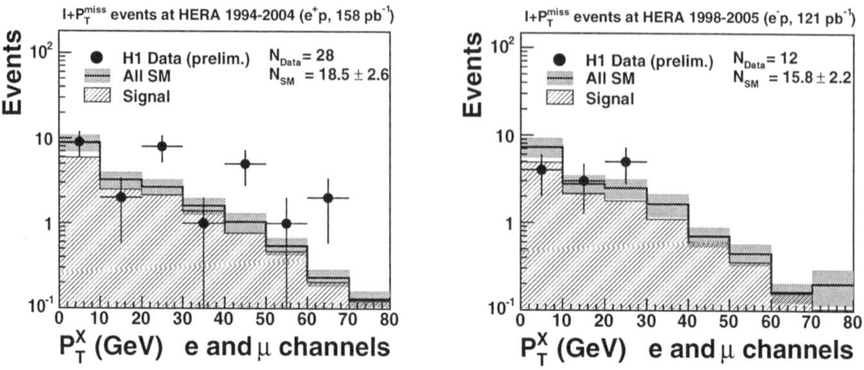

Figure 5. Distribution of the transverse momentum of the hadronic system P_T^X in selected events recorded in (left) the e^+p data sample and (right) the e^-p data sample. The hatched histogram shows the expectation from W production while the total expectation is given by the open histogram.

of $279\,\mathrm{pb}^{-1}$, 17 events are observed at $P_T^X > 25$ GeV for a SM expectation of 9.0 ± 1.5. Amongst the 6 new events observed in HERA II, 5 were recorded during the e^+p running ($53\,\mathrm{pb}^{-1}$) and one during the e^-p running ($107\,\mathrm{pb}^{-1}$). Fig. 5 shows the observed P_T^X distributions separately for the e^+p and e^-p datasets where HERA I and HERA II data are combined, together with the corresponding SM expectations. The observed and expected numbers of events are given in table 1. While the observation in the e^-p data is consistent with the SM expectation, 15 events are observed at $P_T^X > 25$ GeV in the e^+p data for an expectation of 4.6 ± 0.8 events. The probability that the expected yield fluctuates to 15 events or more corresponds to a $3.4\,\sigma$ fluctuation.

The ZEUS experiment has recently carried out a re-analysis of the e channel using the 99-00 e^+p data, resulting in a larger purity in W events. The positron data taken at HERA II have been analysed in the same way, such that the total luminosity amounts to $106\,\mathrm{pb}^{-1}$. The results are also shown in table 1. At $P_T^X > 25$ GeV one event is observed in the data, in agreement with the SM expectation of 1.5 ± 0.18.

Although the rate of events observed in the e channel in the positron data is larger in H1 than in ZEUS, both experiments are compatible with each other within $2.5\,\sigma$ for an average rate of about 4 events per $100\,\mathrm{pb}^{-1}$ in this channel. Assuming that such events are observed in the future H1 and ZEUS data at an average rate of about $7 - 8$ events per $100\,\mathrm{pb}^{-1}$

combining the e and μ channels, a significance of 4σ could be reached[b] from the combined H1 and ZEUS datasets by doubling the e^+p luminosity. It should be noted that new physics scenarios can be found which could explain that such events are observed in e^+p collisions only. For example, in supersymmetry with two R-parity violating couplings involving third generation fields, a top quark could be produced via t-channel sbottom exchange in e^+d collisions. Due to the large value of Bjorken x needed to produce a top quark in the final state, the corresponding process in e^-p collisions would have a much lower cross-section.

Table 1. Summary of the H1 results of searches for events with isolated electrons or muons and missing transverse momentum for the e^+p data (158 pb^{-1}) and the e^-p data (121 pb^{-1}). Data from HERA I and HERA II are combined. The number of events observed by ZEUS in the electron channel, in 106 pb^{-1} of e^+p data, is also shown. The number of observed events at $P_T^X > 25$ GeV is compared to the SM prediction. The quoted errors contain statistical and systematic uncertainties added in quadrature.

$P_T^X > 25$ GeV		Electron obs./exp.	Muon obs./exp.	Combined obs./exp.
H1 Preliminary	1998-2005 e^-p 121 pb^{-1}	2 / 2.4 ± 0.5	0 / 2.0 ± 0.3	2 / 4.4 ± 0.7
H1 Preliminary	1994-2004 e^+p 158 pb^{-1}	9 / 2.3 ± 0.4	6 / 2.3 ± 0.4	15 / 4.6 ± 0.8
ZEUS Preliminary	1999-2004 e^+p 106 pb^{-1}	1 / 1.5 ± 0.18		

4.2. *Multi-lepton events*

If the events reported above were to be explained by some anomalous W production mechanism, an anomalous rate for Z-like events could also be observed. Events with at least two electrons or muons in the final state have been looked for by the H1 collaboration [14]. A slight excess of high mass multi-electron events was observed in the HERA I dataset. The analysis has been repeated using the HERA II data and extended to include other multi-lepton topologies [15]. With a total luminosity of 209 pb^{-1} four events are observed with $\sum_{leptons} P_T > 100$ GeV, three of which being HERA I ee events. This is slightly above the SM prediction of 0.81 ± 0.14.

[b]This estimate is obtained by scaling the SM backgrounds of the H1 analysis in both the e and μ channels.

5. Conclusions

Although some stringent bounds on new physics are set at LEP and the Tevatron, HERA appears to be very well suited to search for new phenomena in some specific cases. In particular, searches for new physics at HERA rarely suffer from huge SM backgrounds. Searches for leptoquarks, for the supersymmetric partners of the top or bottom quarks, and for excited fermions might bring a discovery with the HERA II data. This holds in particular for new physics processes for which the cross-section is larger in e^-p than in e^+p collisions, since the HERA I constraints are not too stringent in such cases. With the future e^+p data to be delivered until mid-2007, the excess of atypical W-like events observed at H1, which corresponds to a $3.4\,\sigma$ deviation, appears to be the best chance for a discovery at HERA II.

References

1. H1 Collab., C. Adloff et al., Phys. Lett. **B568** (2003) 35; ZEUS Collab., S. Chekanov et al., Phys. Lett. **B591** (2004) 23.
2. D0 Collab., D0 Note 4552-CONF.
3. J. Meyer, these proceedings.
4. H1 Collab., Phys. Lett. **B629** (2005) 9; ZEUS Collab., S. Chekanov et al., Phys. Rev. **D68** (2003) 052004.
5. D0 Collab., V.M. Abazov et al., Phys. Rev. **D71** (2005) 071104.
6. H1 Collab., A. Aktas et al., Eur. Phys. J. **C36** (2004) 425.
7. H1 Collab., A. Aktas et al., Phys. Lett. **B616** (2005) 31; ZEUS Collab., contributed paper to EPS'05, abstract #329; *idem*, contributed paper to EPS'05, abstract #330.
8. K. Hagiwara, D. Zeppenfeld and S. Komamiya, Z. Phys. **C29** (1985) 115; F. Boudjema, A. Djouadi and J.L. Kneur, Z. Phys. **C57** (1993) 425.
9. H1 Collab., C. Adloff et al., Phys. Lett. **B548** (2002) 35; ZEUS Collab., S. Chekanov et al., Phys. Lett. **B549** (2002) 32.
10. CDF Collab., D. Acosta et al., Phys. Rev. Lett. **94** (2005) 101802.
11. H1 Collab., V. Andreev et al., Phys. Lett. **B561** (2003) 241.
12. ZEUS Collab., S. Chekanov et al., Phys. Lett. **B559** (2003) 153.
13. H1 Collab., document prepared for the Nov. 2005 DESY-PRC meeting, available at http://www-h1.desy.de/publications/H1preliminary.short_list.html.
14. H1 Collab., A. Aktas et al., Eur. Phys. J. **C31** (2003) 17; H1 Collab., A. Aktas et al., Phys. Lett. **B583** (2004) 28.
15. H1 Collab., contributed to EPS'05, abstract #635.

7 Resonances and Diquarks

NEW RESONANCES IN THE HADRONIC FINAL STATE AT HERA

KATSUO TOKUSHUKU
(on behalf of the H1 and ZEUS collaborations) *
INPS, KEK,
Oho 1-1,
Tsukuba 305-0801, Japan
E-mail: katsuo.tokushuku@kek.jp

HERA-I studies on new resonance searches in the hadronic final state in ep collisions are reviewed. The searches were performed both in meson and baryon resonances. Pentaquark candidates were observed in $K^0_s p$ system in ZEUS and in D^*p system in H1. The two experiments are compared and the compatibility is discussed.

1. Introduction

HERA provides a unique field for new resonance searches. Thanks to the high centre-of-mass energy of HERA, searches can be performed in various kinematic regions. By tagging the scattered electron, the kinematics of the hadronic system can be well defined, primarily by the standard event variables such as Q^2, x, y and W, and then by a particular phase space in the event, such as the target- or current-fragmentation in the Breit frame. Depending on kinematic regions selected by these variables, quite different mechanisms may take place in particle production.

In my talk, the hadron resonance searches in HERA-I are reviewed in the order of the number of quarks in the system; *i.e.* starting from mesons, continuing to baryons including pentaquarks and finally ending with many quark states.

*supported by the Japanese Ministry of Education, Culture, Sports, Science and Technology (MEXT) and its grants for Scientific Research

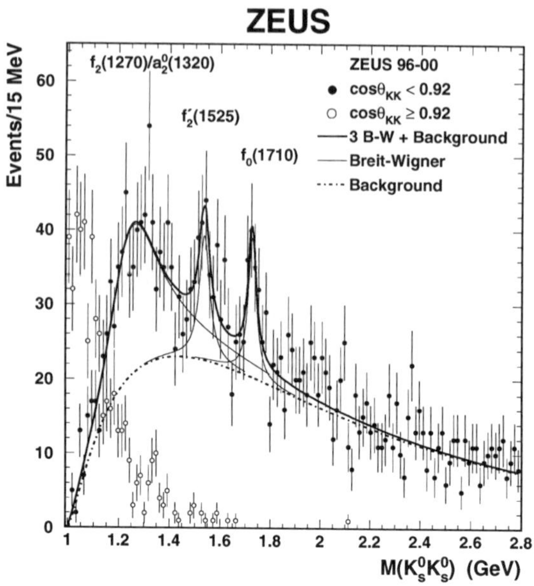

Figure 1. The $K_s^0 K_s^0$ invariant mass spectrum with cuts on the opening angle between the two K_s^0 in the laboratory frame ($\cos\theta_{KK}$). Fitting results using three Breit-Wigners and a background function are also shown.

2. Meson productions at HERA

Meson spectroscopy in the $\pi^+\pi^-$ channel was performed by H1 in photoproduction events[1]. Clear mass peaks from ρ^0, $f_0(980)$ and $f_2(1270)$ were seen. The production cross sections of these resonances and of pions agree well with each other, when they are plotted as a function of its mass plus p_T instead of p_T itself and when the degree of freedom from the spin and isospin is taken into account. Such a thermodynamic behaviour of particle production is often seen in relativistic heavy ion collisions. It is interesting to find a similar behaviour in *ep* collisions.

Figure 1 shows the mass distribution of $K_s^0 K_s^0$ from the ZEUS collaboration[2]. Three resonance structures are observed. The broad peak at low mass may correspond to the $f_2(1270)$ and/or $a_2^0(1320)$ state. The second peak at 1537^{+9}_{-8} MeV is consistent with the $f_2'(1525)$.

The highest mass state has a mass consistent with the glueball candidate $f_0(1710)$. The measured width is 38^{+20}_{-14} MeV, which is narrower than the value by the Particle Data Group of $125{\pm}10$ MeV. But it gives an acceptable chi-square when the data are fitted with a mass width fixed to this value.

Most of the $K_s^0 K_s^0$ pairs (93 %) found in the analysis are in the target region of the Breit frame. If the $f_0(1710)$ is a glueball, it is natural for it to be predominantly seen in the target region, where gluons are abundant. Although this analysis is based on the entire data sample of 1996-2000, it is not enough to make separate plots for various kinematics to see if this is the case. It is important to continue the spectroscopy in the HERA-II period.

3. Baryon productions at HERA

The search for new baryons is a very hot topic since the LEPS experiment found a strange pentaquark candidate Θ^+ in the missing mass distribution in low-energy photon-nucleus interactions[3]. Soon after the observation, many other experiments reported the presence or absence of a signal. The recent high statistic measurements by the CLAS experiment in γp and γd reactions show negative results, while the new measurements from the LEPS experiment indicate an associate production of Θ^+ and $\Lambda(1520)$[4].

As already mentioned, HERA offers a wide range of kinematic region where different hadronisation processes may take place. It is very important to understand how ordinary baryons are produced in various phase spaces, to compare with the pentaquark candidates.

There are still only a few results on stable baryon production at HERA. H1 measured the proton production cross sections in photoproduction. The proton yield is about half of the prediction from the PYTHIA Monte Carlo generator[5]. Lambda production in DIS shown by ZEUS shows a good agreement except for small deviations in the Λ/K^0 ratio at low-x and low-p_T regions[6]. Such a difference may indicate that baryon production in ep collisions is different from e^+e^- collisions. Here again, it is important to study further by specifying the phase space more carefully.

Related to the Θ^+ pentaquark searches shown in the next section, ZEUS studied several strange and charmed baryons which decay to a kaon and a nucleon, such as Λ_c and $\Lambda(1520)$. These baryons are observed in a wide pseudorapidity (η) region, both in photoproduction and in DIS. The behaviour of baryons and anti-baryons is similar; *i.e.* there is no visible baryon/anti-baryon asymmetry[7].

3.1. *Strange pentaquarks*

ZEUS already reported the observation[8] of a peak in the invariant mass distribution of the $K_s^0 p$ system, as shown in figure 2. The peak position is $1521.5 \pm 1.5(stat.)^{+2.8}_{-1.7}(syst.)$ MeV and is consistent with the $\Theta^+(1530)$. The measured Gaussian width is $6.1 \pm 1.6(stat.)^{+2.0}_{-1.4}(syst.)$ MeV, which is above, but consistent with, the experimental resolution of 2.0 ± 0.5 MeV.

On the other hand, the H1 collaboration found negative results[9]. The upper limit on new resonance are shown in figure 3. This can be compared with the visible cross section $\sigma_{vis} = 125 \pm 27^{+36}_{-28}$ pb, reported by ZEUS[10], and corresponding to a slightly larger phase space. In the mass region around 1530 MeV, the H1 upper limit and the ZEUS measurement overlap each other, so that the two measurement are still compatible, with the current statistics.

ZEUS tried to enhance the Θ^+ peak by refining the kinematic cuts[7]. The signal is best seen when the Kp system is going in the proton direction in the laboratory frame and in the high-Q^2 ($Q^2 > 20$ GeV^2) event sample. The $K^0_s p$ peak is more visible than the $K^0_s \bar{p}$ one. This behaviour is contrasted with the case for the normal baryons mentioned in the previous section.

The ratio of the production cross section of Θ^+ to Λ is measured[10] to

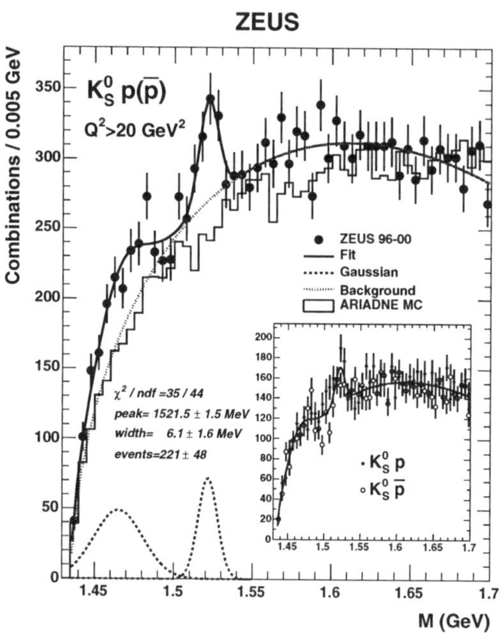

Figure 2. Invariant mass spectrum for the $K^0_s p(\bar{p})$ for $Q^2 > 20$ GeV measured by the ZEUS experiment. Results of a fit with the Gaussians and a background function are also shown. The inset shows the $K^0_s p$ and $K^0_s \bar{p}$ candidates separately.

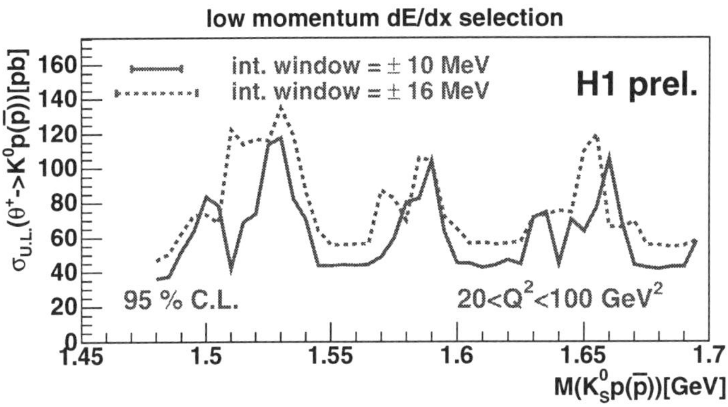

Figure 3. Upper limits set by the H1 experiment on the Θ production cross sections ($\sigma_{U.L.}(e+p \longrightarrow e\Theta^+X \longrightarrow eK^0p(\bar{p})X)$) at 95% confidence level for $20 < Q^2 < 100$ GeV^2.

be about 0.05. This would be large enough for the LEP experiments to see the Θ^+, if the same ratio can be applied to e^+e^- collisions.

All these observations indicate that the production mechanism of the pentaquark is different from ordinary baryons. However, it should be pointed out that these results are from the same event sample and hence the results are highly correlated. It is, therefore, too early to draw any positive/negative conclusions about the strange pentaquark. We need to wait for analyses with the new HERA-II data.

A search of pentaquarks which decay to $\Xi\pi$ was also performed[11]. The narrow peak found by the NA49 experiment[12] at 1862 MeV was not seen in a DIS data sample.

3.2. *Charmed pentaquarks*

The H1 experiment reported a narrow peak in the $D^{*-}p$ mass distribution[13]. Since the D^{*-} contains an anti-c quark, the resonance has an anti-valence quark with a positive baryon number, so that this is another pentaquark candidate. The peak position is $3099 \pm 3 + 5$ MeV and its width is very narrow; 12 ± 3 MeV. As seen in figure 4, the signal is clearly seen both in DIS and in photoproduction, although the background is larger in the photoproduction sample.

The production cross section is estimated based on the events in the peak[14]. Since it is unknown how the $D^{*-}p$ resonance (called $D^*p(3100)$, hereafter) is produced and decays, a model is developed with the RAP-

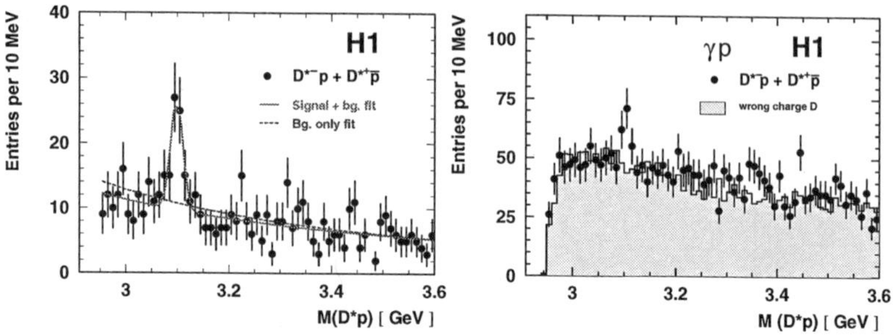

Figure 4. Invariant mass distribution of opposite-charge D^*p combination in DIS and in photoproduction measured by H1.

GAP Monte Carlo, assuming that it is produced in a similar manner to the D mesons such as $D_1(2420)$ and $D_2^*(2460)$. Isotropic decay is also assumed. Cross sections as a function of η and p_T are shown in figure 5. The Monte Carlo predictions with the above assumptions are also shown after the overall visible cross section is normalised to the data. In the model, the η distribution of the resonance is flat but the data have a suppression in the central region. The transverse momentum distribution is slightly softer in the data. These might indicate a different production mechanism of the resonance, but, as already mentioned in the case of the ZEUS studies of Θ^+, the studies done so far are based on a single sample taken in the HERA-I period.

Unfortunately, the resonance structure is not seen in the ZEUS measurements[15]. The observation of H1 implies that the $D^*p(3100)$ contributes roughly 1% of the D^* production rate in DIS with $Q^2 > 1$ GeV^2 although the analysis cuts are different. Figure 6 shows the ZEUS results with such an estimated pentaquark signal. In contrast to the Θ^+ case, the H1 and ZEUS results are already not compatible.

In summary, the pentaquark searches at HERA do not show clear results and it is hard to draw any conclusions at this stage. It is very important for both collaborations to repeat the analyses with the HERA-II data to confirm/reject the candidates.

4. Instantons

To end my talk, I would like to mention searches for the many-quark states induced by QCD instantons[16]. QCD is a non-Abelian gauge theory and

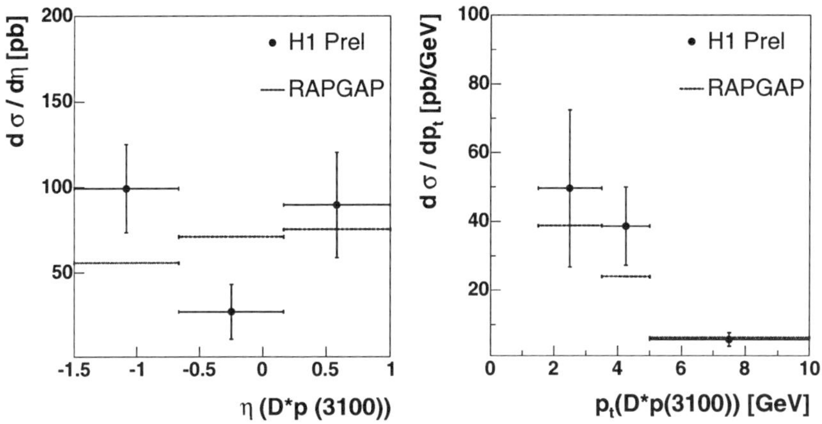

Figure 5. Differential $D^*p(3100)$ cross section as a function of pseudorapidity (η) and transverse momentum (p_T) of the $D^*p(3100)$ system. Data (closed symbols) are compared with the expectation of a model (RAPGAP) assuming a production and decay mechanism as described in the text. Only statistical errors are shown.

has a complicated vacuum structure. Tunnelling processes between topologically different types of vacuum states can occur. These are mediated by instantons.

The cross section for instanton induced processes in DIS is predicted[17] to be $O(10^{-3})$ of the total DIS cross section. This is still small, but by using sophisticated final state selection procedures, it is possible to enhance this fraction.

In (anti-)instanton processes, light quarks and anti-quarks are produced "democratically" and with the same chirality:

$$\gamma^* + g \overset{I}{\Longrightarrow} \sum_{flavours} (\overline{q}_R + q_R) + n_g g, \quad (I \to \overline{I}, R \to L), \tag{1}$$

with $n_g \sim 3$. It is very difficult to measure the handedness in the final state, so the signature used in searches is based on the emission of a large number of partons in a limited rapidity range.

An example[18] of the difference in the final state is shown in Figure 7. The variable used in the plots is the sphericity of particles in the reconstructed instanton rest frame. The left plot is made before enhancement cuts. Instanton events are more circular than "normal" DIS events, but the fraction is very small so that in the plot it is rescaled by a factor of 500.

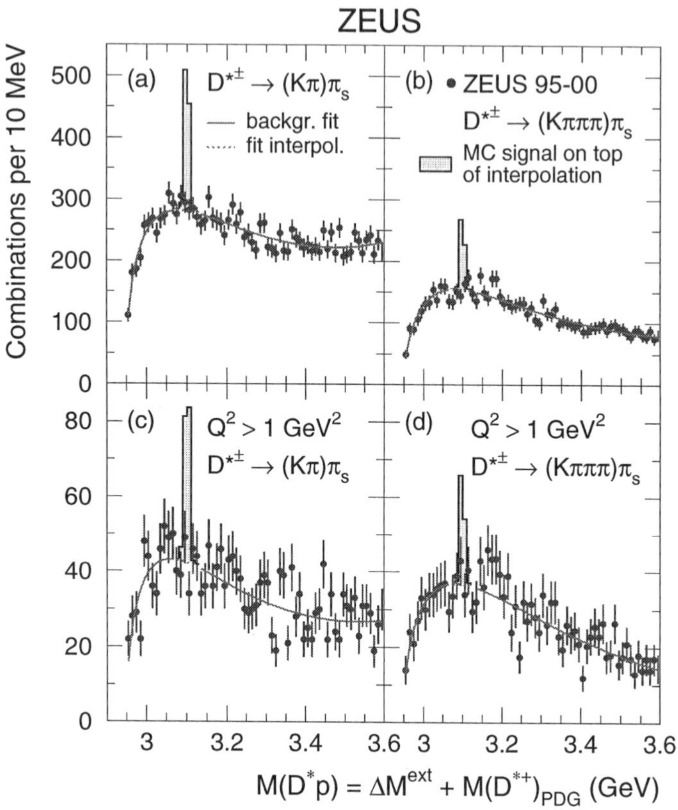

Figure 6. ZEUS measurements on the D^*p mass distributions in two different D^0 decay modes and Q^2 ranges. The shaded histograms show the Monte Carlo $D^*p(3100)$ signal, assuming 1% of D^*'s are the decay product of $D^*p(3100)$.

After enhancement cuts using various kinematic variables[18], the instanton contribution becomes of comparable size to the DIS background (the right plot). The shape of the distributions of the "normal" DIS Monte Carlo predictions is, however, similar to that arising from the instanton process. Moreover, the difference between the two DIS MCs is comparable to the expected instanton contribution. Given these systematic uncertainties, the 95% confidence limit of the cross section for instanton induced events is set to 221 pb, which is about a factor of five above the prediction.

ZEUS made a similar search[19] in a higher Q^2 range ($Q^2 > 120$ GeV^2)

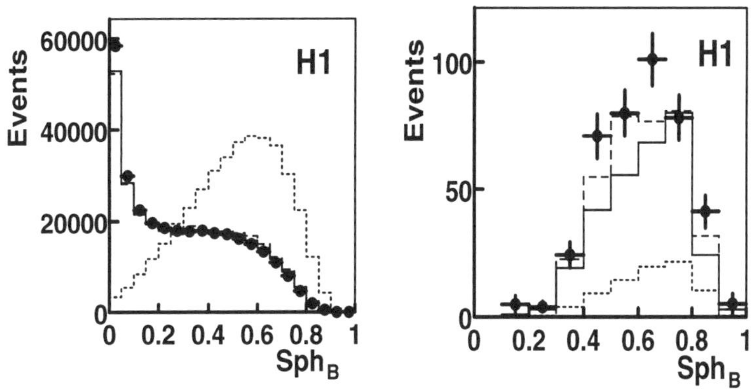

Figure 7. Distributions of the sphericity (Sph_B) in the reconstructed instanton centre-of-mass frame before and after instanton enrichment cuts. Data (filled circles), two QCD model background MCs (solid and dashed line) and the instanton MC (dotted line) are shown. In the left plot, the instanton prediction is scaled up by a factor of 500.

where the theory prediction is expected to be more reliable. The difference of the two DIS MCs again limits making a strong conclusion. Conservatively assuming that all remaining data after the selections belong to an instanton signal, the upper limit on the instanton cross section is set to 26 pb at a 95% confidence level, to be compared with the theoretically predicted cross section of 8.9 pb.

5. Conclusions

Many measurements of meson and baryon production have been performed at HERA. Trying to enhance (exotic) signals, various selections were tried. This capability to control the kinamatic region in the search is one of the advantages at HERA.

Pentaquark results from H1 and ZEUS do not allow us to make a definite conclusion at this stage. We need new results from HERA-II data to answer if a five-quark system exists.

The search for genuine QCD effects such as instanton-induced processes was performed in HERA-I. We need to understand normal DIS better. At the same time we need to find better observables.

All results shown in this talk are from HERA-I data. We expect further progress by using high-statistics HERA-II data, since many results shown are, after various selection cuts, still statistics limited.

References

1. H1 Collaboration, Contributed paper #6-0184 to 32nd International Conference on High Energy Physics, ICHEP04, August 16, 2004, Beijing.
2. S. Chekanov et al. (ZEUS Collaboration), *Phys. Lett.* **B 578**, (2004) 33.
3. T. Nakano et al. (LEPS Collaboration), *Phys. Rev. Lett.* **91**, (2003) 012002.
4. As a recent review, for example, see V. D. Burkert hep-ph/0510309, invited talk at XXII International Symposium on Lepton-Photon Interactions at High Energy, LP2005, June 30, 2005, Uppsala.
5. H1 Collaboration, Contributed paper #1002 to 31st International Conference on High Energy Physics, ICHEP02, July 24, 2002, Amsterdam.
6. ZEUS Collaboration, Contributed paper #275 to XXII International Symposium on Lepton-Photon Interactions at High Energy, LP2005, June 30, 2005, Uppsala.
7. ZEUS Collaboration, Contributed paper #277 to XXII International Symposium on Lepton-Photon Interactions at High Energy, LP2005, June 30, 2005, Uppsala.
8. S. Chekanov et al. (ZEUS Collaboration), *Phys. Lett.* **B 591**, (2004) 7.
9. H1 Collaboration, Contributed paper #400 to XXII International Symposium on Lepton-Photon Interactions at High Energy, LP2005, June 30, 2005, Uppsala.
10. ZEUS Collaboration, Contributed paper #290 to XXII International Symposium on Lepton-Photon Interactions at High Energy, LP2005, June 30, 2005, Uppsala.
11. S. Chekanov et al. (ZEUS Collaboration), *Phys. Lett.* **B 610**, (2005) 212.
12. C. Alt et al. (NA49 Collaboration), *Phys. Rev. Lett.* **92**, (2005) 042003.
13. A. Aktas et al. (H1 Collaboration), *Phys. Lett.* **B 588**, (2004) 17.
14. H1 Collaboration, Contributed paper #401 to XXII International Symposium on Lepton-Photon Interactions at High Energy, LP2005, June 30, 2005, Uppsala.
15. S. Chekanov et al. (ZEUS Collaboration), *Eur. Phys. J.* **C 38**, (2004) 29.
16. G. 't Hooft, *Phys. Rev. Lett.* **37**, (1976) 8;
    G. 't Hooft, *Phys. Rev.* **D 14**, (1976) 3432; Erratum-ibid. **D 18**, (1976) 2199;
    A. A. Belavin et al. *Phys. Lett.* **B 59**, (1975) 85.
17. A. Ringwald and F. Schrempp, *Phys. Lett.* **B 438**, (1998) 217 ;
    ibid. **B 459**, (1999) 249; ibid. **B 503**, (2001) 331.
18. C. Adloff et al. (H1 Collaboration), *Eur. Phys. J.* **C 25**, (2002) 495.
19. S. Chekanov et al. (ZEUS Collaboration), *Eur. Phys. J.* **C 34**, (2004) 255.

HADRON SYSTEMATICS AND EMERGENT DIQUARKS

ALEXANDER SELEM[1,2] AND FRANK WILCZEK[1]

[1] *Center for Theoretical Physics, Laboratory for Nuclear Science, and Department of Physics, Massachusetts Institute of Technology, Cambridge, Massachusetts 02139, USA*

[2] *Department of Physics, University of California, 366 LeConte Hall, MC 7300, Berkeley, CA 94720-0730*

We briefly review a variety of theoretical and phenomenological indications for the probable importance of powerful diquark correlations in hadronic physics. We demonstrate that the bulk of light hadron spectroscopy can be organized using three simple hypotheses: Regge-Chew-Frautschi mass formulae, feebleness of spin-orbit forces, and energetic distinctions among a few different diquark configurations. Those hypotheses can be implemented in a semi-classical model of color flux tubes, extrapolated down from large orbital angular momentum L. We discuss refinements of the model to include the effects of tunneling, mass loading, and internal excitations. We also discern effects of diquark correlations in observed patterns of baryon decays. Many predictions and suggestions for further work appear.

1. Introduction

Quantum chromodynamics (QCD) is established as the fundamental underlying theory of the strong interaction, but its application to low-energy hadron phenomenology is far from a matter of routine deduction. Different idealizations and approximations, whose connection to the underlying theory ranges from tight to tenuous, are presently used to describe different phenomena. Examples include the nonrelativistic quark model, bag model, Skyrme model, large N approaches, Regge phenomenology, and others, not to mention the vast literature of traditional nuclear physics.

With the increasing power of numerical simulation (lattice gauge theory) to provide accurate access to some of what QCD predicts, it becomes an attractive program to build models whose concepts are firmly based in the fundamental theory, and whose underlying parameters can be accessed by numerical simulation.

The concept of diquarks has deep roots in the fundamental theory, and has been invoked to help illuminate several phenomena in the strong interaction. At a crude level, the idea is that pairs of quarks form bound states, which can be treated as (confined) particles, and used as degrees of freedom in parallel with quarks themselves. A more sophisticated discussion should be phrased in terms of strong correlations between pairs of quarks in favorable channels, resulting in a significant energy gap to the appearance of unfavorable channels.

There are simple theoretical arguments that suggest which diquark correlations are favorable. These arguments also suggest that the effect of such correlations should be most important for the lightest quarks, and that the correlations are subject to frustration.

There are several observed phenomena that can be illuminated using these concepts, as we shall review presently. In most of those applications, unfortunately, the diquarks are not well isolated. This complicates their interpretation.

The present investigation was initially motivated as an attempt to address that problem, using the following simple idea. Because diquarks are confined, we cannot study them in perfect isolation. But it is plausible that baryons with large values of the angular momentum L form extended bar-like structures, with quarks pushed to the extremities by centrifugal forces. More specifically, a widely accepted – and, as we shall see, remarkably successful – way to model large L mesons and baryons envisions separated, rotating quark-antiquark or quark-(2 quark) configurations joined by an electric flux tube, or string. Thus we can use large-L baryons to study diquark interactions, and specifically to test whether the quantum numbers that theory predicts to be energetically favorable are in fact so, and by how much.

2. Motivations

2.1. *"Theory"*

Color Cancellation: Since two separated quarks are each in the $\bar{\mathbf{3}}$ representation, while they can combine into a single $\mathbf{3}$ representation, the disturbance produced by the color charges of two quarks in empty space can be halved by bringing them together. Thus, on very general grounds, we should expect an attractive interaction between quarks, when their color state is antisymmetric.

Spin Minimization: Refining this consideration, we can consider the spins of the quarks. Both one-gluon exchange and instanton calculations indicate that the antisymmetric, spin-singlet state is more favorable energetically than the symmetric, spin-triplet state. We might again expect this on very general grounds, in that the total spin is associated with a color magnetic moment, which again disturbs the vacuum. To enforce Fermi statistics overall, then, we must put the quarks in antisymmetric $\bar{\mathbf{3}}$ representation of flavor $SU(3)$. Note that the spin-dependent splitting, being of magnetic origin, is intrinsically relativistic. Formally, at lowest order, it is inversely proportional to the masses of the particles involved. While it would be overly naive, in the context of strong interaction physics, to take that dependence literally – in particular, its suggestion of a divergence for zero mass – we should expect that the splitting will be largest for the u and d quarks, smaller when s is involved, and much smaller for the heavier quarks. We will use the notation $[ud], (ud)$ for the good and bad diquark configurations involving u and d quarks, respectively, and similarly for other flavor combinations.

Repulsion: Since, according to the preceding considerations, energy can be gained by enforcing favorable correlations between quark pairs, we should expect that any effect that disrupts the correlations will induce a repulsive force. The presence of an additional diquark will cause such disruptions, due both to competing interactions and to fermi statistics. So we should expect repulsive forces *between* diquarks that already have favorable correlations.

Emergence: Similarly, the nearby presence of a spectator quark will frustrate favorable diquark correlations. We should expect that the full correlation energy will emerge when the effects of frustration are minimized, for example in baryons at large L (as anticipated above), or in baryons containing a heavy quark (whose spin is ineffective).

2.2. *Phenomenology*

Ground State Spectroscopy: A classic manifestation of energetics that depends on diquark correlations is the $\Sigma - \Lambda$ mass difference. The Λ is isosinglet, so it features the good diquark $[ud]$; while Σ, being isotriplet, features the bad diquark (ud). The splitting

$$\Sigma(1193) - \Lambda(1116) = 76 \text{ MeV} \tag{1}$$

is consistent in sign with the expectations sketched above. Replacing the s

quark with the c quark we find (averaging over the c spin) the much larger splitting

$$\frac{2}{3}\Sigma_c(2520)^{\frac{3}{2}^+} + \frac{1}{3}\Sigma_c(2455)^{\frac{1}{2}^+} - \Lambda_c(2285) = 213 \text{ MeV} \tag{2}$$

Again, this is consistent with expectations: the c quark's spin is much less potent than the s quark's, and frustrates $[ud]$ less.

Structure Functions: One of the oldest observations in deep inelastic scattering is that the ratio of neutron to proton structure functions approaches $\frac{1}{4}$ in the limit $x \to 1$

$$\lim_{x\to 1} \frac{F_2^n(x)}{F_2^p(x)} \to \frac{1}{4} \tag{3}$$

In terms of the twist-two operator matrix elements used in the formal analysis of deep inelastic scattering, this translates into the statement

$$\lim_{n\to\infty} \frac{\langle p|\bar{d}\gamma_{\mu_1}\overleftrightarrow{\nabla}_{\mu_2}\cdots\overleftrightarrow{\nabla}_{\mu_n} d|p\rangle}{\langle p|\bar{u}\gamma_{\mu_1}\overleftrightarrow{\nabla}_{\mu_2}\cdots\overleftrightarrow{\nabla}_{\mu_n} u|p\rangle} \to 0 \tag{4}$$

where spin averaging of forward matrix elements, symmetrization over the μs, and removal of traces is implicit, and a common tensorial form is factored out, together with similar equations where operators with strange quarks, gluons, etc. appear in the numerator. Equation (4) states that in the valence regime $x \to 1$, where the struck parton carries all the longitudinal momentum of the proton, that struck parton must be a u quark. It implies, by isospin symmetry, the corresponding relation for the neutron, namely that in the valence regime within a neutron the parton must be a d quark. Then the ratio of neutron to proton matrix elements will be governed by the ratio of the squares of quark charges, namely $\frac{(-\frac{1}{3})^2}{(\frac{2}{3})^2} = \frac{1}{4}$. Any (isosinglet) contamination from other sources will contribute equally to numerator and denominator, thereby increasing this ratio. Equation (4) is, from the point of view of symmetry, a peculiar relation: it requires an emergent conspiracy between isosinglet and isotriplet operators. It is also, from a general physical point of view, quite remarkable: it is one of the most direct manifestations of the fractional charge on quarks; and it is a sort of hadron = quark identity, closely related to the quark-hadron continuity conjectured to arise in high density QCD. It is an interesting challenge to derive (4) from microscopic QCD, and to estimate the rate of approach to 0.

Fragmentation: A more adventurous application is to fragmentation. One might guess that the formation of baryons in fragmentation of an energetic quark or gluon jet could proceed stepwise, through the formation of diquarks which then fuse with quarks. To the extent this is a tunneling-type process, analogous to pair creation in an electric field, induced by the decay of color flux tubes, one might expect that the good diquark would be significantly more likely to be produced than the bad diquark. This would reflect itself in a large Λ/Σ ratio. And indeed, data from LEP indicates that the value of this ratio is about 10 for leading particles (that is, at large z). In the Particle Data Book one also finds an encouraging ratio for total multiplicities in e^+e^- annihilation: $\Lambda_c : \Sigma_c = .10 \pm .03 : .014 \pm .007$; in this case the c quarks are produced by the initiating current, and we have a pure measure of diquarks.

$\Delta I = \frac{1}{2}$ **Rule:** There are also several indications that diquark correlations have other important dynamical implications. The $\Delta I = \frac{1}{2}$ rule in strangeness-changing nonleptonic decays has also been ascribed to attraction in the diquark channel. The basic operator $\bar{u}\gamma_\mu(1-\gamma_5)d\bar{s}\gamma^\mu(1-\gamma_5)u$ arising from W boson exchange can be analyzed into $[\bar{u}\bar{s}][ud]$, $(\bar{u}\bar{s})(ud)$, and related color-**6** diquark types. Diquark attraction in $[\bar{u}\bar{s}][ud]$ means that there is a larger chance for quarks in this channel to tap into short-distance components of hadronic wavefunctions. This effect is reflected in enhancement of this component of the basic operator as it is renormalized toward small momenta. Such an enhancement is well-known to occur at one-loop order (one gluon exchange).

(Paucity of) Exotics: The basic degrees of freedom in QCD include massless gluons and almost-massless u, d quarks, and the interaction strength, though it "runs" to small coupling at large momentum transfer, is not uniformly small. We might therefore anticipate, heuristically, that low-energy gluons and quark-antiquark pairs are omnipresent, and in particular that the eigenstates of the Hamiltonian – hadrons – will be complicated composites, containing an indefinite number of particles. And indeed, according to the strictest experimental measure of internal structure available, the structure functions of deep inelastic scattering, nucleons do contain an infinite number of soft gluons and quark-antiquark pairs (parton distributions $\sim \frac{dx}{x}$ as $x \to 0$). Yet the main working assumption of the quark model is that hadrons are constructed according to two body plans: mesons, consisting of a quark and an antiquark; and baryons, consisting of three quarks. That paucity of body plans seems to conflict with heuristic expectations.

The puzzles posed by the success of the quark model come into sharp focus in the question of *exotics*. Are there additional body plans in the hadron spectrum, beyond qqq baryons and $\bar{q}q$ mesons (and loose composites thereof)? If not, why not; if so, where are they? As a special case: why don't multi-nucleons merge into single bags, e.g. $qqqqqq$ – or can they? The tension between *a priori* expectations of complex bound states and successful use of simple models defines the main problem of exotics: Why aren't there more of them?

A heuristic explanation can begin along the following lines. Low-energy quark-antiquark pairs are indeed abundant inside hadrons, as are low-energy gluons, but they have (almost) vacuum quantum numbers: they are arranged in flavor and spin singlets. (The "almost" refers to chiral symmetry breaking.) Deviations from the "good" quark-antiquark or gluon-gluon channels, which are color and spin singlets, cost significant energy. States which contain such excitations, above the minimum consistent with their quantum numbers, will tend to be highly unstable. They might be hard to observe as resonances, or become unbound altogether.

The next-best way for extraneous quarks to appear is, according to the preceding considerations, in the form of "good" diquark pairs. Thus a threatening (that is, superficially promising) strategy for constructing low-energy exotics apparently could be based on using those objects as building-blocks. On reflection, there are two reasons "good" diquark correlations help explain the paucity of exotics. First: due to their total antisymmetry, good diquarks conceal spin and flavor. So even if present, they can be hard to discern.

Second, and more important: because of their mutual repulsion, they inhibit mergers. Why do protons and neutrons in close contact retain their integrity, rather than merge into a common bag? A closely related question arises in a sharp form for the H dibaryon that R. Jaffe studied extensively. It has the configuration $uuddss$. In the bag model it appears that a single bag containing these quarks supports a spin-0 state that is quite favorable energetically. A calculation based on quasi-free quarks residing in a common bag, allowing for one-gluon exchange, indicates that H might well be near or even below $\Lambda\Lambda$ threshold, and thus strongly stable; or perhaps even below Λn threshold, and therefore stable even against lowest-order weak interactions. These possibilities appear to be ruled out both experimentally and by numerical solution of QCD (though possibly neither case is completely airtight). Good diquark correlations, together with repulsion between diquarks, suggests a reason why the almost-independent-particle approach fails in this case. Note that this mechanism requires essentially nonperturbative quark interaction effects, beyond one gluon exchange, since one gluon exchange is favorable for H.

3. Spectroscopy

3.1. *Hypotheses in Spectroscopy*

As previewed above, we want to exploit large L baryons to isolate diquarks dynamically. Since L is not a direct observable, but rather a parameter appearing in the context of a spectroscopic model, to access L we must specify the hypotheses of our model with sufficient precision to permit mapping between predicted states and observed mesons and baryons. In practice we did this iteratively: trying out tentative hypotheses, using them to make tentative mappings, and refining or rejecting the hypotheses according to our perception of the overall quality of the fit. We did not attempt to do sophisticated statistical analysis, which given the complex and spotty nature of the data set would be extremely challenging and probably futile. Fortunately the preponderance of the established meson and baryon states can be uniquely fit with appropriate quantum numbers and approximate mass using three simple, physically motivated hypotheses, as follows:

Loaded Flux Tube: The equation $M^2 = \sigma L$ relating mass to orbital angular momentum arises from solving the equations for a spinning relativistic string with tension $\sigma/(2\pi)$, terminated by the boundary condition that both ends move transversely at the speed of light. We might expect it to hold asymptotically for large L in QCD, when an elongated flux tube appears string-like, the rotation is rapid, quark masses are negligible, and semiclassical quantization of its rotation becomes appropriate. The famous Chew-Frautschi formula $M^2 = a + \sigma L$, with simple non-zero values of a (e.g., $a = \frac{1}{2}\sigma$) can result from quantization of an elementary non-interacting string, including zero-point energy for string vibrations.

We can generalize the Chew-Frautschi formula by considering two masses m_1, m_2 connected by a relativistic string with constant tension, T, rotating with angular momentum L. Our general solution naturally arises in a parameterized form in which the energy, E, and L are both expressed in terms of the angular velocity, ω, of the rotating system. In the limit that $m_1, m_2 \to 0$, the usual Chew-Frautschi relationship $E^2 \propto L$ appears.

Consider masses m_1 and m_2 at distances r_1 and r_2 away from the center of rotation respectively. The whole system spins with angular velocity ω. It is also useful to define:

$$\gamma_i = \frac{1}{\sqrt{1-(\omega r_i)^2}} \tag{5}$$

where the subscript i can be 1 or 2 (for the mentioned masses). It is straightforward to write the energy of the system:

$$E = m_1\gamma_1 + m_2\gamma_2 + \frac{T}{\omega}\int_0^{\omega r_1}\frac{1}{\sqrt{1-u^2}}du + \frac{T}{\omega}\int_0^{\omega r_2}\frac{1}{\sqrt{1-u^2}}du. \tag{6}$$

The last two terms are associated with the energy of the string. Similarly, the angular momentum can be written as:

$$L = m_1\omega r_1^2\gamma_1 + m_2\omega r_2^2\gamma_2 + \frac{T}{\omega^2}\int_0^{\omega r_1}\frac{u^2}{\sqrt{1-u^2}}du + \frac{T}{\omega^2}\int_0^{\omega r_2}\frac{u^2}{\sqrt{1-u^2}}du. \tag{7}$$

Carrying out the integrals gives:

$$E = m_1\gamma_1 + m_2\gamma_2 + \frac{T}{\omega}(\arcsin[\omega r_1] + \arcsin[\omega r_2]), \tag{8a}$$

$$L = m_1\omega r_1^2\gamma_1 + m_2\omega r_2^2\gamma_2 + \frac{T}{\omega^2}\frac{1}{2}\Big(-\omega r_1\sqrt{1-(\omega r_1)^2} \tag{8b}$$
$$+ \arcsin[\omega r_1] - \omega r_2\sqrt{1-(\omega r_2)^2} + \arcsin[\omega r_2]\Big)\,.$$

Furthermore the following relationship between the tension and angular acceleration holds for each mass:

$$m_i\omega^2 r_i = \frac{T}{\gamma_i^2}. \tag{9}$$

The limit $L \to 0$ is singular: it requires $\omega \to \infty$ and $r \to 0$, and is quite sensitive to the presence of small quark masses or interactions between the ends. Thus we do not expect our simple semi-classical approach to be accurate in this sector. The situation is reminiscent of the relation between vortices and rotons in superfluid He^4.

The resulting parametric expressions for E and L are opaque to humans, but easy for a computer to handle. For the first corrections at small m_1, m_2 (and $L \neq 0$) we find, after some algebra,

$$E \approx \sqrt{\sigma L} + \kappa L^{-\frac{1}{4}}\mu^{\frac{3}{2}} \tag{10}$$

with

$$\kappa \equiv \frac{2}{3}\frac{\pi^{\frac{1}{2}}}{\sigma^{\frac{1}{4}}} \tag{11}$$

and

$$\mu^{\frac{3}{2}} \equiv m_1^{\frac{3}{2}} + m_2^{\frac{3}{2}} \tag{12}$$

This is a useful expression, since it allows us to extract expressions for quark and diquark mass differences from the observed values of baryon and meson mass differences. Numerically, $\kappa \approx 1.15\ \text{GeV}^{-\frac{1}{2}}$ for $\sigma \approx 1.1\ \text{GeV}^2$.

Note that the usual correction ascribed to zero-point vibrations, i.e. a classic intercept of the type $E^2 = a + (2\pi T)L$, yields corrections of the form $E \to \sqrt{\sigma L} + \frac{a}{2\sqrt{\sigma L}}$. It becomes subdominant to mass corrections at large L.

For heavy-light systems the corresponding formula is

$$E - M = \sqrt{\frac{\sigma L}{2}} + 2^{\frac{1}{4}}\kappa L^{-\frac{1}{4}}\mu^{\frac{3}{2}} \tag{13}$$

where M is the heavy quark mass and μ is the light quark mass. The effective tension is halved.

Emergent Diquarks: We allow diquarks with different quantum numbers and masses at the end of the flux tubes.

Smallness of Spin-Orbit Forces: If one inserts a linear confining potential between quark and antiquark into the Dirac equation, the spin-orbit forces depend on whether the potential is taken as a scalar or the fourth component of a vector. These two options give answers of equal magnitude but opposite sign. The magnitude is much too large for the data to accommodate. We obtain a good fit by simply ignoring spin-orbit forces.

The nature of the fits is best conveyed by reference to the enclosed Tables. The next subsection provides a running commentary upon them.

3.2. *Assignments*

In Table 1, we display assignments for non-strange baryons. The rules of the game are as follows. Good diquarks have spin 0 and isospin 0. Thus when they are assembled into non-strange baryons the total isospin is $\frac{1}{2}$ (i.e., they are nucleons N) and the spin-parity J^P is $(L + \frac{1}{2})^L$ or $(L - \frac{1}{2})^L$, each represented once (except for $L = 0$, of course). In general, we put a dash $--$ in boxes where no state is expected). We expect these states to be approximately degenerate, reflecting the smallness of spin-orbit forces. Bad diquarks have spin 1 and isospin 1. They can be joined with the other quark into either isospin $\frac{3}{2}$ Δs or isospin $\frac{1}{2}$ nucleons. They

$(2J)^0$	$(2L+1)^L$	$(2L-1)^L$	$(2L+3)^L$	$(2L+3)^L$	$(2L+1)^L$	$(2L+1)^L$	$(2L-1)^L$	$(2L-1)^L$	$(2L-3)^L$	$(2L-3)^L$
L	good 0	good 1	bad 0	bad 0	bad 1	bad 1	bad 2	bad 2	bad 3	bad 3
0	N (930)	–	–	Δ (1232)	–	–	–	–	–	–
1	N (1520)	N (1535)	N (1675)		N (1700)	Δ (1700)	N (1650)	Δ (1620)	–	–
2	N (1680)	N (1720)	N (1990)	Δ (1950)	N (2000)	Δ(1905) Δ(2000)	N (1900)	Δ (1920)		Δ (1910)
3			N (2250)	Δ (2400)	N (2190)		N (2200)		N (2080)	
4	N (2220)			Δ (2420)		Δ (2300)				
5	N (2600)			Δ (2750)						
6	N (2700)			Δ (2950)						

Table 1. Classification of non-strange baryons. See text for explanation.

can also be joined into various spin-parities, as indicated in Table 1. The integer that appears below "good" or "bad" in the second row indicates the misalignment, that is how much the total angular momentum differs from its maxiumum for the given spin and orbital angular momenta. The boxes in gray indicate quantum numbers that can be reached in two ways. In one case – $\Delta(1905)$ and $\Delta(2000)$ – two states have been resolved; we predict that better measurements will reveal widespread doubling. (There are several additional known cases of doublings of this type elsewhere in the baryon and meson spectrum, as we shall see.) All the "bad" diquark states in a given row should be approximately degenerate, according to the hypotheses of small coupling between the ends of the flux tubes (including weak flavor-dependent interactions) and weak spin-orbit forces. The L assignments are of course constrained by the Chew–Frautschi-Regge mass formula as well as by quantum numbers.

The approximate degeneracies our model predicts are well represented in the data, as is the systematic splitting between good and bad diquark configurations, which saturates around 200 MeV at $L = 2$.

After the preceding discussion, only a few comments need be added. The $\Lambda(1405)$ is very light for its position in Table 2, presumably reflecting the influence of the nearby KN threshold. The favorable energetics of good versus bad diquark configurations is apparent in the comparison of different columns. The quantum numbers of columns 4 and 5 are consistent with the body-plan $[us] - d$. The energy difference between this and the $[ud] - s$ body plan saturates at about 100 MeV. We can reach additional spin-parities with the "sbad" body plan $(us) - d$ or with $(ud) - s$. Only the former supports isospin-$\frac{1}{2}$ Λs, so the appearance of near-degenerate Λs

$(2J)^0$ L	$(2L+1)^L$ good 0	$(2L-1)^L$ good 1	$(2L+1)^L$ sgood 0	$(2L-1)^L$ sgood 1	$(2L+3)^L$ sbad 0	$(2L+1)^L$ sbad 1	$(2L-1)^L$ sbad 2
0	Λ (1116)	–	Σ (1193)	–	Σ (1385)	–	–
1	Λ (1520)	Λ (1405)	Σ(1670) Λ(1690)	Σ(1620) Λ(1670)	Σ(1775) Λ(1830)		Σ(1690) Σ(1750) Λ(1800)
2	Λ (1820)		Σ (1915)	Λ (1890)	Σ (2030)		Σ (2080)
3	Λ (2100)		Σ (2250)				
4	Λ (2350)		Σ (2455)				
5	Λ (2585)		Σ (2620)				

Table 2. Classification of S=1 baryons. See text for explanation.

and Σs is evidence for that body plan (though the Σs could be mixtures).

$(2J)^0$ L	$(2L+1)^L$ sgood 0	$(2L+3)^L$ sbad 0	$(2L+1)^L$ sbad 1
0	Ξ (1313)	Ξ (1530)	–
1	Ξ (1690)		Ξ (1820)
2	Ξ (1950)		Ξ (2030)
3	Ξ (2255)		
4	Ξ (2370)		

Table 3. Classification of S=2 baryons. See text for explanation.

The sparse Table 3 provides some additional support for the diquark distinctions just discussed.

These baryon tables include most of the relevant resonances classified 2∗ or better by the Particle Data Book. The few exceptions will be discussed below (Section 4.3).

The systematic approximate degeneracies across the rows in Table 4, ranging over different ways of arranging the spins relative to each other and to the orbit, and the family indices, are striking. It supports the hypotheses of weak spin-orbit forces and dynamical independence between objects on the ends. Note that the "doubling" phenomenon we mentioned as (mostly) an unfulfilled prediction for baryons is here reflected in the many cases where the same spin-parity can be reached in two ways (columns 2/3, 4/5, 6/7, and 8/9).

$(2J)^0$	$(2L+2)^{L+1}$	$(2L+2)^{L+1}$	$(2L)^{L+1}$	$(2L)^{L+1}$	$(2L)^{L+1}$	$(2L)^{L+1}$	$(2L-2)^{L+1}$	$(2L-2)^{L+1}$
L	I=S=1 0	I=0, S=1 0	I=S=1 1	I=0, S=1 1	I=1, S=1 1	I=S=0 1	I=S=1 2	I=0, S=1 2
0	$\rho(770)$	$\omega(783)$	–	–	$\pi(140)$	$\eta(550)$	–	–
1	a(1320)	f(1270)	a(1260)	f(1285)	b(1235)	$\eta(1170)$	a(1450)	f(1370)
2	$\rho(1690)$	$\omega(1670)$			$\pi(1670)$	$\eta(1645)$	$\rho(1700)$	$\omega(1650)$
3	a(2040)	f(2050)						f(2010)
4	$\rho(2350)$							
5	a(2450)	f(2510)						

Table 4. Classification of nonstrange mesons. See text for explanation.

$(2J)^0$	$(2L+2)^{L+1}$	$(2L)^{L+1}$	$(2L)^{L+1}$	$(2L-2)^{L+1}$
L	ss, S=1	ss, S=1	ss, S=0	ss, S=1
0	$\varphi(1020)$	–	$\eta'(960)$	–
1	f'(1525)	f(1420)	$\eta'(1380)$	f(1500)
2	$\varphi(1850)$			$\varphi(1680)$
3				f(2300) f(2340)

Table 5. Classification of hidden strangeness mesons. See text for explanation.

The $\varphi(1680)$ in Table 5 is uncomfortably light. There are two states $f(2300), f(2340)$ where the $\bar{s} - s$ body plan allows only one. The extra state is plausibly ascribed to pure glue (see Section 4.3).

Still more approximate degeneracies and doubling are found in Table 6.

These meson tables contain most of the relevant resonances classified awarded a bullet • by the Particle Data Group. The few exceptions will be discussed below (Section 4.3).

3.3. *Regge-Chew-Frautschi Fits*

Figure 1 shows some of the more prominent Regge trajectories. The near-linearity down to $L = 0$, and the universal slope, are evident.

Table 7 shows all the trajectories that contain 3 or more points, with their slopes and intercepts. The bottom four series would appear to favor a larger slope, but in each case there are extenuating circumstances. The

$(2J)^0$	$(2L+2)^{L+1}$	$(2L)^{L+1}$	$(2L-2)^{L+1}$
L	0	1	2
0	$K^*(892)$	$K(495)$	–
1	$K^*(1430)$	$K(1270)$ $K(1400)$	$K^*(1430)$
2	$K^*(1780)$	$K(1770)$ $K(1820)$	$K^*(1680)$
3	$K^*(2045)$	$K(2320)$	
4	$K^*(2380)$	$K(2500)$	

Table 6. Classification of strange mesons. See text for explanation.

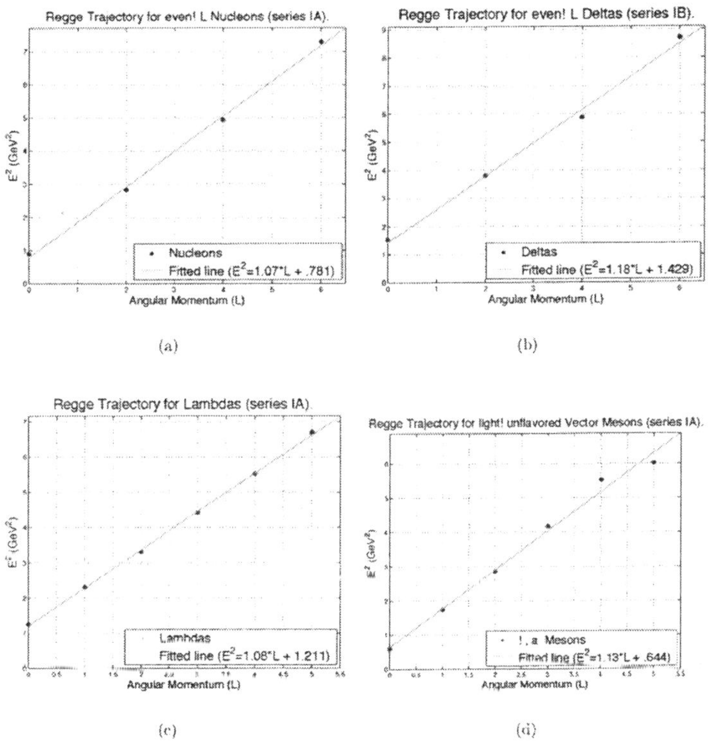

Figure 1. Fits to some prominent Regge trajectories. See text for explanation.

$\eta, \pi/b$ and K series are affected by chiral symmetry breaking. Their $L = 0$ members are approximate Nambu-Goldstone bosons, i.e. largely collective

Series	# Points	σ (GeV^2)	a(GeV^2)
nucleons, [ud], even L, 0	4	1.07	.781
deltas, (ud), even L, 0	3	1.18	1.43
lambdas, all L, [ud], 0	6	1.08	1.21
sigmas, [us], even L, 0	3	1.15	1.40
sigmas, [us], odd L, 0	3	1.09	1.85
cascades, [us], even L, 0	3	.97	1.78
ρ/a, all L, 0	6	1.13	.64
ω/f, all L, 0	5	1.16	.55
φ/f, all L, 0	3	1.19	1.07
K*, all L, 0	5	1.19	.80
η, all L, 1	3	1.20	.26
π/b, all L, 1	3	1.38	.06
K, all L, 1	5	1.56	.23
φ/f, all L, 1	3	1.52	.41

Table 7. Parameters extracted from prominent Regge trajectories. See text for explanation.

states, and it would be surprising if simple quark-model ideas described them accurately. Apart from that, the $K(2320)$ is too heavy for its location in Table 6: we'd be happy to see it migrate down in mass. The other bad actor is $\varphi(1680)$, mentioned previously. It infects the bottom row.

For purposes of Figure 1 we have not attempted to separate mass effects from an intrinsic intercept.

3.4. *The $L = 2$ Slice*

For low L the hypotheses of our model become dubious, and for very large L the data becomes sparse. Fortunately, $L = 2$ (see Table 8) appears to be large enough for simple dynamics to apply, and yet has enough data to make the case powerfully.

Diquark Distinctions: In the first row, the first two N entries (light gray) contain the good diquark, while the remainder contains the bad diquark. The two sets are clearly distinguished. In the fourth row, the good diquark Λ is clearly split below the others; the rest come in two tiers, plausibly representing $[us] - d$ and $(us) - d$, as discussed previously. Similarly for the final column.

Clear good/bad distinction	2	N (1680)	N (1720)	N (1990)	Δ (1950)	N (2000)	Δ(1905) Δ(2000)	N (1900)	Δ (1920)		Δ (1910)
$m_{good} \sim m_q$	2	ρ(1690)	ω(1670)			π(1670)	η(1645)	ρ(1700)	ω(1650)		
$m_{good} \sim m_q$	2	K*(1780)	K(1770) K(1820)	K*(1680)							
anomalous state	2	φ(1850)			φ(1680)						
	2	Λ(1820)		Σ (1915)	Λ(1890)	Σ (2030)		Σ (2080)			
	2	Ξ(1950)		Ξ(2030)							

Table 8. Close-up on the $L = 2$ sector. See text for explanation.

Dynamical Independence: The masses of the mesons and baryons are little affected by how we combine the spin and flavor of the two ends. This, together with the following point, is shown by the near-degeneracy we find among states in each row, once diquark splittings are taken into account.

Feeble Spin-Orbit Forces: As just mentioned.

Good Diquark - Antiquark Degeneracy: The near-degeneracy between the baryon and meson states in light gray, and between the $\Lambda(1820)$ and K mesons states in dark gray, exhibits the near-degeneracy between the good diquark and a light antiquark.

Anomalous State: All this striking success highlights the anomalous nature of the one thing that doesn't fit: $\varphi(1680)$ is too light.

4. Supplements

4.1. *Even-Odd Effect*

While the Λ trajectory for M^2 against L is remarkably straight (see Figure 1c), passing through all integers from 0 to 5, this is not always the case.

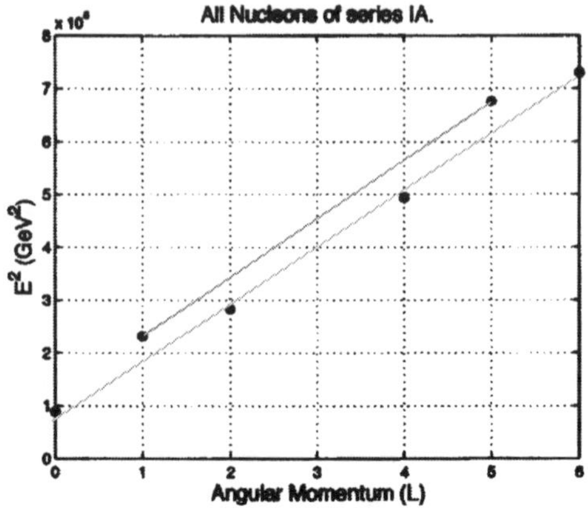

Figure 2. Splitting of even from odd L trajectories. See text for explanation.

For example, as shown in Figure 2, the series of nucleons with different values of L, to which we ascribe the body plan $[ud]-d$, do not quite lie on a straight-line trajectory. Rather, they seem to split into two trajectories, one for even L, and a higher one for odd L. There appear to be similar effects in other sectors (though the data is poor).

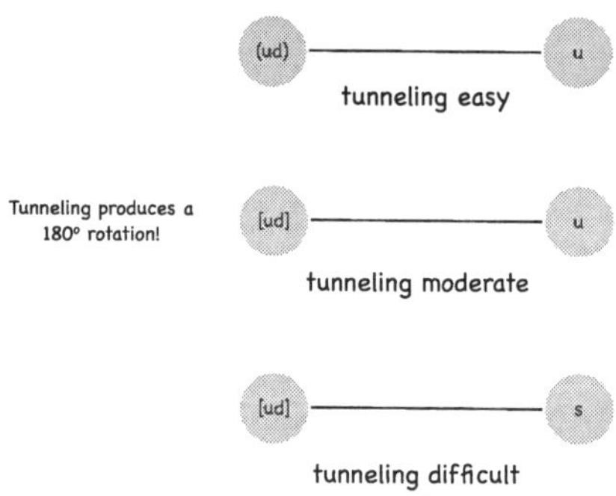

Figure 3. Possibility of tunneling in different configurations. See text for explanation.

There is a nice explanation for the difference, shown in Figure 3. Before considering rotation levels, we should fix the internal wave function. Now if tunneling is possible, we should take symmetric or antisymmetric superpositions between the two permutations of the ends. Since interchange of the ends has the same effect as a 180° rotation, symmetric wave functions must be associated to even L, antisymmetric to odd L. Since a node in the wave function is costly, the odd L should be higher in mass. This is what we see in the nucleons. On the other hand, for the good-diquark Λ to permute its two ends, all three quarks would have to tunnel. So in that case the even-odd splitting should be very small, as is observed.

4.2. *Charmed Mesons and Baryons*

We will discuss this subject in more detail elsewhere. Existing data is sparse, but certainly consistent with the ideas discussed here. For example, the series $\Lambda_c(2285), \Lambda_c(2625), \Lambda_c(2880)$ at $J^P = \frac{1}{2}^+, \frac{3}{2}^-, \frac{5}{2}^+$ is well fit using the mass-loaded string formulae with $M_c = 1600, m_{[ud]} = 180$ MeV, and $\sigma = .974$ GeV^2.

4.3. *Additional States*

There are a few resonances that don't fit into our Tables naturally. We'll briefly discuss those now.

Glueball Candidates: The $f(1710)$, with $J^{PC} = 0^{++}$ does not fit, and there are two 2^{++} states $f(2300), f(2340)$ where only one belongs. These extra states have the right quantum numbers to be glueballs, and their masses are in broad agreement with expectations from lattice gauge theory.

Daughters? $\pi(1300), \eta(1295), \eta(1440)$ are extra 0^{-+} states, as is $\pi(1800)$. $K^*(1410)$, $\rho(1450), \omega(1420)$ are extra vector states. Formally, they could be accommodated as daughters of the corresponding ground-state mesons, i.e. as internal excitations of the flux tube. Of course the classical flux tube picture is badly strained at this L, but we note with a smile that the masses of these "daughters" differ approximately by $\delta M^2 = \sigma$ from the corresponding mesons (2σ in the case of $\pi(1800)$). At this point, we've mentioned all the relevant meson candidates.

Pentaquarks?? and Baryon Daughters: We assume θ^+ is gone for good. $N(1440), \Lambda(1600), N^*(1710)$ could be daughters of the usual octet, and $\Sigma(1940) - \frac{3}{2}^+, \Lambda(2110) - \frac{5}{2}^-$ could be $L = 1, 2$ versions. At this point, we've mentioned all the relevant baryon candidates.

Diquark-Antidiquark: Given the emerging near-degeneracy of $[ud] - q$

and $\bar{u} - q$ at large L, it is hard to resist the inference that $[\bar{u}\bar{d}] - qq$ states should exist in the same mass range as $u - qq$. Could there, for example, be a a set of doubly-strange positive parity $J = 1, 2, 3$ mesons with mass close to 2 GeV?

4.4. *Diquarks in Baryon Decays*

Finally, let us mention an interesting dynamical application of our classification. Since the good $[ud]$ diquark is so favorable, we might expect it to retain its integrity even as the baryon containing it decays. Specifically, we might expect that Λ baryons with the body plan $[ud] - s$ prefer to decay into (generalized) nucleons containing good diquarks and $\bar{K}$ mesons, as opposed to strange baryons and π mesons.

The $\Lambda(1520) - \frac{3}{2}^-$ is a remarkable case. It decays to $N\bar{K}$ 45 % of the time, $\Sigma\pi$ 42 %, and $\Lambda\pi\pi$ 10 % (taking the data perhaps too literally). At first sight these numbers may not appear impressive, but upon reflection they are startling. Parity and angular momentum conservation bump up the NK channel to d-wave, and it is not very far from threshold. $\Lambda\pi\pi$ is also phase-space challenged! So these channels, in which the good diquark retains its integrity, are working against considerable handicaps, yet they hold their own.

$\Lambda(1820)$ decays 55-65 % into $N\bar{K}$, versus 8-14 % into $\Sigma\pi$.

$\Lambda(2100)$ decays 25-35 % into $N\bar{K}$, and another 10-20 % into N plus excited $\bar{K}$s, versus 5 % into $\Sigma\pi$.

Moving to strange baryons with the body plan $[us] - d$, we expect the opposite effect, that is preference for strange baryon channels. Unfortunately the data is very sparse, basically restricting us to $\Sigma(1670)$. That particle decays 30-60 % into $\Sigma\pi$ and 5-15 % into $\Lambda\pi$, but only 7-13 % into $N\bar{K}$. So our expectation is vindicated.

If we take $[ud]$ integrity at face value, and apply it to our conjectured $[\bar{u}\bar{d}] - ss$ meson, we are led to expect that $\bar{N}\Xi$ is a prominent decay channel, if it is kinematically allowed.

5. Discussion

Our hope that large L spectroscopy would give convincing evidence for energetically significant diquark correlations is amply fulfilled. In conjunction with the loaded flux-tube model, assuming negligible spin-orbit forces, this idea forms the basis of a simple and surprisingly successful account of the preponderance of hadron spectroscopy.

Our hypotheses are quite different in spirit from those adopted in the nonrelativistic quark model, or in any simple potential model. The observation that pairs of quarks in the good diquark configuration have mass comparable to that of a single quark, which appears quite directly and strikingly in the $L = 2$ data, is a qualitative challenge to the foundation of such models. For relativistic potential or bag models, the smallness of spin-orbit forces poses a qualitative challenge.

The phenomenological importance of diquarks as building blocks of baryons is difficult to accommodate within large N_c approaches, since diquark phenomenology relies on $N_c = 3 \not> 2 + 1$. Specifically, the approximate degeneracy between mesons and baryons with good diquarks, which is a striking feature of the (moderately) large L data, seems hard to reconcile with the radically different nature of mesons and baryons at large N_c.

Our fits, and the hypotheses that underlie them, suggest that the dynamics of light-quark QCD simplifies at large L. That dynamics appears to be well described by a flux tube or string with universal tension, with quarks or diquarks at the ends. Elementary ideas about confinement of color suggest that dynamics of this sort should emerge asymptotically at large L, but several details are surprising and important. Semiclassical quantization of rigidly rotating configurations gives a remarkably successful picture of the spectrum down to small L, with broad brush even to $L = 0$. There appears to be little interaction between the two objects at the ends, even at $L = 1$, and spin-orbit forces are quite small. Evidence for internal excitations of the flux tube is at best equivocal.

Challenges to theory: Can these regularities be derived from fundamental QCD? Can they provide the basis of a systematic approximation scheme? Can we get good numerical predictions for large L resonances from fundamental theory? Can we get good numerical data on emergent diquark splittings from fundamental theory? Apart from spectroscopy, can we make better contact between the dynamical applications of diquark ideas mentioned in "Motivations" and fundamental QCD?

Challenges to experiment: Fill in the tables, or demonstrate convincingly that they have holes! Check whether the "anomalous" states are real. Find the predicted diquark-antidiquark states, or rule them out. Determine whether the heavy-light spectroscopy predicted for mesons and baryons containing c and b together with light quarks matches reality.

Finally, a common challenge: How important is "intrinsic noise" in the hadronic spectrum? In discerning our simplicities and broad trends, we did not allow ourselves to be discouraged by occasional discrepancies at the level of 50 or even 100 MeV. Should we be? Putting it another way: must we predict that more accurate measurements will bring all the masses into line? Or does the presence of many nearby states (including continuum states) with degenerate quantum numbers imply some intrinsic scatter into the spectrum, in the spirit of random matrix theory? If so, can we quantify – and aspire to predict – the statistics of the residuals?

Acknowledgments

The work of FW is supported in part by funds provided by the U.S. Department of Energy under cooperative research agreement DE-FC02-94ER40818.

8 Future Projects

IMPORTANCE OF A MEASUREMENT OF $F_L(X, Q^2)$ AT HERA

R. S. THORNE*

Cavendish Laboratory, University of Cambridge, J. J. Thomson Avenue, Cambridge, CB3 0HE

E-mail: thorne@hep.phy.cam.ac.uk

I investigate what a direct measurement of the longitudinal structure function $F_L(x, Q^2)$ could teach us about the structure of the proton and the best way in which to use perturbative QCD for structure functions. I assume HERA running at a lowered beam energy for approximately 4-5 months and examine how well the measurement could distinguish between different theoretical approaches. I conclude that such a measurement would provide useful information on how to calculate structure functions and parton distributions at small x.

1. Introduction

It would be vital to have a real accurate measurement of $F_L(x, Q^2)$ at HERA since this gives an independent test of the gluon distribution at low x to accompany that determined from $dF_2(x, Q^2)/d\ln Q^2$ [1–5]. At present the fits to $F_2(x, Q^2)$ at low x are reasonably good, but the gluon is free to vary in order to make them as successful as possible. It is essential to have a cross-check. (It is important to note that $F_L(x, Q^2)$ is a much better discriminator of the gluon distribution, and/or of different theories, for given $F_2(x, Q^2)$ than the charm contribution. $F_2^c(x, Q^2)$ is constrained to evolve in exactly the same way as $F_2^{tot}(x, Q^2)$ (with appropriate charge weighting) for $W^2 \gg m_c^2$, so is hardly independent. At lower W^2 the suppression is determined mainly by kinematics.) Currently there is a consistency check on the relationship between $F_2(x, Q^2)$ and $F_L(x, Q^2)$ at high y since both contribute to the total cross-section measured at HERA. Hence, there are effective "determinations" of $F_L(x, Q^2)$ obtained by extrapolating to high y using either NLO perturbative QCD or using $(d\sigma/d\ln y)_{Q^2}$ whilst making assumptions about $(dF_2(x, Q^2)/d\ln y)_{Q^2}$ [6]. This is a good consistency test

*Royal Society University Research Fellow.

of a given theory, usually NLO QCD, and could show up major flaws. However, it relies on small differences between two large quantities, so its accuracy is limited. Also, for an extraction of $F_L(x, Q^2)$ it has model-dependent uncertainties which are difficult to quantify fully [7]. A real measurement would be a much more direct test of the success of different theories in QCD.

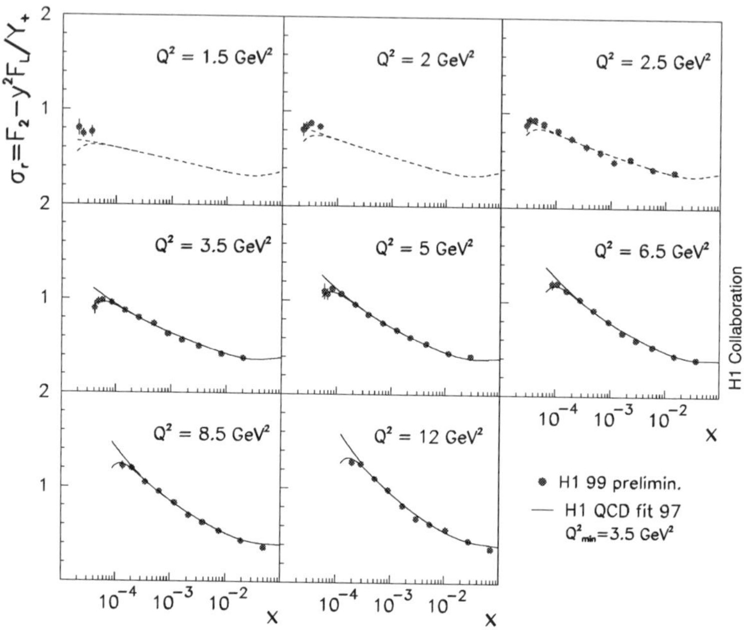

Figure 1. The NLO consistency check of $F_L(x, Q^2)$ for the H1 fit.

The consistency check of $F_L(x, Q^2)$ extracted by H1 versus $F_L(x, Q^2)$ predicted from their QCD fit [5] is shown in Figure 9 in the second of [6]. However, due to the potentially large, strongly correlated, model-dependent errors in the "measured" $F_L(x, Q^2)$ it is far more revealing to see plots like Figure 1 [6]. The turn-over in $\tilde{\sigma}(x, Q^2) = F_2(x, Q^2) - y^2/(1 + (1 - y)^2) F_L(x, Q^2)$ is clearly matched by the $F_L(x, Q^2)$ contribution. However, the same consistency check for the fit of $\tilde{\sigma}(x, Q^2)$ for MRST partons at NLO fails at the lower Q^2 values, as seen on the left-hand side of Figure 2. This is because of the different gluon obtained from a full global fit. Hence, the consistency check is not universally successful at NLO.[a]

[a] Additionally, Alekhin performed fits to DIS data, using the reduced cross-section for HERA data, and allowed higher-twist corrections to be determined phenomenologically.

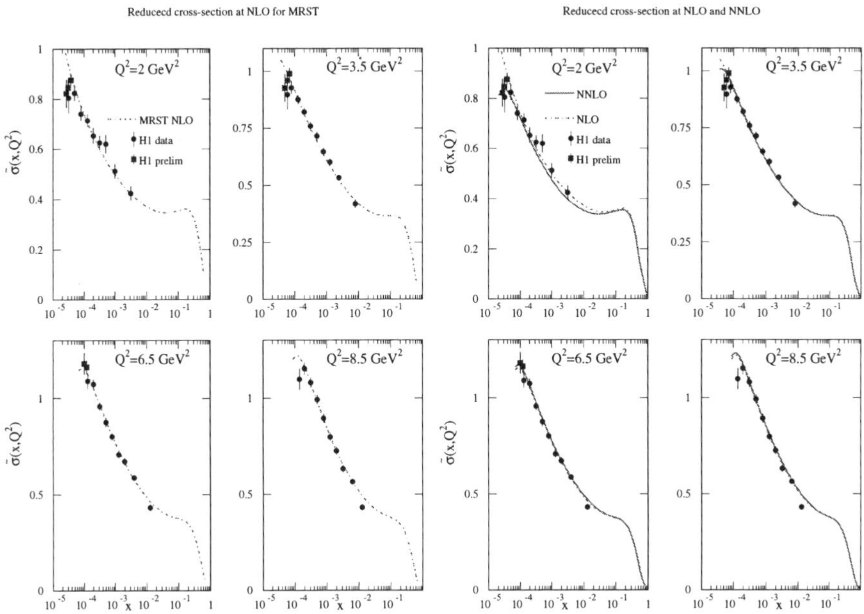

Figure 2. The NLO consistency check of $F_L(x,Q^2)$ for the H1 fit (left). The consistency check of $F_L(x,Q^2)$ for the NLO and NNLO MRST fits (right).

Hence, many current NLO global fits show problems regarding $F_L(x,Q^2)$ at high y. In general they provide a good fit to HERA data, but there are some problems in $dF_2/d\ln Q^2$, e.g. Figure 17 in [8]. However, standard perturbation theory is not necessarily reliable in general because of increasing logs at higher orders, e.g. at small x

$$P^1_{qg} \sim \alpha_S(\mu^2) \qquad P^2_{qg} \sim \frac{\alpha_s^2(\mu^2)}{x} \qquad P^n_{qg} \sim \frac{\alpha_s^n(\mu^2)\ln^{n-2}(1/x)}{x} \tag{1}$$

and similarly

$$C^1_{Lg} \sim \alpha_S(\mu^2) \qquad C^2_{Lg} \sim \frac{\alpha_s^2(\mu^2)}{x} \qquad C^n_{Lg} \sim \frac{\alpha_s^n(\mu^2)\ln^{n-2}(1/x)}{x}, \tag{2}$$

and hence enhancements at higher orders are possible.

However, we can already see precisely what happens at NNLO. The splitting functions have been calculated at NNLO [9], and recently the coefficient functions for $F_L(x,Q^2)$ have been finished [10]. The gluon extracted from the MRST global fit at LO, NLO and NNLO is shown in Figure 3.

He found an unambiguous positive correction for $F_L(x,Q^2)$, i.e. the consistency check fails for the purely perturbative fit [3].

Additional positive small-x contributions in P_{qg} at each order lead to a smaller low-x gluon at each order.[b]

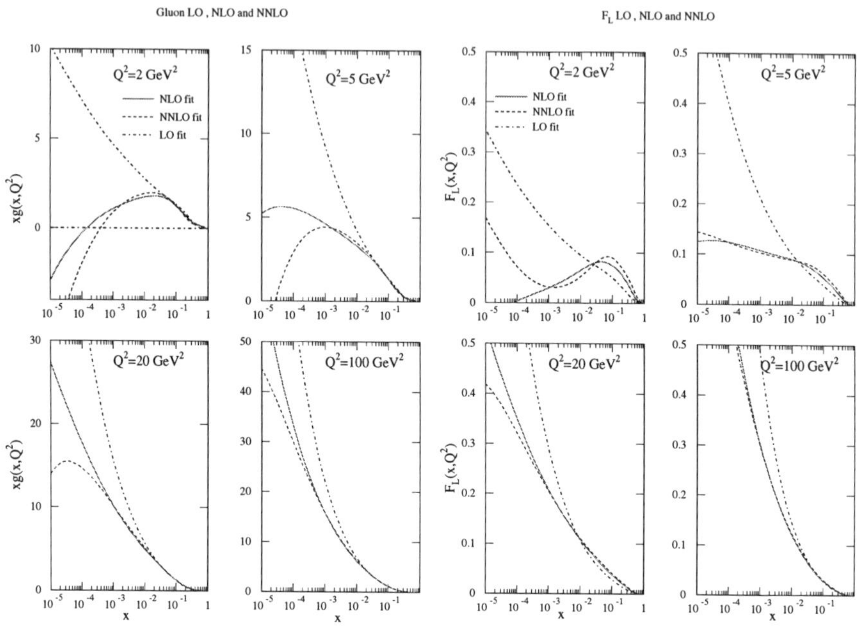

Figure 3. The gluon extracted from the global fit at LO, NLO and NNLO (left). $F_L(x,Q^2)$ predicted from the global fit at LO, NLO and NNLO (right).

The NNLO $O(\alpha_s^3)$ longitudinal coefficient function $C^3_{Lg}(x)$ given by

$$C^3_{Lg}(x) = n_f \left(\frac{\alpha_S}{4\pi}\right)^3 \left(\frac{409.5\ln(1/x)}{x} - \frac{2044.7}{x} - \cdots\right). \tag{3}$$

There is clearly a significant positive contribution at small x, and this counters the decrease in small-x gluon. $F_L(x,Q^2)$ predicted from the global fit at LO, NLO and NNLO is shown in Figure 3. The NNLO coefficient function more than compensates for the decrease in the NNLO gluon.

Without considering the high-y HERA data, the NNLO fit is not much better than the NLO fit, though it is a slight improvement [2]. However, the NNLO contribution to $F_L(x,Q^2)$ largely solves the previous high-y problem

[b]This conclusion relies on a correct application of flavour thresholds in a General Variable Flavour Number Scheme at NNLO [11], not present in earlier approximate NNLO MRST fits. The correct treatment of flavour is particularly important at NNLO because discontinuities in unphysical quantities appear at this order.

with $\tilde{\sigma}(x, Q^2)$, as seen on the right-hand side of Figure 2.[c] But these data are not very precise, the effective error on $F_L(x, Q^2)$ being $\sim 30 - 40\%$. It is of real importance to have some accurate measurement of $F_L(x, Q^2)$ at small x.

HERA has proposed some running at lower beam energy before finishing in order to make a direct measurement of $F_L(x, Q^2)$. The expectation is to measure data from $Q^2 = 5 - 40\text{GeV}^2$ and $x = 0.0001 - 0.003$ with a typical error of at best $12 - 15\%$ [12]. How important would this be in distinguishing between different theoretical approaches to structure functions? There has been a study by ZEUS [13] on the impact of such data on the accuracy with which $g(x, Q^2)$ is determined if $F_L(x, Q^2)$ is roughly as expected from a NLO fit. There is a significant although not enormous improvement in the gluon uncertainty. However, this is not, in my view, the most interesting question. Rather, it is important to see if the potential measurement could tell apart different theoretical treatments, e.g. whether we need go beyond the standard fixed-order perturbation theory approach. There has also been a study of this by ZEUS [14], with *extreme* theoretical predictions, and the discriminating power is obvious. However, in this case the extremes are based on unrealistic models (out-of-date partons and partons from one order used with coefficient functions from another). Furthermore, all data points are assumed to line up, i.e. the χ^2 for the correct theory would be 0. A more sophisticated approach is needed.

2. Test of Theoretical Models

I consider a variety of more plausible theoretical variations. A fit that performs a double resummation of leading $\ln(1/x)$ and β_0 terms leads to a better fit to small-x data than a conventional perturbative fit [15]. The resummation also seems to stabilize $F_L(x, Q^2)$ at small x and Q^2. The fit has some problems at higher x (particularly for Drell-Yan data), and NLO contributions to resummation are needed for precision [16], hence the prediction is somewhat approximate, but it has the correct trend.[d] Alternatively, a dipole-motivated fit [19–25] contains higher terms in $\ln(1/x)$ and higher twists. It also guarantees reasonable behaviour for $F_L(x, Q^2)$ at low Q^2 due to the form of wavefunction. In a quantitative comparison I use

[c] The high-y fit would fail with gluons that are positive at small x and Q^2 – $F_L(x, Q^2)$ would be too big and the turnover too great.

[d] Similar results would be likely from the approaches in [17,18] since the resummations, though different in detail, have the same qualitative features.

my own dipole-motivated fit [26] in order to avoid problems in the heavy flavour treatment in some other approaches. The evolution of various predictions for $F_L(x, Q^2)$ at $x = 0.0001$ and $x = 0.001$ is seen in Figure 4. The resummation and dipole predictions are behaving sensibly at low Q^2. The NLO prediction is becoming negative at the lowest values, while the NNLO prediction is becoming flat at $Q^2 \sim 2\text{GeV}^2$ for $x = 0.0001$. It has a slight turn up at even smaller x, implying the necessity for even further corrections. The results are shown for various values of Q^2 on the left-hand side of Figure 5. They suggest that a measurement of $F_L(x, Q^2)$ over as wide a range of x and Q^2 as possible would be very useful.

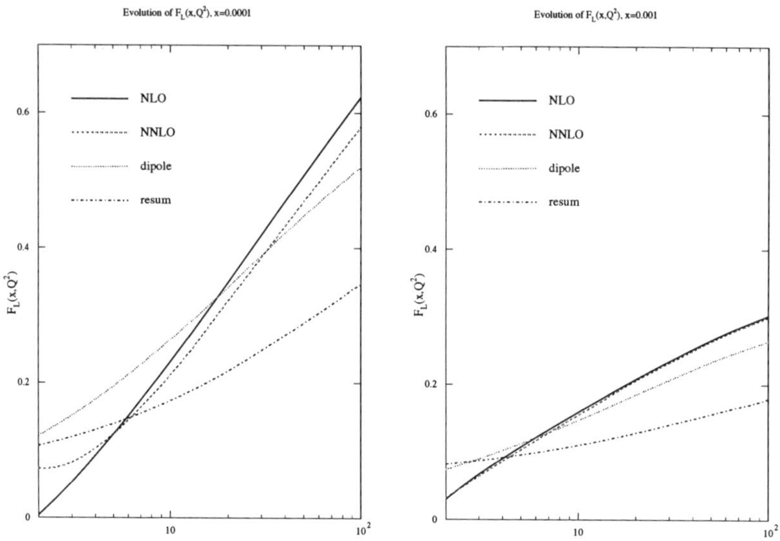

Figure 4. Evolution of various predictions for $F_L(x, Q^2)$ at $x = 0.0001$ (left) and $x = 0.001$ (right).

In particular, the dipole fit produces a rather different shape and size prediction for $F_L(x, Q^2)$ from that at NLO and NNLO. Hence I generate a set of data based on the central dipole prediction but with a random scatter ($\chi^2 = 20/18$ for the dipole prediction). The comparison of the pseudo-data to other predictions is shown on the right-hand side of Figure 5, where I also show points at $Q^2 = 2\text{GeV}^2$ that might have been measured at HERA III and might be at eRHIC [27]. Points at 40GeV^2 are not as useful, as the errors are bigger and the theoretical curves are converging. From Figure 5 it is clear that there is some reasonable differentiating power, but this is

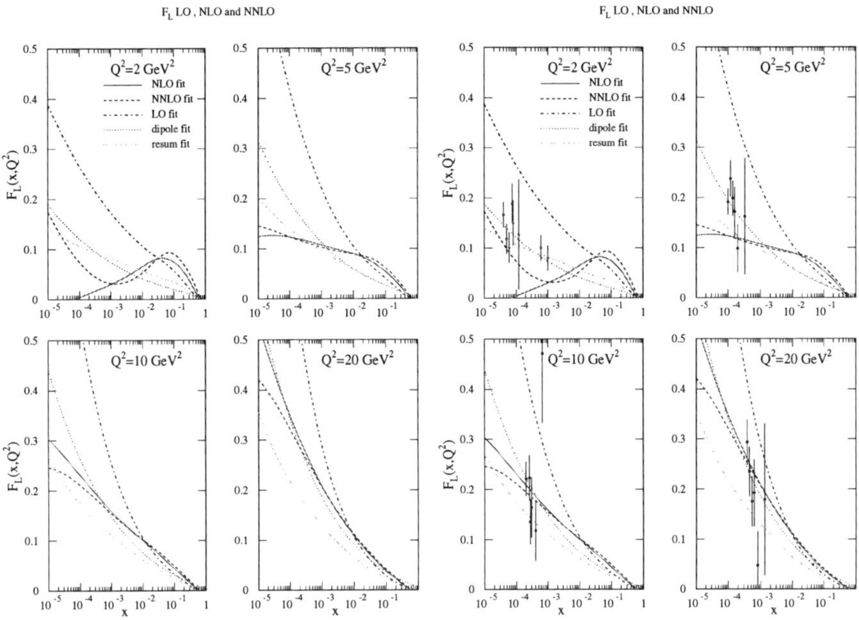

Figure 5. $F_L(x, Q^2)$ predicted from the global fit at LO, NLO and NNLO, from a fit which performs a double resummation of leading $\ln(1/x)$ and β_0 terms, and from a dipole model type fit (left). Comparison of data to various predictions. Also shown are points at $Q^2 = 2\text{GeV}^2$ that might have been measured at HERA III and for the two highest x values might be at eRHIC (right).

comparing the central predictions for a given theoretical framework only. We must also consider the uncertainties.

This is shown in Figure 6, where the left-hand side shows the comparison at NLO as the weight of the $F_L(x, Q^2)$ *data* is increased in the fit. The best fit results in $\chi^2 = 27/18$ for the $F_L(x, Q^2)$ *data* but this is becoming an unacceptable global fit. The next-best fit is an acceptable global fit, and $\chi^2 = 29/18$ for the $F_L(x, Q^2)$ *data*. The NLO fit to the $F_L(x, Q^2)$ *data* is never particularly good because the shape in Q^2 is never quite correct. The comparison at NNLO as the weight of the $F_L(x, Q^2)$ *data* is increased in the fit is similar. The best fit results in $\chi^2 = 26/18$ for the $F_L(x, Q^2)$ *data* but is becoming an unacceptable global fit. The next-best fit is an acceptable global fit, and $\chi^2 = 31/18$ for the $F_L(x, Q^2)$ *data*. Again the NNLO fit to $F_L(x, Q^2)$ *data* always gets the shape in Q^2 slightly wrong.

As well as the resummation and dipole hypotheses we can also look at explicit higher twist possibilities, in particular the renormalon correction due to the nonsinglet quark sector. This is a different picture from the case

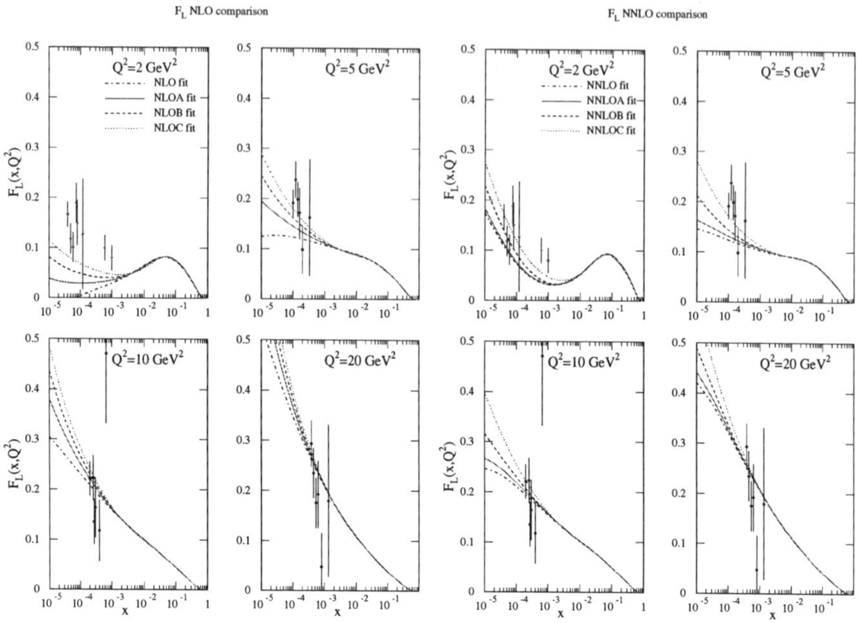

Figure 6. Comparison at NLO (left) and at NNLO (right) as the weight of $F_L(x, Q^2)$ *data* is increased in the fit.

for $F_2(x, Q^2)$, where the renormalon calculation of higher twist dies away at small x due to satisfying the Adler sum rule. It is a completely different picture for $F_L(x, Q^2)$ – at small x $F_L^{HT}(x, Q^2) \propto F_2(x, Q^2)$. The explicit renormalon calculation [28] gives

$$F_L^{HT}(x, Q^2) = \frac{A}{Q^2} \otimes F_2(x, Q^2) \tag{4}$$

where the estimate for A for the first moment of the structure function is

$$A = \frac{8C_f \exp(5/3)}{3\beta_0} \Lambda_{QCD}^2 \approx 0.4\,\mathrm{GeV}^2. \tag{5}$$

This effect has nothing to do with the gluon distribution, and is not part of the higher twist contribution in the dipole approach. The higher twist does mix with higher orders though. I add it to the NLO prediction. The renormalon correction could be a rather significant effect, as seen in Figure 7, where I generate a new set of data based on the central higher twist prediction. (The *data* at $Q^2 = 40\mathrm{GeV}^2$ are shown. All predictions give $\chi^2 =\sim 6/6$ for the six points at $Q^2 = 40\mathrm{GeV}^2$ (except LO)). It is most similar to the dipole prediction but the data give $\chi^2 = 25/18$ for the dipole prediction curve – perhaps at the edge of distinguishability. The renormalon-based data are clearly able to rule out the central NLO and NNLO curves, but one must repeat the study done for *dipole data.*

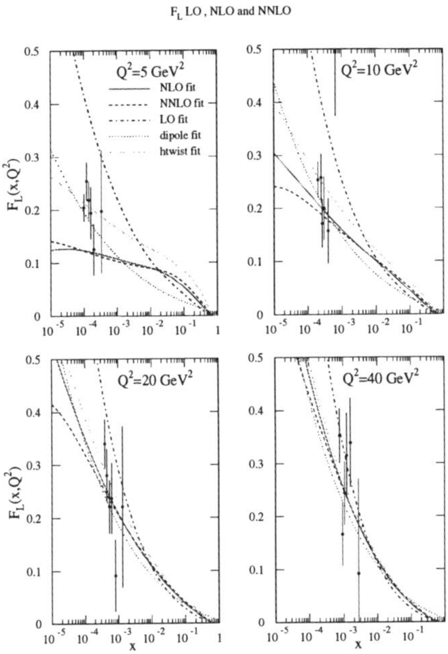

Figure 7. Comparison of data to predictions with bin at $Q^2 = 40\text{GeV}^2$ shown.

Again I look at the NLO fit as the weight of the $F_L(x, Q^2)$ *data* is increased in the fit. The best fit results in $\chi^2 = 22/18$ for the $F_L(x, Q^2)$ *data*, but is an unacceptable global fit – $\Delta\chi^2 > 60$. The next-best fit is a marginally acceptable global fit and $\chi^2 = 27/18$ for the $F_L(x, Q^2)$ *data*. Hence, in this case the NLO fit to $F_L(x, Q^2)$ *data* can get the shape in Q^2 (for $Q^2 \geq 5\text{GeV}^2$) more-or-less right, but the deterioration in the global fit required to do so is worse than for the *dipole data*. The comparison at NNLO as the weight of $F_L(x, Q^2)$ *data* is increased in the fit is again similar to that at NLO. The best fit results in $\chi^2 = 23/18$ for $F_L(x, Q^2)$ *data* but is a poor global fit – $\Delta\chi^2 > 50$. The next-best fit is a moderately acceptable global fit, and $\chi^2 = 29/18$ for $F_L(x, Q^2)$ *data*.

3. Conclusions

The measurement of $F_L(x, Q^2)$ seems to be the best way to determine reliably the gluon distribution at low x, particularly at low Q^2, and to determine whether fixed-order calculations are sufficient or whether resummations, or other theoretical extensions may be needed. Currently we can perform global fits to all up-to-date data over a wide range of parameter space, and the fit quality is fairly good, but there are some minor problems. We could require higher orders, higher twist and/or some type of resummation, all of which might have a potentially large impact on the predicted $F_L(x, Q^2)$ and other quantities. Hence, $F_L(x, Q^2)$ is a vital measurement for our understanding of precisely how best to use perturbative QCD to describe the structure of the proton and also for making really reliable predictions and comparisons at the LHC. The lowest Q^2 possible would be useful. The proposed measurement at HERA would have a reasonable ability to distinguish between different theoretical approaches, due to both the inability to fit $F_L(x, Q^2)$ because of the shape and the deterioration in global fits needed in order to match the general features of $F_L(x, Q^2)$ data, and would play a central role in determining the best way to use QCD.

Acknowledgments

I would like to thank Max Klein for supplying me with the simulated H1 data for $F_L(x, Q^2)$, and him and Mandy Cooper-Sarkar, Claire Gwenlan, Alan Martin and James Stirling for numerous discussions on the subject of the longitudinal structure function.

References

1. CTEQ Collaboration: J. Pumplin *et al.*, JHEP **0207** (2002) 012.
2. A.D. Martin, R.G. Roberts, W.J. Stirling and R.S. Thorne, Phys. Lett. **B604** (2004) 61.
3. S.I. Alekhin, Phys. Rev. **D68** (2003) 014002.
4. ZEUS Collaboration: S. Chekanov *et al.*, Eur. Phys. J. **C42** (2005) 1.
5. H1 Collaboration: C. Adloff *et al.*, Eur. Phys. J. **C21** (2001) 33.
6. H1 Collaboration: C. Adloff *et al.*, Phys. Lett. **B393** (1997) 452; N Gogitidze, J. Phys. **G28** (2002) 751, `hep-ph/0201047`; E.W. Lobodzinska, proceedings of 11th International Workshop on Deep Inelastic Scattering (DIS 2003), St. Petersburg, Russia, 23-27 Apr 2003, `hep-ph/0311180`.
7. R.S. Thorne, Phys. Lett. **B418** (1998) 371.
8. R.S. Thorne, invited talk at 21st International Symposium on Lepton and Photon Interactions at High Energies (LP 03), Batavia, Illinois, 11-16 Aug 2003, Int. J. Mod. Phys. **A19** (2004) 1074.
9. A. Vogt, S. Moch and J.A.M. Vermaseren, Nucl. Phys. **B688** (2004) 101; Nucl. Phys. **B691** (2004) 129.

10. A. Vogt, S. Moch and J.A.M. Vermaseren Phys. Lett **B606** (2005) 123; hep-ph/0504242.
11. R.S. Thorne, proceedings of 13th International Workshop on Deep Inelastic Scattering (DIS 05), Madison, Wisconsin, 27 Apr - 1 May 2005, p. 847, hep-ph/0506251.
12. M. Klein, proceedings of 12th International Workshop on Deep Inelastic Scattering (DIS 2004), Strbske Pleso, Slovakia, 14-18 Apr 2004, p. 309; J. Feltesse, these proceedings.
13. C. Gwenlan, proceedings of 13th International Workshop on Deep Inelastic Scattering (DIS 05), Madison, Wisconsin, 27 Apr - 1 May 2005, p. 396, hep-ex/0507032.
14. C. Gwenlan, A.M. Cooper-Sarkar and C. Targett-Adams, HERA and the LHC: A Workshop on the Implications of HERA for LHC Physics: CERN - DESY Workshop 2004/2005 hep-ph/0509220.
15. R.S. Thorne, Phys. Rev. **D60** (1999) 054031; Phys. Rev. **D64** (2001) 074005.
16. C.D. White and R.S. Thorne, in preparation.
17. M. Ciafaloni, D. Colferai, G.P. Salam and A.M. Stasto, Phys. Lett. **B587** (2004) 87.
18. G. Altarelli, R.D. Ball and S. Forte, Nucl. Phys. **B674** (2003) 459.
19. K. Golec-Biernat and M. Wüsthoff, Phys. Rev. **D59** (1999) 1999, etc.
20. M. McDermott, L. Frankfurt, V. Guzey and M. Strikman, Eur. Phys. J. **C16**, (2000) 641.
21. J. Bartels, K. Golec-Biernat and H. Kowalski, Phys. Rev. **D66** (2002) 014001.
22. E. Gotsman, E. Levin, M. Lublinsky and U. Maor, Eur. Phys. J. **C27** (2003) 411.
23. E. Iancu, K. Itakaru and S. Munier, Phys. Lett. **B590** (2004) 199.
24. J.R. Forshaw, and G. Shaw, JHEP **0412** (2004) 052.
25. H. Kowalski and D. Teaney, Phys. Rev. **D68** (2003) 114005.
26. R.S. Thorne, Phys. Rev. **D71** (2005) 054024.
27. I. Abt, A. Caldwell, X. Liu and J. Sutiak hep-ex/0407053.
28. E. Stein *et al.*, Phys.Lett. **B376** (1996) 177; M. Dasgupta and B.R. Webber, Phys. Lett. **B382** (1996) 273.

MEASUREMENT OF THE LONGITUDINAL PROTON STRUCTURE FUNCTION AT LOW X AT HERA

JOËL FELTESSE
DSM/DAPNIA, CEA/Saclay
F-91191 Gif-sur-Yvette, France

The theoretical interest in the longitudinal inclusive $F_L(x, Q^2)$ and diffractive $F_L^D(x_{I\!P}, x, Q^2)$ Structure Functions are briefly mentioned. A simulation based on running HERA for three months with a reduced proton beam energy shows that measurements are possible with five and three sigma significance for F_L and F_L^D , respectively.

1. Motivations

Since 1992 the H1 and ZEUS experiments have been central in testing the properties of QCD and discovered surprising behaviours in unexplored regions. Principle among these is the rise of the Structure Function and the strong gluon density at very low x. The HERA facility is scheduled to end operation by mid-2007 and an important piece of the program is still missing: the measurement of the longitudinal Structure Function $F_L(x, Q^2)$. It is simply related to the double inclusive differential cross section

$$\frac{d^2\sigma}{dxdQ^2} = \frac{2\pi\alpha^2 Y_+}{Q^4 x} \cdot [F_2(x, Q^2) - f(y) \cdot F_L(x, Q^2)] \tag{1}$$

with $y = Q^2/sx$, $Y_+ = 1 + (1 - y)^2$, $f(y) = y^2/Y_+$ and s the centre of mass system energy squared. The functions $F_2(x, Q^2)$ and $F_L(x, Q^2)$ are basic Structure Functions to study the proton's interior structure down to the smallest reachable dimensions. The extraction of $F_2(x, Q^2)$ from the measured differential cross section has only been possible at low y, where the contribution of $F_L(x, Q^2)$ can be neglected or in making assumptions on $F_L(x, Q^2)$ at large y. The measurement of $F_L(x, Q^2)$ is difficult and has never been performed at very low x, the domain of excellence of HERA. After more than ten years of in-depth study of systematic effects in the measurement of the inclusive cross-sections, the HERA experiments are probably the best experiments that ever were to perform an accurate measurement of the longitudinal Structure Function of the proton. In the Quark Parton Model, due to helicity conservation [1], $F_L=0$. In QCD, the presence of gluons interacting with quarks leads to a sizeable longitudinal Structure Function $F_L(x,Q^2)$ at low x, which is directly related to the gluon distribution $xg(x,Q^2)$ at Leading Order [2]

$$F_L = \frac{\alpha_s(Q^2)}{4\pi}x^2 \int_x^1 \frac{dz}{z^3}\left[\frac{16}{3}F_2(z,Q^2) + 8(\Sigma e_q^2)\left(1-\frac{x}{z}\right)zg(z,Q^2)\right] \quad (2)$$

At small x, the right-hand side is dominated by the gluon distribution $g(x,Q^2)$. There is a very direct relation between the gluon distribution and F_L which persists when b-quark threshold effects or terms at order α_s^2 (see reference [3]) and even α_s^3 (see reference [4]) are included, at least above $Q^2 \approx 10\ GeV^2$. The longitudinal Structure Function would be a clean probe of the gluon distribution in the x domain $(10^{-4}, 10^{-3})$.

At present the gluon density at low x is poorly constrained by the data . It has only been indirectly determined from a fit of the scaling violations of the $F_2(x,Q^2)$ Structure Function. Good fits of the inclusive cross sections from MRST [5] and CTEQ [6] deviate in the relative contributions of sea quarks and gluons at low x to F_L by almost a factor two (see Figure 1). Moreover, the fits rely on the NLO DGLAP evolution equations without resummation which become questionable at the lowest x values [7]. There are

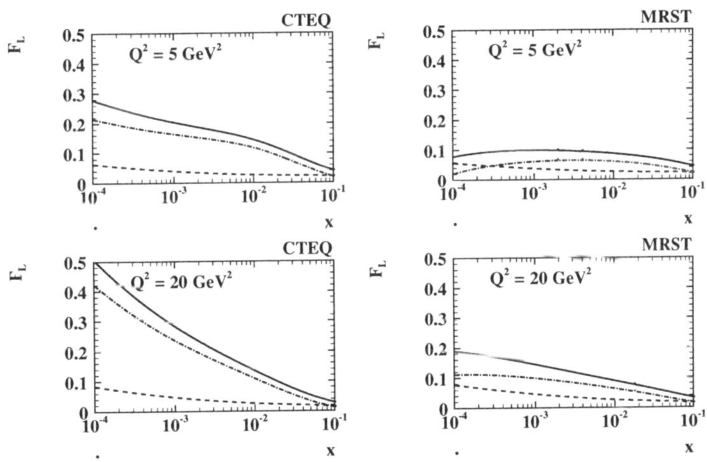

Figure 1. F_L prediction (solid line) from parametrisations of MRST and CTEQ. The dashed-dotted lines and the dashed lines represent the gluon and the F_2 components of F_L respectively.

suspicions that higher twist effects would cancel in the Structure Function F_2 but would generate a large difference in F_L at low x [8].

The dynamics of large density of partons at low x is still not well understood. A direct measurement of F_L at HERA would be a powerful consistency check of the theory at small x [4,9]. In addition, it would provide a constraint on the determination of the gluon density in the x range of W and Higgs production at LHC [10].

2. Direct measurement of F_L

2.1. *Method*

The method to measure $F_L(x, Q^2)$is simply to measure the differential cross section at fixed Q^2 and x at different y, by varying the beam energies (see Eq. (1)) and to perform a straight line fit of the reduced cross section

$$\sigma_r = [F_2(x, Q^2) - f(y) \cdot F_L(x, Q^2)] \tag{3}$$

as a function of $f(y)$ to extract F_2 and F_L .

To illustrate the method, the reduced cross section, at $x = 2\ 10^{-4}$ and $Q^2 = 10\ GeV^2$, is plotted in Figure 2 for three proton beam energies. One can notice that:

- F_L is very sensitive to small relative shifts of the cross sections and not to absolute errors.
- The precision on F_L requires the difference between the $f(y)$ values at low and high beam energies, $f(y_{low}) - f(y_{high})$, to be large. So the difference between the two centre of mass energies should be large.
- The precision varies as $1/f(y_{low}) \approx 1/y^2_{low}$. So y_{low} should be as large as possible.
- For a relative systematic shift of 3.5 % between the cross sections at $E_p = 920\ GeV$ and $E_p = 460\ GeV$, the systematic error on F_L would be $\approx 0.035/0.18 = 20\%$.
- A measurement at an intermediate beam energy of 575 GeV would not improve significantly the error on F_L but would be an excellent cross check of the systematics.

2.2. *Reducing proton or electron beam energy ?*

To reduce energy in the centre of mass, the proton beam energy E_p or the electron beam energy E_e or both can be reduced. However, in HERA

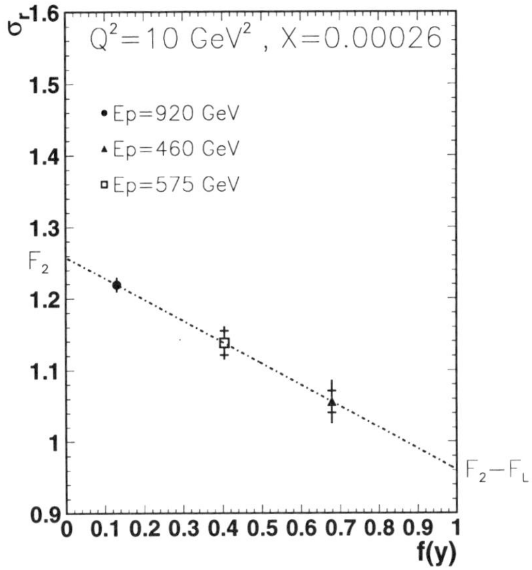

Figure 2. Simulated measurement of the reduced cross section for data at 920 GeV (30 pb^{-1}), 575 GeV (3.5 pb^{-1}) and 460 GeV (5 pb^{-1}). The inner error bar represents the statistical error. The full error bar denotes the statistical and systematic uncertainty added in quadrature.

electron-proton kinematics it is preferable to only reduce the proton energy to optimise the experimental systematics by comparing cross sections mainly in the same part of the apparatus [11].

The choice of the reduced proton beam energy is a compromise between the requirement to have the lowest proton beam energy and the luminosity which varies as the square of the beam energy. A reasonable guess for the lowest beam energy is 460 GeV, a value which could be tuned when the luminosity would be better known.

2.3. *Using Initial State Radiation events ?*

To access different y values at fixed Q^2 and x a clever method based on using initial state radiation events in DIS events has been tried. The emission of hard photons in a direction close to the incoming electron

$$e + p \longrightarrow e + \gamma + X \tag{4}$$

can be interpreted as leading to a reduced effective beam energy. The hard photon can be measured in the photon detectors which are part of the luminosity systems of H1 or ZEUS [12]. Unfortunately, an irreducible pile-up of Bethe-Heitler quasi-elastic electron proton scattering events

$$e + p \longrightarrow e + \gamma + p \tag{5}$$

has prevented a clean reconstruction of the photon energy in the photon detector. The Q^2, x and y variables of the radiative DIS events have been reconstructed using the electron energy and angle combined with the total momentum of the hadronic final state in the main detector, yielding disappointing results for F_L [13,14].

2.4. *Indirect determination*

The H1 Collaboration has invented methods to determine F_L from the measurement of the differential cross section at standard beam energies [15]. These methods rely on extrapolating fits of data obtained at relatively low y into the region of high y, where the longitudinal Structure Function starts to contribute significantly to the cross section. The extrapolation is not model independent, different theoretical models will lead to different extrapolations [16]. These methods are not fully satisfying and cannot replace a true deconvolution of the two Structure Functions F_2 and F_L by varying the beam energies.

3. Simulation of a direct measurement

3.1. *Parameters*

In the simulation we have considered two beam energy settings : i) the standard beam energies of $E_e = 27.6\ GeV$ and $E_p = 920$, ii) a reduced proton beam energy $E_p = 460\ GeV$ combined to the standard electron beam energy, with luminosities of 30 and 10 pb^{-1} respectively. We have used the H1 parameterisation of F_2 and F_L [17], which provides F_L values around 0.3. In a case study for the workshop we have assumed systematic errors based on the state of analysis reached so far in inclusive low Q^2 cross section of H1. With the H1 detector, at large y and low Q^2, the kinematics of the scattered electron can be best reconstructed using the backward high resolution Spacal calorimeter for an energy E'_e of the scattered electron as low as 3 GeV and the Backward Silicon tracker BST or the Central Jet Chamber for a precise angle measurement. A cut on the sum $\sum E - p_z$ of the hadronic final state and the scattered electron reduces the size of the radiative corrections and the residual photoproduction background. The following sources of systematic errors have been considered:

- Relative shift on the energy of the scattered electron rising from 0.2% at 30 GeV to 2% at 3 GeV.
- Relative shift on the electron angle of 0.2 mrad for $175^0 > \theta_e > 165^0$ in the Backward Silicon Tracker and of 1 mrad for $\theta_e < 165^0$.
- Uncertainty on the photoproduction background subtraction varying from 0 % at $y = 0.65$ to 4 % at $y = 0.9$.
- 0.5 % uncertainty on the residual radiative corrections.
- An overall uncertainty of 1 % on the measurement of the cross section at low beam energy settings covers relative uncertainties on electron identification, trigger efficiency, vertex efficiency, and relative luminosity.

3.2. *Results*

To evaluate the errors two independent methods have been considered: an analytic calculation [18] and a fast Monte-Carlo simulation. They provide statistical and systematic errors which are in excellent agreement. In the following we refer to detailed numbers obtained by the fast simulation. As

Table 1. Statistical and systematic errors in percent in the range $7 < Q^2 < 13\ GeV^2$ based on data at 920 GeV (30 pb^{-1}) and 460 GeV (10 pb^{-1}). The first three lines contain the y range and the average x and F_L values for each bin. These are followed by the statistical uncertainty (δ_{stat}), the uncorrelated (δ_{unc}) and the radiative correction (δ_{rad}) uncertainties and the correlated systematics due to the photoproduction background ($\delta_{\gamma p}$), the electron energy ($\delta_{E'_e}$) and the angle ($\delta_{\theta'_e}$). The last two lines summarise the systematic (δ_{syst}) and total (δ_{tot}) uncertainties.

y *at* 460 *GeV*	0.90-0.78	0.78-0.68	0.68-0.58	0.58-0.40
$< x >$	0.00023	0.00026	0.00030	0.00040
$< F_L >$	0.303	0.293	0.283	0.266
δ_{stat}	7.2	10.0	13.3	16.3
δ_{unc}	6.4	9.3	13.8	28.0
δ_{rad}	3.2	4.7	7.0	13.9
$\delta_{\gamma p}$	13.0	4.6	1.0	0.0
$\delta_{E'_e}$	5.7	2.5	0.2	5.5
$\delta_{\theta'_e}$	0.7	1.9	6.2	13.5
δ_{syst}	17.2	11.8	16.7	34.5
δ_{tot}	18.6	15.5	21.3	38.2

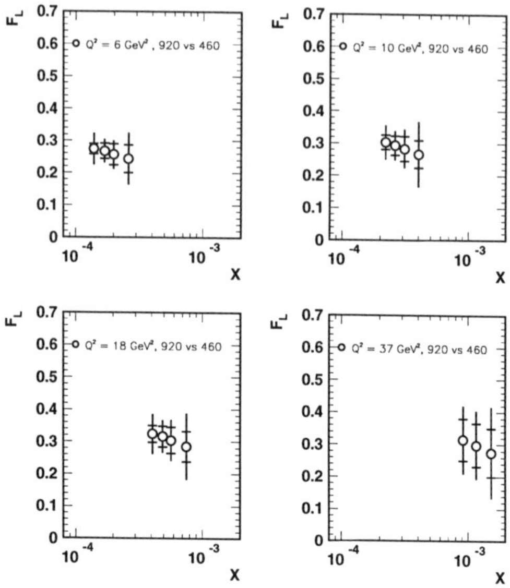

Figure 3. Simulated measurement of the longitudinal Structure Function $F_L(x, Q^2)$ for data at 920 GeV (30 pb^{-1}) and 460 GeV (10 pb^{-1}). The inner error bar is the statistical error. The full error bar denotes the statistical and systematic uncertainty added in quadrature.

an example, the statistical error and the errors on F_L due to each source of systematic errors are given in Table 1 in the range $7 < Q^2 < 13\ GeV^2$ and in four intervals in y. In the first 3 bins in y the overall precision on F_L is around 20 %. It is remarkable that, had the statistical error or the largest source of systematic error turned out to be twice larger, the overall precision would stay below 25 %. The result of simulated F_L is displayed in Figure 3 showing that with the above assumptions F_L can be measured, at the three to five sigma accuracy level, in four bins of Q^2 at very low x values.

4. Direct measurement of the longitudinal diffractive Structure Function F_L^D

In diffractive processes $ep \to eXp$, a reduced cross section can be defined by analogy with inclusive scattering [19]. It is related to the triple differential cross section by

$$\frac{d^3\sigma}{dx_{I\!P}dxdQ^2} = \frac{2\pi\alpha^2 Y_+}{Q^4 x} \cdot \sigma_r^D(x_{I\!P}, x, Q^2), \tag{6}$$

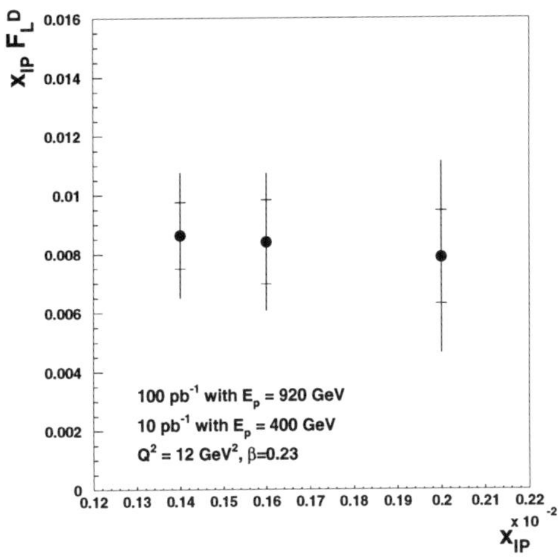

Figure 4. Illustration of the simulated result for F_L^D with statistical (inner bars) and total (outer bars) errors.

where

$$\sigma_r^D(x_{\mathbb{P}}, x, Q^2) = [F_2^D(x_{\mathbb{P}}, x, Q^2) - f(y) \cdot F_L^D(x_{\mathbb{P}}, x, Q^2)] \tag{7}$$

and $x_{\mathbb{P}}$ is the fraction of the longitudinal momentum lost by the proton. Nothing is experimentally known about $F_L^D(x_{\mathbb{P}}, x, Q^2)$. The dominant role played by gluons in the diffractive parton densities [19] implies that F_L^D must be relatively large. A measurement of F_L^D to even modest precision would provide a very powerful independent tool to verify our understanding of the underlying dynamics of deep inelastic diffractive scattering. As in the inclusive measurements the most promising possibility to extract F_L^D is to compare data at fixed Q^2, $x_{\mathbb{P}}$ and x, but from different centre of mass energies s and hence from different y values. The expected uncertainties on the measurement are estimated using the RAPGAP generator [20] and values of F_2^D and F_L^D from the H1 NLO QCD fit to diffractive DIS data [19]. The expected sources of uncertainties are the same as in the inclusive measurement, combined with an overall uncorrelated error of 2.4 % for luminosities, efficiencies and acceptance corrections and with a 4 % uncertainty on the energy scale of the hadronic final state (to reconstruct $x_{\mathbb{P}}$). An illustration of the expected measurement, in the range $7 < Q^2 < 30\ GeV^2$ at the average $\beta = 0.23$ (where $\beta = x/x_{\mathbb{P}}$), is shown in Figure 4. In this first simulation [21] the assumed luminosity is 10 pb^{-1}, as for the inclusive measurement, but at a proton beam energy of 400 GeV, which is of no significant influence for the estimation. The results are encouraging. The diffractive longitudinal Structure Function , based on the H1 2002 fit, would be measured to a three sigma accuracy. The dominant errors come from uncertainties which are uncorrelated between the two beam energies and which can still be improved.

5. Possible Scenarios

Based on a recent estimate of the luminosity at reduced proton energy [22], it is possible to consider two possible scenarios:

(i) *One reduced proton energy at* $E_p = 460\ GeV$. It would need 3 weeks to change the configuration of the machine and to tune the luminosity plus 10 weeks to record 10 pb^{-1} of good data with High Voltage of trackers on.

(ii) *Two reduced proton energies at* $E_p = 460\ GeV$ *and* $E_p = 575\ GeV$: 3 weeks to prepare the machine at $E_p = 460\ GeV$ plus 5 weeks of data taking to get 5 pb^{-1} of good data, plus 3 weeks to prepare the machine at $E_p = 575\ GeV$ followed by 2 weeks to record 3.5 pb^{-1} of good data.

The first scenario would have the advantage to provide more statistics to reduce the error on F_L at x around 10^{-3} and to provide enough luminosity to perform a 3 sigma measurement of F_L^D . The second scenario would provide an excellent check of the systematic errors. The two scenarios require the same total amount of time. At the present level of uncertainty on the luminosity at reduced proton energy the scenario (i) looks rather safer and preferable.

6. Conclusion

A direct measurement of the longitudinal Structure Function $F_L(x, Q^2)$with five sigma significance, in the x range from 10^{-4} to 10^{-3} in four bins of Q^2, can be performed at HERA. It requires about three months at reduced proton beam energy. In addition it would provide the first measurement of the diffractive Structure Function $F_L^D(x_{I\!P}, x, Q^2)$ at the three sigma level. Due to possible conflict with alternative scenarios where the increase of recorded luminosity at normal beam settings would be privileged, it would be wise to do the measurement in the last three months of HERA running in spring 2007 and to wait on the last development on searches for physics beyond the standard model [23] to take a decision by the end of 2006.

Acknowledgments

I would like to thank Laurent Schoeffel for useful discussions and help in preparing the manuscript.

References

1. C. Callan and D. Gross, *Phys. Rev. Lett.*. **22** (1969) 145.
2. G. Altarelli and G. Martinelli, *Phys. Lett.* **B 76** (1978) 89.
3. L.H. Orr and W J. Stirling Phys. Rev. Lett. **66** (1991) 1673.
4. S. Moch, J.A.M. Vermaseren and A. Vogt *Phys. Lett.* **B 606** (2005) 123.
5. A. D. Martin et al., *Eur. Phys. J.* **C 23** (2002) 73.
6. J. Pumplin et al., *JHEP* **0207** (2002) 012.
7. R. Ball, these proceedings.
8. J. Bartels, K. Golec-Biernat and K. Peters, *Eur. Phys. J.* **C 17** (2000) 121.
9. R.S. Thorne, these proceedings [hep-ph/0511351].
10. V. Lendermann, these proceedings.
11. M. Klein, "On the future measurement of the longitudinal Structure Function", Proc. Workshop DIS 2004, Strbske Pleso, Slovakia, Eds. D. Bruncko, J. Ferencei and P. Strizenec, Vol. 1, p. 309. [http://www.saske.sk/UEF/OSF/DIS].
12. M. W. Krasny, W. Placzek and H. Spiesberger, *Z. Phys.* **C 53** (1992) 687.
13. C. Issever, Ph.D. thesis, Dortmund University, Germany, 2000 [http://www.h1-desy.de].
14. J. Cole, "Structure Function Measurements using Radiative Events", Proc. Workshop DIS 2003, St. Petersburg, Russia , Eds. V. Kim and L. Lipatov; A. Bornheim, Ph. D. thesis, Bonn University, Germany, 1999 [http://www.zeus-desy.de].
15. H1 Coll., C. Adloff et al., *Eur. Phys. J* **C 21** (2001) 33 [hep-ex/0012053]; H1 Coll.," Determination of the longitudinal proton Structure Function $F_L(x,Q^2)$ at low Q^2 at HERA", Proc. ICHEP04, Beijing, China,[http://www.h1-desy.de].
16. R.S. Thorne, *Phys. Lett.* **B 418** (1998) 371.
17. H1 Coll., C. Adloff et al., *Eur. Phys. J.* **C 30** (2003) 1.
18. M. Klein, private communication.
19. H1 Coll., Paper 980 contributed to ICHEP 2002, Amsterdam, H1 prelim-02-012) [http://www.h1-desy.de].
20. H. Jung, *Comp. Phys. Commun.* **86** (1995) 147.
21. P. Newman, private communication.
22. F. Willeke, "Prospects for operating HERA with lower proton energy", informal memo, 15 Sept. 2005, unpublished.
23. E. Perez, these proceedings [hep-ex/0601028].

HERA AND THE LHC

ALBERT DE ROECK

CERN, 1211 Geneva 23

This report summarizes some of the main results of the one year long workshop on HERA and the LHC.

1. Introduction

In roughly two years time from this writing, i.e. in the second half of the year 2007, the Large Hadron Collider (LHC), presently under construction at CERN, Geneva, will come into operation. This collider will produce proton-proton interactions at a centre of mass system (CMS) energy of 14 TeV. Present experimental and theoretical indications are that this energy range, also called the TeV-scale or Terascale, will break new ground in the understanding of particle physics and even the Universe [1].

Specifically, the LHC will unveil the mystery of electro-weak symmetry breaking, either by discovering the Higgs bosons, or otherwhise. Furthermore the chances are extremely high that new physics will be discovered, such as supersymmetry, extra dimensions or other. The LHC will also be a precision instrument[2], allowing for a measurement of the masses of the top quark and W boson to respectively 1 GeV and 15 MeV.

However the LHC will also allow for e.g. new measurements in the field of QCD, b and c physics, diffraction etc., in this new energy regime. Many of these measurements will need to be made and understood early on, in order to allow to estimate backgrounds correctly for searches of new phenomena. Precision measurements will also need a good understanding of QCD, both in the perturbative range (parton showering, jets,...) and the nonperturbative range (fragmentation, underlying events, minimum bias event cross sections,...). Since the protons are composite particles, consisting of gluons and quarks, the *pp* cross sections of hard scattering processes depend on the parton distributions in the proton. The LHC can make some measurements of these quantities, but will rely to a large extend on precision data collected at other colliders, in particular data from HERA.

2. The HERA/LHC Workshop

The goals of the HERA-LHC workshop have been defined as follows:

- To identify and prioritize those measurements to be made at HERA which have an impact on the physics reach of the LHC.
- To encourage and stimulate the transfer of knowledge between the HERA and LHC communities and establish an ongoing interaction.
- To encourage and stimulate theory and phenomenological efforts related to the above goals.
- To examine and improve theoretical and experimental tools related to the above goals.
- To increase the quantitative understanding of the implications of HERA measurements on LHC physics.

Five working groups have been formed: (WG1) Parton Densities; (WG2) Multi-jet Final States; (WG3) Heavy Quarks; (WG4) Diffraction; (WG5) MC-Tools. The first meeting took place at CERN in March '04 (250 participants), and the final meeting was held at DESY in April '05 (150 participants). More information can be found in Ref. 3.

3. WG1: Parton Distributions

Parton distribution functions are the prime measurements that are made at HERA. The charge weighted quark distributions are measured directly via the structure function F_2. The gluon distributions can be measured indirectly via QCD evolution fits of F_2 or semi-directly in e.g. jet and charm cross section measurements.

The F_2 structure functions at HERA are now measured with a precision of typically 2% or better in large kinematic regions, and are basically limited by systematics. The Run-II high statistics HERA data is expected in particular to improve the region of large x and Q^2 which is still statistically limited.

Taking naively the simple spread of the existing PDFs gives up to a 10% uncertainty in the Standard Model (SM) Higgs cross section, as demonstrated in Fig. 1 [7]. The message from the workshop is clear: ultimately we have to do better than that.

The working group has defined the following program to be studied:

- Study and document the potential experimental and theoretical accuracy for various LHC processes (Drell-Yan, W, Z, WW, γ+ jet

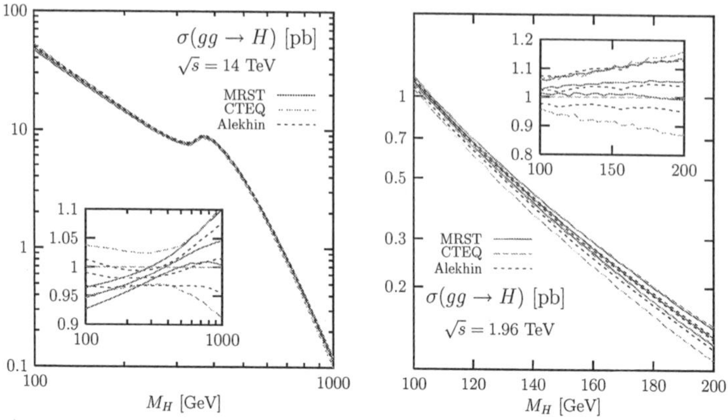

Figure 1. The CTEQ[4], MRST[5] and Alekhin[6] PDF uncertainty bands for the NLO cross sections for the production of the Higgs boson at the LHC (left) and Tevatron (right) for the process $gg \to$ Higgs[7]. The insert shows the spread of the predictions when the NLO cross sections are normalized to the prediction of the reference CTEQ6M set.

production...) How can these be used for precision measurements at the LHC and e.g for luminosity determination? Cross sections and distributions will be studied and benchmarked with LHC detector simulation.

- Study of the impact of PDFs on LHC measurements. Here one will try to make the most of the HERA data. Is there a need for F_L and/or eD scattering? Can one judge which PDF is preferred? If so, what are the most precise PDFs and their errors?
- On the more theoretical side: what is the impact of small x and large x resummation and saturation corrections on PDFs? How well is the QCD evolution validated in the different kinematic regimes? How can we verify this at HERA and what is the impact on the LHC?

The systematic study of well measurable LHC final states is ongoing. As an example the uncertainties for W, Z and di-boson production with experimental cuts, for the parton distributions and perturbative scales, are 4-5% and 4-9% respectively [8].

Many of the processes in this study can be used for the extraction of in-

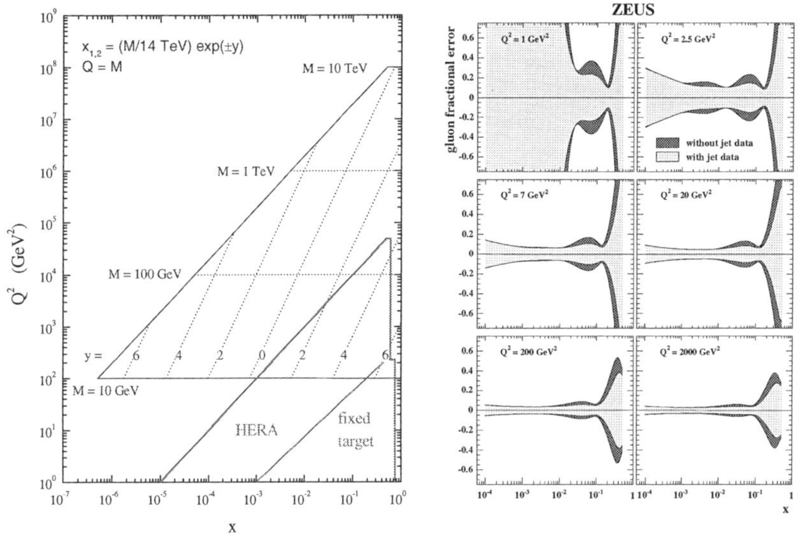

Figure 2. (left) The kinematic plane (x, Q^2) and the reach of the LHC, together with that of the existing data (HERA, fixed target). Lines of constant pseudo-rapidity are shown to indicate the kinematics of the produced objects in the LHC centre of mass frame [9]. (right) The total experimental uncertainty on the gluon PDF for a fit including jets, compared to a fit not including jet data (outer error bands). The uncertainties are shown as fractional differences from the central values of the fits[10], for several values of Q^2.

formation on the PDFs, but it needs still to be quantified to what precision this can be done.

Fig. 2 shows the plane in x, Q^2 covered presently by HERA and the part that will be covered by the LHC [9]. Extrapolation or rather QCD evolution of the PDFs will be required over about 3 orders of magnitude. Clearly we need to understand as good as we can the evolution in the region where we have precise data at present, to check the uncertainty which is 'tolerated' by these data (e.g. the amount of non-linear effects). In the course of this workshop the NNLO splitting functions for the DGLAP evolution became available [11], so full NNLO fits can be made soon. Low-x resummation is important, and it was shown that it can lead to differences of about 20% at $x = 10^{-3}$ and low Q^2 for the gluon distribution extracted by global fits [12]. On the high x side, $x > 0.7$, resummations can lead to 15% changes in the quark distributions [13].

The key issues nowadays for the global fits are the selection of data, a consistent treatment of errors and calculation of error bands. There are

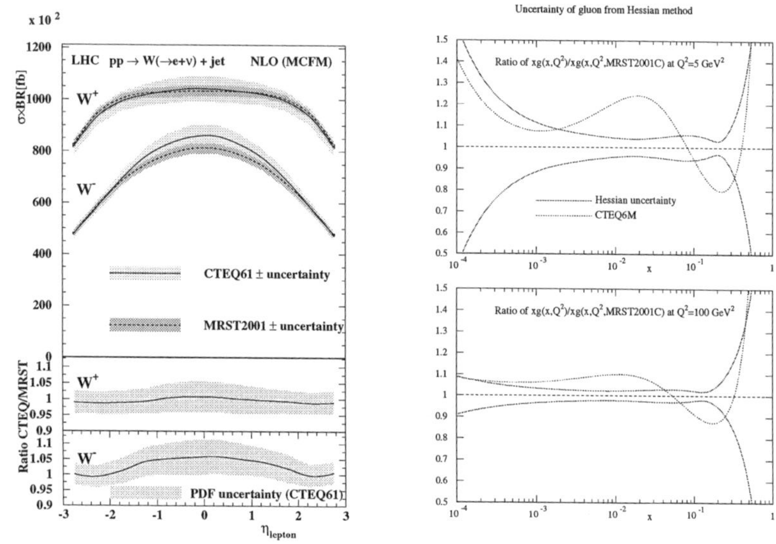

Figure 3. (Left) The PDF uncertainties for W^+ and W^- production[8]. (Right) The gluon distribution uncertainty[12] from MRST, compared with the CTEQ central values.

some tensions observed between data sets which need to be understood. While several prescriptions are being tried out for the error treatment, one radical way to approach this is to take data of one experiment only, but try to include as much as possible information. ZEUS presented an encouraging study on a combined PDF study using F_2 data and jet cross sections. Fig. 2 shows the potential gain in the uncertainty of the gluon distribution. Particularly at medium-x one can gain of order of 30% in precision in the gluon determination.

A new initiative that started during this workshop are the first steps towards a creation of combined data sets from HERA, i.e. really combining the experimental data points, rather than using the sets as two independent ones in the fit. The first results are very encouraging: they show that the extracted PDF fit from the combined data set can be much better than the fit to the sum of all the data points. What happens in practice is that one experiment 'calibrates' the other during the combining procedure. Similar improvements have been noted at LEP in combining measurements.

Turning back for a moment to the present PDF uncertainty: Fig. 3 shows the PDF error bands one gets using the present prescriptions of the PDF uncertainties, for W+jet production at the LHC. One notes that the error band of one PDF does not cover the central value of the other. One of the main reasons is the low-x behaviour of the parton distributions which is presently very different for the two sets of PDFs shown in Fig. 3. Both PDFs however are consistent with the HERA low-x data. Clearly nature may have chosen one or the other way, so how can one make progress here? What is needed are measurements that are more directly sensitive to the gluon in that region. The measurement of the longitudinal structure function F_L could do the trick, if it can reach the necessary precision. Better than F_2^{charm}, F_L is as fundamental as F_2 with little theoretical ambiguity. To make a clean measurement of F_L HERA will have to operate some time at lower energies, and this is not yet on the program. Similarly for a good flavour separation and non-singlet structure function extraction, electron scattering on Deuterons would be needed. HERA is a unique machine and if these measurements do NOT happen at HERA, they won't happen for at least a very long time to come.

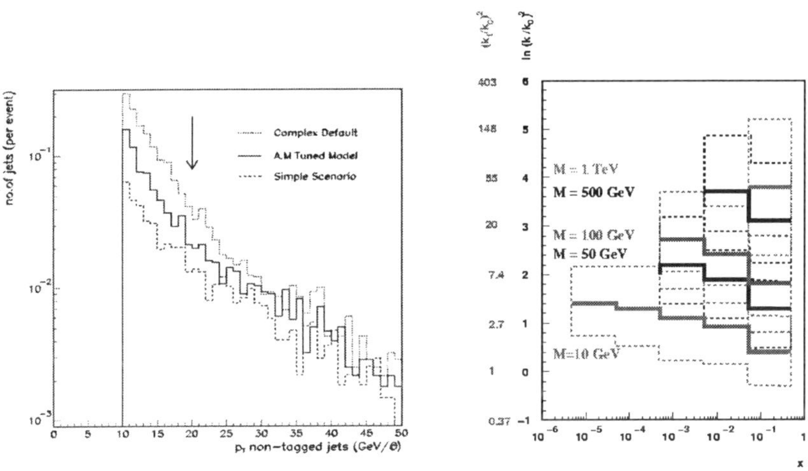

Figure 4. (left) Number of central jets per event in an analysis of $H \to WW^* \to 2l$ for different models/assumptions of the underlying event. The study was performed with ATLFAST. (Right) the k_T from QCD evolution for different values of the mass of a produced system M in $gg \to M$.

4. WG2: Multi-Jet Final States and Energy Flow

The following topics were studied by WG2:

- The study of the structure of the underlying event, and of minimum

bias events. New models were proposed and tested during the workshop. Tunes to existing data were discussed. A task force was installed to study similar observables in ep as done in pp for the tuning.
- The gap survival probability. The dynamics of gaps void of particles in pp and the consequences for the LHC are still poorly understood. New measurements were suggested to make further progress.
- A study of the phenomenology related to the CASCADE Monte Carlo, which shows differences with other QCD generators at the LHC at low-x.
- Unintegrated PDFs and their importance e.g. on p_T distributions of the Higgs particle.
- Issues connected with Matrix Element/Parton Shower matching.
- Resummation of event shape variables.
- Future parton shower developments, such as unintegrated parton correlation functions and QEDxQCD exponentation.

Certainly one of the unknowns for studies at the LHC at present is the control of the underlying event and the event shape and number of minimum bias events which will be added to hard scattering event as pile-up: we expect about 4 interactions per bunch crossing on average at the first years luminosity of $2.10^{33}\text{cm}^{-2}\text{s}^{-1}$. Studies of tunes of PYTHIA and to some extend also HERWIG have been made using Tevatron and even lower energy data. These tunes should be validated next with the plethora of available HERA data. New models are now available: a new PYTHIA version, Jimmy for HERWIG, and the SHERPA underlying event. All these need tuning and validating. The effect of the (importance of) underlying event was demonstrated with the vector boson fusion channel for Higgs production. In this process two forward jets are produced, plus the Higgs, chosen to decay in two W's, which in turn decay leptonically. Hence there is no color flow and hadron activity in the central region, except from the underlying event. To select these events over background a central jet veto is introduced, the efficiency of which will be affected by the underlying event model. Results in Fig. 4(left) show that there is a 10% variation in the selection efficiency, depending on the model chosen for the underlying event.

A challenge for final state studies will be to predict cross sections and topologies for many-jet events at the LHC, e.g. 8-jets or more. Certain SUSY cascades can lead to such number of jets, and a pure event counting technique will need a solid prediction of the QCD background. This needs good matching between matrix elements and parton showers. Such matching algorithms have been developed over the past year, in particular for *ee* and *pp* scattering, and are now being extended to *ep* such that these can be used to test on HERA multi-jet data.

A very important aspect is the initial k_T in the hard scattering, built up during the parton evolution before, say, the gluon enters in the hard scattering to produce a Higgs in the process $gg \to$ Higgs. The growth in k_T can be large as shown in Fig. 4(right) for a CASCADE calculation, for massive systems, thus affecting the p_T distribution of the produced particle. This means that for such production processes the unintegrated partons will be needed to correctly follow this evolution and provide the expected k_T in the scattering. HERA can test these k_T predictions and their effects with its data, and will allow to measure the unintegrated PDFs via final state measurements.

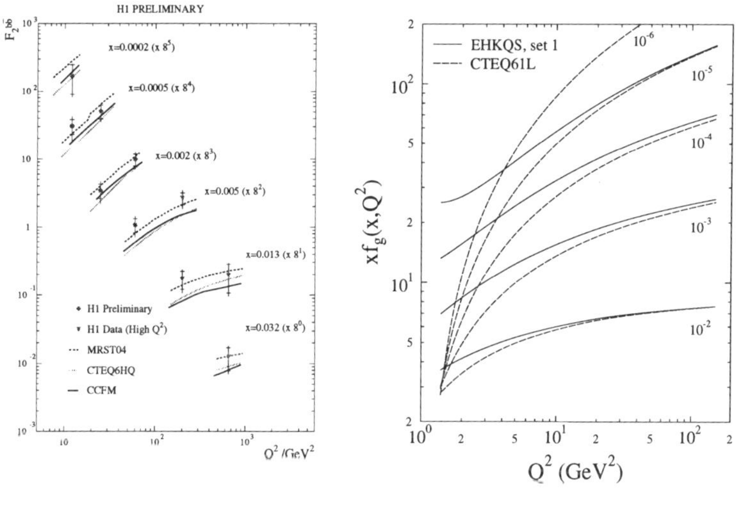

Figure 5. (Left) Data on $F_2^b(x, Q^2)$ from the H1 experiment, compared to QCD predictions. (Right) Comparison of EHKQS set 1 (solid line) and CTEQ611 (dashed line) gluon distributions as a function of Q^2 for various x values.

5. WG3: Heavy Flavours

A list of measurements to be done at HERA has been proposed by WG3:

- The charm and bottom structure functions F_2^c and F_2^b.

- Charm exclusive final states in γp and DIS: cross sections, fragmentation universality, contributions from higher charm resonances.
- Charm exclusive final states with jets.
- Bottom exclusive final states.
- Double quark tags.
- Charm and bottom in charged current events.
- Quarkonia.
- Diffractive production of charm.

To have significant impact and improve the already available data, at least 400 pb^{-1} will be needed at HERA-II. The topics listed are of general interest for the study of heavy flavour physics, but several have direct impact on the LHC. A clear case is the measurement of F_2^b, which is important for $bb \to$ Higgs production contribution. This needs a measurement of F_2^b at a scale of $m_H/2$. Fig. 5 shows recent results of a measurement of F_2^b from H1 based on HERA-I data [14]. The HERA-II data could reduce the errors by a factor of 4.

Heavy flavour measurements are also very sensitive to non-linear QCD evolution effects in the parton distributions. Fits to the HERA F_2 data at small x and small Q^2 improve by adding non-linear terms to the gluon evolution, see Fig. 5 [15]. This will lead to more charm production at low p_T [16]. The effects will become visible at the LHC for p_T values below about 2 GeV. ALICE will be best placed to measure these effects in the LHC data, since they can measure p_T values down to almost zero.

6. WG4: Diffraction

This working group studied the following topics:

- Diffractive Higgs production.
- Backgrounds to diffractive Higgs.
- Diffractive factorization breaking in di-jet, charm and leading neutron production.
- Rapidity gap survival.
- New measurements e.g. F_L^D.
- Exclusive diffractive di-jets.
- Saturation effects and relation to multiple interactions and the gap survival.

A large part of the activities was the transfer of experience, knowledge, design and operation of the detectors for forward physics from HERA to the LHC.

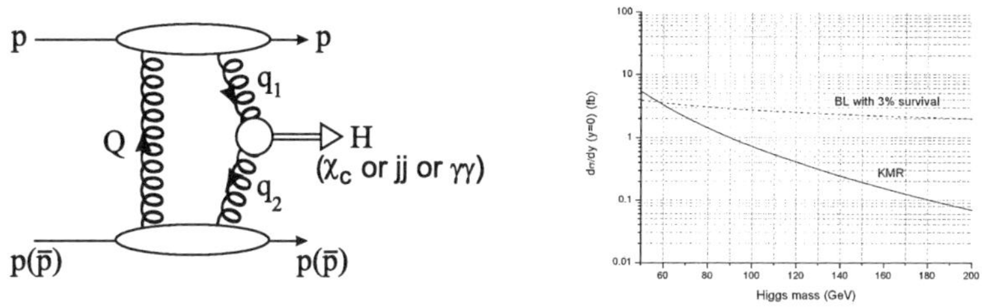

Figure 6. (Left) Diagram for exclusive Higgs production. (Right) Evolution of the cross section for KMR[19] and the model proposed in[20] as function of mass.

A topic of recent strong interest is the possibility to produce central diffractive Higgs particles in pp collisions, see Fig. 6. The advantages of this channel are [17]: a good missing mass resolution, of order 1-2 GeV via the protons for the Higgs, and low backgrounds. The cross sections are generally of the order of femtobarns and there has been quite some discussion on the validity of certain calculations. Also Monte Carlo models have been compared with one another in detail. The differences are basically understood as due to Sudakov suppression factors and parton distributions. In particular the Exhume[18] program is considered to give the more natural expected η behaviour. The KMR[19] calculation has been checked by independent groups and found to be ok. In all it means that the perturbative cross section for the Standard Model exclusive Higgs production is likely to stay below 10 fb. There are however alternative model predictions, based on non-perturbative calculations. Fig. 5(right) shows the different energy dependence in the KMR and the model proposed in [20]. It is not excluded that the total exclusive cross section could be larger than the one calculated in [19] if an additional soft component would be present.

It will be important in the coming year to test and measure the ingredients that go in that calculation. An example is the rescattering effects in collisions. It has been suggested to look into events with jets and a leading neutron at HERA [21] and study eg. $x - p_T$ correlations.

An input used in the exclusive Higgs cross section calculations are the generalized unintegrated parton distributions. HERA can measure these distributions in exclusive J/ψ production. The double pomeron process itself can be measured at HERA in the reaction $\gamma p \rightarrow V + X + p$ with V

a vector meson and X the centrally produced system. Finally, the leading proton spectra as measured at HERA are found not to be described with standard Monte Carlo generators. This has an effect on the background studies to diffractive processes at the LHC, and some tuning based on the HERA leading baryon measurements will be essential.

Diffraction and low-x is part of the LHC physics program and there are plans to equip the central detectors with detectors in the forward region, which also offers new opportunities for groups to join in this activity.

7. WG5: Tools

WG5 had the following program:

- Parton distribution library: LHAPDF is now the official carrier of the PDFs. It is used by the LHC experiments in generators. The HERA PDFs have been added recently. LHAPDF allows for uncertainty estimates. The Pion and Photon PDFs have been added to the library. Should the F_2^D parametrizations also be added?
- NLOLIB framework for NLO QCD programs. A uniform user interface is being developed, as well as an interface to HZTOOL. e^+e^-/ep have been included but pp still needs to be added.
- HZTOOL/JetWeb/RunMC/Cedar tools for Monte Carlo tuning. All HERA results have been included, some e^+e^- results. Include more pp data?
- Discussions on RAPGAP and CASCADE Monte Carlo programs for inclusive and diffractive pp.
- Plenty of exchange on other MC tools, leading to new MC tools and comparisons with ep where possible.
- Continuation of the MC@LHC workshop, concerning validation of MC programs.

8. The Verdict and Outlook

Coming back to the goals that were set at the start of the workshop, one can say items (1) $\rightarrow$ (4) have been achieved. For item (5) many studies are still ongoing, and more quantitative examples/results are expected for the proceedings end '05.

The final meeting is not the end of the workshop, however. The link between the communities is now strong and should not fade away. Therefore it was unanimously decided to continue the workshop but on a "one meeting per year basis". The next meeting will be in June 2006 at CERN. Everybody is invited to continue (or start) participating in the workshop.

Acknowledgments

It is a pleasure to thank all participants of the workshop for their work, and especially Hannes Jung for the co-organization. My thanks go also to the organizers of Ringberg 05 for this kind invitation.

References

1. J. G. Branson et al., hep-ph/0110021.
2. Fabiola Gianotti, Monica Pepe-Altarelli, Nucl. Phys. Proc. Suppl. 89 (2000) 189, hep-ex/0006016.
3. http://www.desy.de/~heralhc/proceedings/proceedings.html, hep-ph/0601012, hep-ph/0601013.
4. J. Pumplin et al., JHEP **0207** (2002) 012, hep-ph/0201195
5. A.D. Martin, R.G. Roberts, W.J. Stirling, R.S. Thorne, Eur. Phys. J. **C23** (2002) 73 hep-ph/0110215
6. Sergey Alekhin, Phys. Rev. **D68** (2003) 014002, hep-ph/0211096
7. Abdelhak Djouadi, Samir Ferrag, Phys. Lett. **B586** (2004) 352, hep-ph/0310209.
8. H. Stenzel, contribution to this workshop.
9. A.D. Martin, R.G. Roberts, W.J. Stirling, R.S. Thorne, Eur. Phys. J. **C14** 2000, 133.
10. ZEUS Collaboration, S. Chekanov et al., hep-ph/0503274.
11. S. Moch, J.A.M. Vermaseren, A. Vogt, Nucl. Phys. **B691** (2004) 129, hep-ph/0404111
12. A.D. Martin, R.G. Roberts, W.J. Stirling, R.S. Thorne, Eur. Phys. J. **C35** (2004) 325, hep-ph/0308087
13. G. Corcella and L. Magnea, hep-ph/0507042
14. H1 Collaboration, A. Aktas et al., Eur. Phys. J. **C41** (2005) 453, hep-ex/0502010.
15. K.J. Eskola, V.J. Kolhinen, R. Vogt, Phys. Lett. **B582** (2004) 157, hep-ph/031011.
16. A Dainese, J.Phys. **G30** (2004) 1787-1799, hep-ph/040309.
17. A. de Roeck et al., Eur. Phys. J. **C25** (2002) 391, hep-ph/0207042.
18. J. Monk, A. Pilkington, hep-ph/0502077.
19. Valery A. Khoze, Alan D. Martin, M.G. Ryskin, Eur. Phys. J. **C14** (2000) 525, hep-ph/0002072.
20. M. Boonekamp, R. Peschanski and C. Royon, Nucl. Phys. **B669** (2003) 277, Erratum-ibid. **B676** (2004) 493, hep-ph/0301244.
21. A. Kaidalov and V. Khoze, private communication.

List of participants

Name	Institution	E-mail
Simon Albino	Würzburg U	Simon.Albino@desy.de
Richard Ball	Edinburgh U	rdb@ph.ed.ac.uk
Andrea Banfi	Cambridge U	abanfi@hep.phy.cam.ac.uk
Olaf Behnke	Heidelberg U	obehnke@mail.desy.de
Gerd Buschhorn	MPI Munich	gwb@mppmu.mpg.de
Allen Caldwell	MPI Munich	caldwell@mppmu.mpg.de
Amanda Cooper-Sarkar	Oxford U	a.cooper-sarkar@physics.ox.ac.uk
Albert De Roeck	CERN	deroeck@mail.cern.ch
Frank Ellinghaus	Colorado U	Frank.Ellinghaus@desy.de
Laurent Favart	IIHE Bruxelles	lfavart@ulb.ac.be
Joel Feltesse	CEA Saclay	joel.feltesse@cea.fr
Sean Fleming	UC San Diego	spfleming@yahoo.com
Frank Fugel	Hamburg U	Frank.Fugel@desy.de
Lidia Goerlich	PAN Krakow	goerlich@mail.desy.de
Günter Grindhammer	MPI Munich	guenterg@mail.desy.de
Rosita Jurgeleit	MPI Munich	roj@mppmu.mpg.de
Bernd Kniehl	Hamburg U	kniehl@desy.de
Michael Klasen	LPSC Grenoble	klasen@lpsc.in2p3.fr
Henri Kowalski	DESY Hamburg	henri.kowalski@desy.de
Gustav Kramer	Hamburg U	kramer@mail.desy.de
Victor Lendermann	Heidelberg U	victor@mail.desy.de
Andrej Liptaj	MPI Munich	liptaj@mail.desy.de
John Loizides	UC London	john.harry.loizides@desy.de
Luminita Mihaila	Karlsruhe U	luminita.mihaila@desy.de
Andreas Metz	Bochum U	Andreas.Metz@tp2.ruhr-uni-bochum.de
Joachim Meyer	DESY Hamburg	jmeyer@mail.desy.de
Zoltan Nagy	Zurich U	nagyz@physik.unizh.ch
Andrei Nikiforov	MPI Munich	nikifor@mail.desy.de
Wolfgang Ochs	MPI Munich	wwo@mppmu.mpg.de
Emmanuelle Perez	DESY Hamburg	eperez@hep.saclay.cea.fr
David Saxon	Glasgow U	d.saxon@physics.gla.ac.uk
Ingo Schienbein	DESY Hamburg	Ingo.Schienbein@desy.de
Graham Shaw	Manchester U	graham.shaw@manchester.ac.uk
Robert Thorne	Cambridge U	thorne@hep.phy.cam.ac.uk
Thomas Teubner	Liverpool U	Thomas.Teubner@liverpool.ac.uk
Katsuo Tokushuku	KEK Tsukuba	katsuo.tokushuku@desy.de
Daniel Traynor	QM London	d.traynor@qmw.ac.uk
Zoltan Trocsanyi	Debrecen U	Zoltan.Trocsany@cern.ch
Graeme Watt	DESY Hamburg	Graeme.Watt@desy.de
Katarzyna Wichmann	Hamburg U	kklimek@mail.desy.de
Frank Wilczek	MIT	wilczek@mit.edu
Benedikt Zihlmann	Gent U	zihlmann@mail.desy.de